Decoding Euroc

Decoding Eurocode 7

Andrew Bond and Andrew Harris

CRC Press
Taylor & Francis Group
Boca Raton London New York

CRC Press is an imprint of the
Taylor & Francis Group, an **informa** business
A TAYLOR & FRANCIS BOOK

CRC Press
Taylor & Francis Group
6000 Broken Sound Parkway NW, Suite 300
Boca Raton, FL 33487-2742

First issued in paperback 2021

ISBN 13: 978-1-03-209961-3 (pbk)
ISBN 13: 978-0-415-40948-3 (hbk)

British Library Cataloguing in Publication Data
A catalogue record for this book is available from the British Library

Library of Congress Cataloging in Publication Data
Bond, Andrew, 1959-
  Decoding Eurocode 7 / Andrew Bond and Andrew Harris. -- 1st ed. p. cm.
  Includes bibliographical references and index.
  ISBN 978-0-415-40948-3 (hardback : alk. paper) 1. Engineering
geology—Standards—Europe. 2. Structural design—Standards—Europe. 3. Standards, Engineering—Europe. I. Harris, Andrew, 1955- II. Title. III. Title: Decoding Eurocode seven.

  TA705.4.E85B66 2008
  624.102'184--dc22   2008018633

Visit the Taylor & Francis Web site at
http://www.taylorandfrancis.com

and the CRC Press Web site at
http://www.crcpress.com

Contents

About the authors

Andrew Bond (MA MSc PhD DIC MICE CEng) is a UK delegate on the Eurocode 7 committee, a former member of the UK's National Strategy Committee, and co-author of BSI's *Extracts from the Structural Eurocodes for students of structural design*. He is a former editor of *Geotechnical Engineering*.

Andrew gained first class honours from Cambridge University in 1981, before working for WS Atkins and Partners on a variety of civil, structural, and geotechnical engineering projects. He obtained his MSc from Imperial College in 1984 and his PhD in 1989, for pioneering research into the behaviour of driven piles and design of the Imperial College Pile. He joined Geotechnical Consulting Group in 1989, becoming a Director in 1995. While at GCG, he developed the computer programs ReWaRD® (for embedded retaining wall design) and ReActiv® (for reinforced slope design).

Andrew set up his own company, Geocentrix, in 1999, for whom he has developed the pile design program Repute® and delivered a wide range of Eurocode training courses and lectures, both publically and privately. In 2006, he set up Geomantix (with Andrew Harris), to provide specialist geotechnical consultancy services.

Andrew Harris (MSc DIC MICE CEng FGS) is co-author of BSI's *Extracts from the Structural Eurocodes for students of structural design* and a frequent speaker on Eurocode 7 and the design of pile foundations.

Andrew gained first class honours from Kingston Polytechnic in 1977, before working for Rendel Palmer and Tritton on a variety of civil and geotechnical engineering projects. He obtained his MSc from Imperial College in 1980, after which he joined Peter Fraenkel and Partners, including three years on overseas placement in Hong Kong and Indonesia. In 1985, Andrew joined Kingston University as a senior lecturer, rising to Associate Dean before joining C L Associates as regional manager in 2004.

In 2006, Andrew set up his own company, Geomantix (with Andrew Bond), to provide specialist geotechnical consultancy services. He also continues at Kingston University as a part-time lecturer in geotechnical engineering.

Acknowledgements

We would like to thank the following friends and colleagues who kindly reviewed parts of the manuscript for us:

Tony Barley, David Beadman, Eddie Bromhead, Owen Brooker, Richard Driscoll, Derek Egan, Chris Hendy, Angela Manby, Devon Mothersill, David Norbury, Trevor Orr, David Rowbottom, Giuseppe Scarpelli, Hans Schneider, Ian Smith, Tony Suckling, Viv Troughton, Austin Weltman, Shon Williams, and Hugo Wood.

Any errors that remain in the book are entirely our fault not theirs.

We would also like to thank the following people and organizations for providing project information for inclusion in our worked examples: Chris Hendy and Claire Seward, Atkins (§2); Viv Troughton, Stent Foundations (§5 and §13); Tony Suckling, Stent Foundations, formerly Cementation Foundations Skanska (§5 and §13); the Singapore Building and Construction Authority (§5); Donald Cook and Chris Hoy, Donaldson Associates (§7); CL Associates (§13); and Bob Handley, Aarsleff Piling (§13).

Andrew Bond would like to thank his colleagues on CEN Technical Committee 250/SC7, with whom he has hade many fruitful discussions during the development of Eurocode 7:

Christophe Bauduin, Richard Driscoll, Eric Farrell, Roger Frank, the late Niels Krebs-Oveson, Jean-Pierre Magnan, Trevor Orr, John Powell, Giuseppe Scarpelli, Hans Schneider, Bernd Schuppener, Brian Simpson, and other members of SC7 too numerous to mention.

Permission to reproduce extracts from British Standards was granted by the British Standards Institution (BSI). British Standards can be obtained in PDF format from the BSI online shop www.bsi-global.com/en/Shop/ or by contacting BSI Customer Services for hard copies, telephone +44 (0)20 8996 9001 or email cservices@bsi-global.com.

Thanks go to Simon Bates (Editorial Assistant) and Faith McDonald (Production Editor) of Spon for answering the multitude of questions we had during the 'creative process' and for their contribution to the book's production. We are grateful to Tony Moore (Senior Editor) for not giving up on this commission, despite the many years it took us to get started.

We would particularly like to thank Jack Offord for creating most of the illustrations that appear in the book, including the front cover. Jack's ability to realize highly original and attractive artwork from our (often vague) initial instructions has greatly added to the book's appeal. His pastiche of Munch's 'The Scream' (which appears in the Epilogue) is a particularly good example of Jack's ability to create pictures that communicate their messages with stunning clarity.

Thanks also to Val Harris for proof-reading the final manuscript and helping remove a multitude of errors invisible to us; and to Jenny Bond for helping assemble the table of contents and index. We are grateful to Andrew Shackleton of Asgard Publishing for providing a thorough review of the text and its layout, resulting in significant improvements to the book.

Finally, we would both like to thank our families for their forbearance while we were writing this book. Their active support and encouragement were essential to the (hopefully) successful outcome of our endeavours.

We end with the words of Winston Churchill (British Prime Minister, 1940–45 and 1951–55):

'Writing a book is an adventure. To begin with, it is a toy and an amusement; then it becomes a mistress, ... a master, and then a tyrant ... just as you are about to be reconciled to your servitude, you kill the monster, and fling him out to the public.'

Andrew Bond
Andrew Harris
June 2008

Prologue

'The Structural Eurocodes are a European suite of codes for structural design ... developed over ... more than twenty-five years. By 2010 they will have effectively replaced the current British Standards as the primary basis for designing buildings and civil engineering structures in the UK.'[1]

Construction Products Directive 89/106/CE

In December 1988, the Commission of the European Community issued Council Directive 89/106/EEC (aka the 'Construction Products Directive'), which sets out essential requirements for all construction works:

'... products must be suitable for construction works which ... are fit for their intended use ... [and] satisfy the following essential requirements ... for an economically reasonable working life...
> *1. Mechanical resistance and stability*
> *2. Safety in the case of fire*
> *3. Hygiene, health and the environment*
> *4. Safety in use'*[2]

The Commission directed the European Committee for Standardization (CEN, see Chapter 1) to prepare a series of European Standards (ENs) containing common unified calculation methods for assessing the mechanical resistance of structures or parts thereof.[3] These standards, or 'Euronorms', would be used to check conformity with Essential Requirement No 1 of the Construction Products Directive and aspects of Essential Requirements Nos 2 and 4. Additionally, the ENs would provide a basis for specifying contracts for construction works and serve as a framework for future harmonized technical specifications for construction products.

CEN established a technical committee (TC 250) to oversee development of these European Standards, which are collectively known as the 'Structural Eurocodes' (but which have recently been renamed the 'EN Eurocodes'). The countries who are members of CEN, and hence will be adopting the Eurocodes into their engineering practice, are shown on **Plate 1** (which can be found in the book's colour section).

Scope of the book

Decoding Eurocode 7 provides a detailed examination of Eurocode 7 Parts 1 and 2 (ENs 1997-1 and -2), together with an overview of more than a hundred associated European and International standards that this major new geotechnical standard depends upon.

The associated documents include another fifty-six standards in the Structural Eurocodes suite (ENs 1990 to 1999); sixty-one geotechnical investigation and testing standards (EN ISOs 14688, 14689, 17892, 22282, 22475, 22476, and 22477); and eleven standards covering the execution of special geotechnical works (ENs 1536, 1537, 12063, 12715, 12716, 12699, 14199, 14475, 14679, 14731, and 15237).

The topics covered by the book (see **Figure 0.1**) include the Structural Eurocodes (Chapter 1); the basis of structural design and general rules for geotechnical design (Chapters 2–3); ground investigation, testing, and

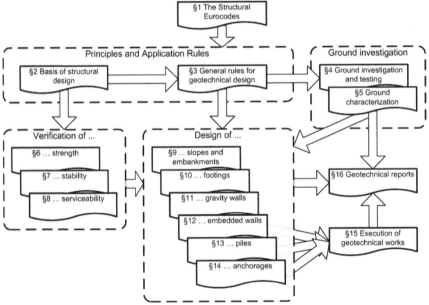

Figure 0.1. Topics covered by Decoding Eurocode 7

characterization (Chapters 4–5); verification of strength, stability, and serviceability (Chapters 6–8); design of slopes and embankments, footings, gravity walls, embedded walls, piles, and anchorages (Chapters 9–14); execution of geotechnical works (Chapter 15); and geotechnical reports (Chapter 16).

Key features of the book

A distinctive feature of *Decoding Eurocode 7* is its extensive use of flow diagrams (e.g. **Figure 0.1**), which help explain how reliability is introduced into the design process, and mind maps (e.g. **Figure 0.2**), which bring together a mixture of information into a coherent framework. The book is enhanced by a colour section between its middle pages.

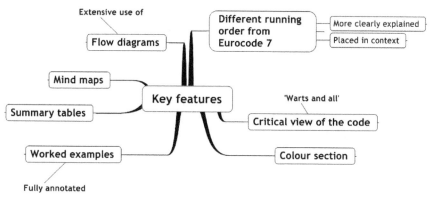

Figure 0.2. *Key features of Decoding Eurocode 7*

The book deliberately presents Principles and Application Rules in a completely different running order from the Eurocodes, so they can be explained more clearly and placed more sensibly in the context of existing practice. Information that is scattered in Eurocode 7 is brought together in helpful summary tables and fully annotated worked examples are included in several chapters to demonstrate how the Eurocodes are used in practice.

Although one of the authors of this book has been involved in the development of Eurocode 7, this has not prevented us taking a critical view of the code, presenting it 'warts and all'. This will allow engineers reading this book to come to an informed opinion about the merits of Eurocode 7, which we are confident will generally be very positive.

Outline of the book

Chapter 1 introduces the various Structural Eurocodes, the links between them, and their timetable for publication. This chapter places the Eurocodes into the wider landscape of standards development not only by the European Committee for Standardization (CEN), but also by the International Standards Organization (ISO) and various national standards bodies, such as the British Standards Institution (BSI) and its German counterpart, Deutsches Institut für Normung (DIN).

Chapter 2 discusses the basis of structural design, as set out in the Structural Eurocodes. It reviews the contents of the 'head' Eurocode (EN 1990) in some detail, including topics such as: requirements; assumptions; Principles and Application Rules; principles of limit state design; design situations; ultimate limit states EQU, STR, and FAT; serviceability limit states; actions, combinations, and effects; material properties and resistance; geometrical data; structural analysis and design by testing; and verification by the partial factor method. The chapter concludes with worked examples involving a shear wall under combined loading, loads on a pile group supporting an elevated bridge deck, and the results of compression tests on concrete cylinders.

Chapter 3 presents the general rules for geotechnical design that are set out in Eurocode 7 Part 1, including: design requirements, complexity of design, and geotechnical categories; limit states; actions and design situations; design and construction considerations; geotechnical design by calculation, prescriptive measures, testing, and observation; supervision, monitoring, and maintenance; and the Geotechnical Design Report.

Chapter 4 discusses ground investigation and testing as described in Eurocode 7 Part 2, including: planning ground investigations; spacing and depth of investigation points; identification and classification of soil and rock; soil and rock sampling; groundwater measurements; field tests in soil and rock; laboratory tests in soil and rock; and testing of geotechnical structures. The chapter concludes with worked examples covering ground investigation for a hotel complex, including specification of field work, borehole logging, and specification of laboratory tests.

Chapter 5 reviews the immensely important topic of ground characterization, including: deriving geotechnical parameters from correlations, theory, and empiricism; obtaining the characteristic value; and statistical methods for ground characterization (pros and cons). The chapter concludes with worked examples that illustrate the statistical determination of parameters for Thames Gravel, Singapore Marine Clay, and London and Lambeth Clays.

Chapter 6 to 8 discuss the verification of strength (ultimate limit states GEO and STR), stability (EQU, UPL, and HYD), and serviceability (serviceability limit states). After reviewing the basis of design, each chapter focuses on how Eurocode 7 introduces reliability into the design and then discusses particular aspects of that verification. Chapter 6 explains the three Design Approaches introduced in Eurocode 7; Chapter 7 demonstrates the similarity of the EQU, UPL, and HYD limit states; and Chapter 8 reviews ways of determining settlement. Chapter 7 concludes with worked examples that illustrate the subjects covered (worked examples for strength and serviceability appear later, in Chapters 9–14).

Chapters 9 to 14 deal with the design of particular geotechnical structures: 9, slopes and embankments; 10, footings; 11, gravity walls; 12, embedded walls; 13, piles; and 14, anchorages. Each chapter deals with the topic in a similar manner, discussing the required levels of ground investigation; design situations and limit states; basis of design; design methods and the application of partial factors; serviceability; and issues related to supervision, monitoring and maintenance. A key feature of all these chapters is a set of worked examples that illustrate the application of Eurocode 7 to common problems and highlights some of the ambiguities in the use of partial factors.

Chapter 15 describes the contents of the series of European Standards that have been published under the generic title *Execution of special geotechnical works*, covering bored piles, displacement piles, micropiles, sheet pile walls, diaphragm walls, ground anchors, reinforced fill, soil nailing, grouting, jet grouting, deep mixing, deep vibration, and vertical drainage.

Chapter 16 introduces the two key geotechnical reports that are defined in Eurocode 7: the Ground Investigation Report (GIR) and the Geotechnical Design Report (GDR). Also in this chapter is a discussion of other reports that contribute to the GIR, such as drilling and sampling records, field investigation reports, and laboratory test reports. The chapter also includes a comparison with existing practice.

The final chapter, Epilogue, summarizes the impact that the EN Eurocodes – and Eurocode 7 in particular – will have on existing design practice.

The book concludes with three Appendices that give charts for slope stability design (Appendix 1), earth pressure coefficients (Appendix 2), and notes on the worked examples (Appendix 3).

Included at the end of a number of chapters in this book are a series of worked examples that illustrate the application of Eurocode 7 to common geotechnical design situations. To help you work through these examples, we have provided in Appendix 3 some notes explaining their format and notation. We recommend that you read the notes before studying the worked examples in detail.

Further information

We have established a website to support the book at:

www.decodingeurocode7.com

Here you will find further worked examples, discussion of topics that wouldn't fit within these pages, and corrections of any errors in the manuscript that we become aware of. For ongoing news and views about the Structural Eurocodes, you can also visit Andrew Bond's blog, which he has been writing since May 2006, at:

www.eurocode7.com

Notes and references

1. Institution of Structural Engineers (2004), *National Strategy for Implementation of the Structural Eurocodes*, Institution of Structural Engineers. Quotation taken from p. 9. The report can be downloaded from: www.istructe.org.uk/technical/files/eurocodes.pdf.

2. The text of the Construction Products Directive is published on the European Commission's website (http://ec.europa.eu). Follow the links to Enterprise > Industry Sectors > Construction > Directive 89/106/CE or go directly to the following web address: ec.europa.eu/enterprise/construction/internal/cpd/cpd_en.htm.

3. See Europa website: www.cenorm.be.

Chapter 1

The Structural Eurocodes

'The Eurocodes will become the Europe wide means of designing Civil and Structural engineering works and so ... they are of vital importance to both the design and Construction sectors of the Civil and Building Industries.'[1]

1.1   The Structural Eurocode programme

The Structural Eurocodes are a suite of ten standards for the design of buildings and civil engineering works, as illustrated in **Figure 1.1** and **Plate 2** (in the book's colour section). These standards are divided into fifty-eight parts and are accompanied by National Annexes issued by the various European countries that have introduced the Eurocodes into their design practice.

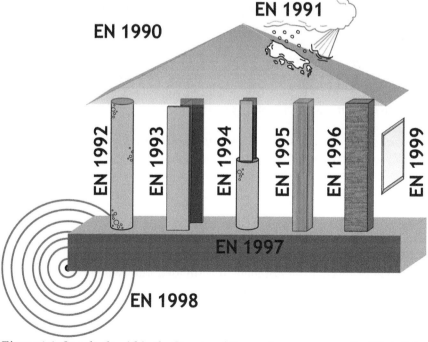

Figure 1.1. Standards within the Structural Eurocodes programme. See Plate 2 for colour version.

Outside and above the building in **Figure 1.1** are ENs 1990 and 1991, key aspects of which are discussed in **Chapter 2, Basis of structural design**.

Eurocode – Basis of structural design (EN 1990) establishes principles and requirements for the safety, serviceability, and durability of structures; describes the basis for their design and verification; and gives guidelines for related aspects of structural reliability. This code – occasionally and mistakenly referred to as 'Eurocode 0' – explains the fundamental engineering approach that underlies the entire suite of Eurocodes.[2]

Eurocode 1 – Actions on structures (EN 1991) gives design guidance and actions for the structural design of buildings and civil engineering works, including some geotechnical aspects. It is divided into four parts, with Part 1 further divided into seven sub-parts.[3]

The pillars of the building in **Figure 1.1** are the 'resistance' codes (ENs 1992 to 1996 and 1999), each of which provides detailed rules for the design of structures built in a particular material type.

Eurocode 2 – Design of concrete structures (EN 1992) covers the design of buildings and civil engineering works in plain, reinforced, and prestressed concrete. It is divided into three parts, with Part 1 further divided into two sub-parts.[4]

Eurocode 3 – Design of steel structures (EN 1993) covers the design of buildings and civil engineering works in steel. It is divided into six parts, with Parts 1, 3, and 4 further divided into twelve, two, and three sub-parts, respectively (for a total of twenty documents).[5]

Eurocode 4 – Design of composite steel and concrete structures (EN 1994) covers the design of composite structures and members for buildings and civil engineering works. It is divided into two parts, with Part 1 further divided into two sub-parts.[6]

Eurocode 5 – Design of timber structures (EN 1995) covers the design of buildings and civil engineering works in solid, sawn, planed, pole, or glue-laminated timber or in wood-based structural products or panels joined with adhesives or mechanical fasteners. It is divided into two parts, with Part 1 further divided into two sub-parts.[7]

Eurocode 6 – Design of masonry structures (EN 1996) covers the design of unreinforced, reinforced, prestressed, and confined masonry for buildings works. It is divided into three parts, with Part 1 further divided into two sub-parts.[8]

Eurocode 9 – Design of aluminium structures (EN 1999) covers the design of buildings and civil and structural engineering works in aluminium. It is published in one part which is divided into five sub-parts.[9]

Finally, supporting the building in **Figure 1.1** are Eurocodes 7 and 8.

Eurocode 7 – Geotechnical design (EN 1997) covers geotechnical aspects of the design of buildings and civil engineering works. It is divided into two parts (with no sub-parts).[10]

Eurocode 8 – Design of structures for earthquake resistance (EN 1998) covers the design and construction of buildings and civil engineering works in seismic regions. It is divided into six parts (with no sub-parts). EN 1998 provides additional rules for design that supplement those given in the resistance codes for concrete, steel, and other materials.[11]

Thus, there are a total of 58 parts and sub-parts that constitute the ten European standards in the EN Eurocode suite.

1.1.1   Links between the Eurocodes

Figure 1.2 and **Plate 3** (in the colour section) show the connections between the main parts of the Structural Eurocodes (in the style of the London Underground tube map). Only the main parts (not the sub-parts) of the Eurocodes are shown.

Along the 'Central Line' (in red on **Plate 3**) are the ten parts (Parts 1) giving general rules and rules for buildings. For example, EN 1992-1 provides general rules for concrete structures.

Around the 'Circle Line' (in yellow on **Plate 3**) are the six parts (Parts 2) giving rules for bridges. For example, EN 1992-2 provides design and detailing rules for concrete bridges.

The dashed line running north to south links ENs 1992-3 and 1993-3, which deal with liquid retaining and containing structures; the grey line running west to east beneath the Central Line links ENs 1991-4, 1993-4, and 1998-4, which cover silos and tanks; and the dashed lines in the south-east quadrant link ENs 1993-5, 1997-1 and -2, and 1998-5, which are concerned with foundations.

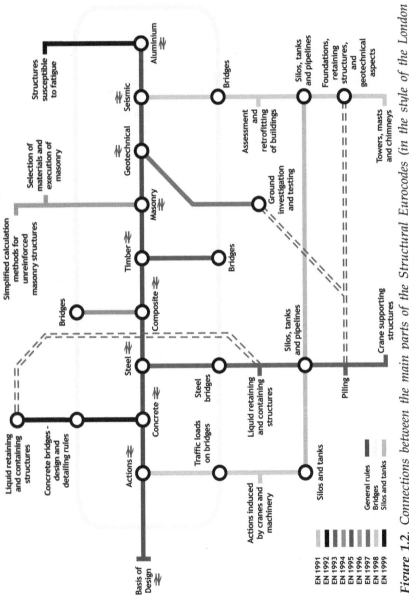

Figure 1.2. *Connections between the main parts of the Structural Eurocodes (in the style of the London Underground tube map)* ©*Transport for London). See* **Plate 3** *for colour version.*

Structures susceptible to fatigue

Aluminium

Selection of materials and execution of masonry

Bridges

Simplified calculation methods for unreinforced masonry structures

Seismic

Geotechnical

Silos, tanks and pipelines

Foundations, retaining structures, and geotechnical aspects

Masonry

Assessment and retrofitting of buildings

Ground investigation and testing

Towers, masts and chimneys

Timber

Bridges

Liquid retaining and containing structures

Composite

Bridges

Concrete bridges - design and detailing rules

Steel

Silos, tanks and pipelines

Crane supporting structures

Steel bridges

Piling

Concrete

Traffic loads on bridges

Liquid retaining and containing structures

Actions

Silos and tanks

Actions induced by cranes and machinery

Silos and tanks

Basis of Design

EN 1991
EN 1992
EN 1993
EN 1994
EN 1995
EN 1996
EN 1997
EN 1998
EN 1999

General rules
Bridges
Silos and tanks

1.1.2 Timetable for publication

The first generation of Eurocodes were developed in the 1980s, under the direction of the Commission of the European Community. The Commission's aim was to establish harmonized technical rules that would eventually replace national standards for the design of construction works across Europe.[12]

In 1989, the Commission and countries in the European Union (EU) and the European Free Trade Association (EFTA) transferred responsibility for the Structural Eurocodes to the European Committee for Standardization (CEN, see Section 1.3.2).

Between 1991 and 1999, trial versions of the Structural Eurocodes were published as pre-standards (ENVs or 'EuroNorm Vornorm'). The intended lifetime for ENVs is three years, during which time they can be used provisionally but without the status of a fully agreed European standard (EN or 'European Norm'). The experience gained during this period was used to modify the ENVs so they could be approved as ENs. Many of the pre-standard Eurocodes underwent significant revision before publication as fully-fledged Euro-norms.

As **Figure 1.3** shows, work on the final (EN) versions of the Structural Eurocodes began in July 1998 and was not completed until November 2006, when the final parts of Eurocode 9 were ratified by the relevant drafting committee (the 'date of ratification'). Each standard was translated into the three official CEN languages – English, French, and German – and then made available to the national standards bodies (see Section 1.3.3). The dates of availability of the various Eurocodes varied from April 2002 for EN 1990 to May 2007 for EN 1999-1-3.

The final step in the implementation of the Structural Eurocodes is the publication of each EN as a national standard in each country affiliated to CEN (see Sections 1.3.2 and 1.3.3). By the end of 2007, more than three quarters of the Eurocodes parts had appeared in some (but not all) countries and the final quarter are expected in the first half of 2008.

1.2 The wider landscape

Not only is Eurocode 7 an integral part of the EN Eurocodes, it also forms the hub of a large number of European and international standards covering site investigation, construction, and geotechnical testing.

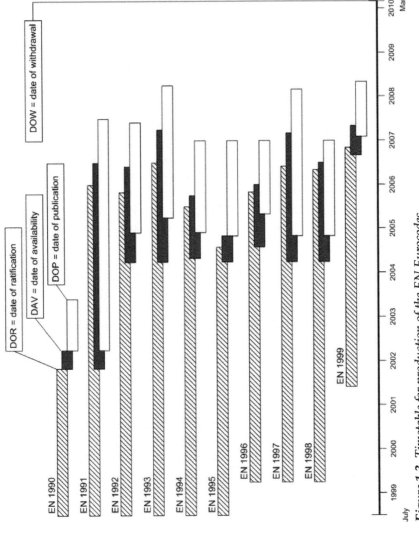

Figure 1.3. Timetable for production of the EN Eurocodes

Figure 1.4 and **Plate 4** (in the colour section) show the connections between Eurocode 7 and its associated standards, overlaid on the mainline railway map for Great Britain. Running east to west (in green on **Plate 4**) are the ten EN Eurocodes developed by CEN's TC (Technical Committee) 250, as discussed earlier in Section 1.1. (See Section 1.3.2 for discussion of CEN.)

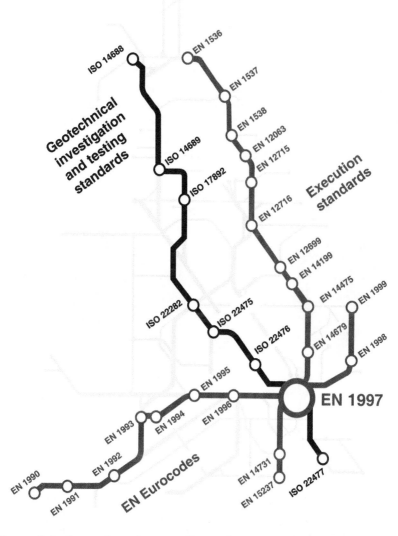

*Figure 1.4. Connections between Eurocode 7 and associated European and international standards (based on the National Rail schematic map, ©Association of Train Operating Companies). See **Plate 4** for colour version.*

Running north-west to south-east (in blue on **Plate 4**) are the seven geotechnical investigation and testing standards developed jointly by ISO's TC 182 and CEN's TC 341, which are discussed in detail in Chapter 4. (See Section 1.3.1 for discussion of ISO.)

Finally, travelling north to south along the east coast of Britain (in red on **Plate 4**) are the twelve execution standards developed by CEN's TC 288, which are discussed in detail in Chapter 15.

1.3   Standards organizations

1.3.1   International Standardization Organization (ISO)

The International Organization for Standardization (known as ISO, after the Greek word 'isos' meaning 'equal') was founded in 1947 to 'facilitate the international coordination and unification of industrial standards'.[13] ISO is a network of national standards bodies from 158 countries (comprising 103 member bodies, 46 correspondent members, and 9 subscriber members). **Figure 1.5** illustrates the current membership of ISO.

Figure 1.5. Members of ISO

Based in Geneva, ISO has almost 200 technical committees (TCs) which are broken down into approximately 500 subcommittees, 2000 working groups, and 60 ad hoc study groups. By the end of 2007, ISO had published over 16,500 international standards and standards-type documents, with 1,250 new standards being published each year. A quarter of ISO's standards are in the engineering technologies sector, with a further quarter covering materials technologies.[14]

'ISO standards avoid having to reinvent the wheel. They distil knowledge and make it available to all. In this way, they propagate new advances and transfer technology, making them a valuable source of knowledge.'[15]

ISO publishes standards on topics ranging from company organization, management, and quality to telecommunications, including audio and video engineering; from textile and leather technology to agriculture and food technology; from mining and minerals to construction materials; and finally from railway engineering, shipbuilding, and marine structures to building and civil engineering. These standards can be purchased directly from ISO, through its website www.iso.org.

1.3.2 European Committee for Standardization (CEN)

The European Committee for Standardization (known as CEN, after its French name Committé Européen de Normalisation) was founded in 1961 by the national standards bodies in the European Economic Community (EEC) and the European Free Trade Association (EFTA).[16]

Based in Brussels, CEN currently comprises thirty national members, seven associates (for example, the European Construction Industry Federation, FIEC), and two counsellors (representing the EEC and EFTA); four affiliates (mainly central and eastern European countries); and nine partners (countries outside Europe such as Australia, Egypt, and Russia), which have committed to implementing certain European standards as their own national standards. **Figure 1.6** and **Plate 1** (in the colour section) illustrate the current membership of CEN.

CEN has over 250 technical committees (TCs), numbered from TC 10 (covering lifts, escalators, and moving walks) to TC 353 (covering information and communication technologies for learning, education, and training). The technical committees that are active in the building and civil engineering sector include:
  TC 124 Timber structures
  TC 127 Fire safety in buildings
  TC 135 Execution of steel structures and aluminium structures
  TC 151 Construction equipment and building material machinery
  TC 189 Geosynthetics
  TC 250 Structural Eurocodes
  TC 288 Execution of special geotechnical works
  TC 341 Geotechnical investigation and testing

The committees of interest to us in this book are TCs 250, 288, and 341.

By the end of 2007, CEN had published nearly 13,000 European Standards. Approximately 16% of CEN's standards are in the building and civil engineering sector, with a further 14% covering materials. The only sectors which are larger are mechanical engineering and transport-and-packaging.

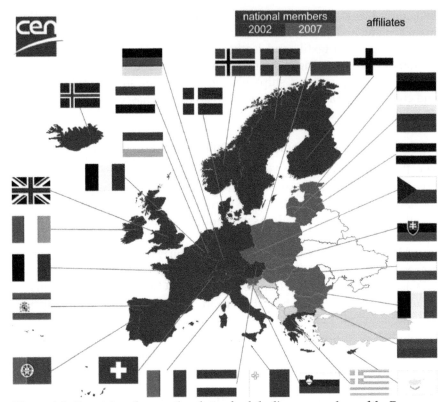

*Figure 1.6. Countries whose national standards bodies are members of the European Committee for Standardization. See **Plate 1** for colour version.*

CEN does not itself sell European Standards, but instead makes them available through its national members (such as the British Standards Institution, BSI).

1.3.3   National standards bodies (NSBs)

Each country in ISO and in CEN is represented by its national standards body (NSB), which is usually the organization responsible for setting standards in its country. NSBs play an important role in the implementation of the Eurocodes.

Each country's standards body is responsible for publishing European standards as national standards with the same legal standing. NSBs may translate European standards from the official CEN text (which is produced in English, French, and German), but they must not deviate from or alter any part of that text.

The NSB must retain the prefix EN in the national designation for the relevant European Standard. For example, EN 1990 becomes:

> BS EN 1990 in the UK
> NF EN 1990 in France
> DIN EN 1990 in Germany

The EN designation signifies that the standard's technical content is exactly the same in all countries throughout Europe.

Each country may add a National Title Page, a National Foreword, and a National Annex to each Eurocode part.[†] The National Annex provides the missing information in the Eurocode 'jigsaw', as illustrated in **Figure 1.7** and discussed below.

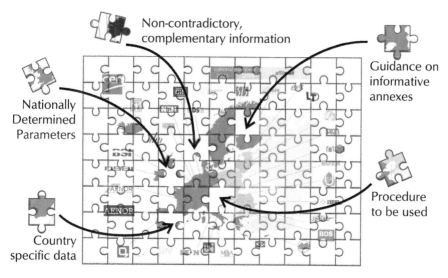

Figure 1.7. Information provided in the National Annex to an EN Eurocode

---

[†]National Annexes are only required for 57 of the 58 EN parts. The exception is EN 1998-3 for assessment and retrofitting of buildings.

Nationally Determined Parameters
Nationally Determined Parameters (NDPs) are parameters that are 'left open in the Eurocode for national choice'.[17] NDPs include partial factors, correlation factors, combination factors, model factors, and such like.

Country-specific data
Country-specific data refers to information of a geographical nature that only has relevance in a specific country — for example, charts showing wind pressure.

The procedure to be used where a choice is allowed
Some Eurocodes allow a choice of procedures where no single procedure could be agreed by the drafting committee. Eurocode 6 allows a choice between six classes of execution control and Eurocode 7 a choice between three different Design Approaches for the STR and GEO limit states (see Chapter 6).

Guidance on the informative annexes
A country's National Annex may offer guidance on the status of any informative annexes in the relevant Eurocode. Informative annexes may be 'promoted' to normative (i.e. mandatory) status, remain as purely informative, or be dismissed as unsuited for use in that country.

Reference to non-contradictory, complementary information
A National Annex can provide references to what is termed 'non-contradictory, complementary information' (NCCI), that is other design guidance that supports the relevant Eurocode without contradicting it. Where contradictions occur, Eurocode provisions take precedence.

1.3.4   Role of Eurocode 7 in UK practice

A significant challenge facing each national standards body over the coming years is how to accommodate the Structural Eurocodes into its existing national standards.

Figure 1.8 illustrates how Eurocode 7 interacts with various documents commonly used in UK geotechnical practice. These include so-called 'residual standards', i.e. existing British Standards that do not conflict with Eurocode 7; *de facto* standards such as Highways Agency and CIRIA documents; and 'Published Documents' (PDs), which are 'supporting documents produced by committee for information only [and] includes guidance, reports, and recommendations'.[18]

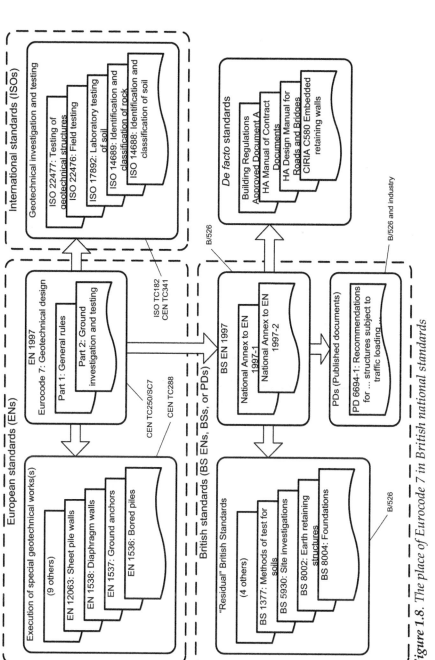

International standards (ISOs)

Geotechnical investigation and testing

- ISO 22477: Testing of geotechnical structures
- ISO 22476: Field testing
- ISO 17892: Laboratory testing of soil
- ISO 14689: Identification and classification of rock
- ISO 14688: Identification and classification of soil

ISO TC182
CEN TC341

European standards (ENs)

EN 1997
Eurocode 7: Geotechnical design
- Part 1: General rules
- Part 2: Ground investigation and testing

CEN TC250/SC7

CEN TC288

Execution of special geotechnical works(s)
- (9 others)
- EN 12063: Sheet pile walls
- EN 1538: Diaphragm walls
- EN 1537: Ground anchors
- EN 1536: Bored piles

De facto standards

- Building Regulations Approved Document A
- HA Manual of Contract Documents
- HA Design Manual for Roads and Bridges
- CIRIA C580 Embedded retaining walls

B/526

British standards (BS ENs, BSs, or PDs)

BS EN 1997
- National Annex to EN 1997-1
- National Annex to EN 1997-2

PDs (Published documents)
- PD 6694-1: Recommendations for ... structures subject to traffic loading ...

"Residual" British Standards
- (4 others)
- BS 1377: Methods of test for soils
- BS 5930: Site investigations
- BS 8002: Earth retaining structures
- BS 8004: Foundations

B/526

B/526 and industry

Figure 1.8. The place of Eurocode 7 in British national standards

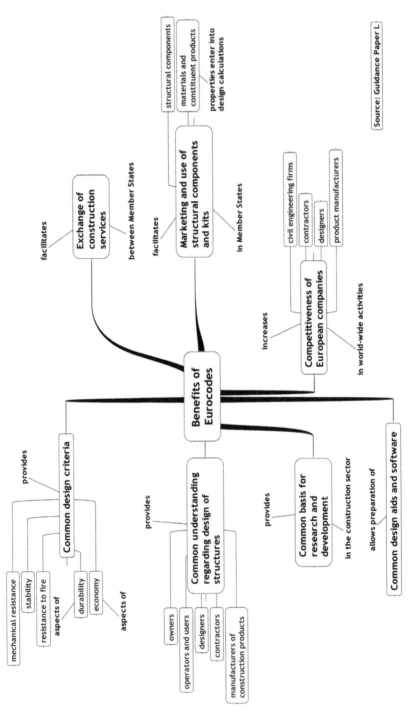

Figure 1.9. Benefits of the Eurocodes

1.4   Summary of key points

The importance of the Structural Eurocodes to design practice across Europe during the first part of the 21st century cannot be overstated.

Never before have design standards for all the major construction materials (steel, concrete, timber, masonry, and aluminium) been changed at virtually the same time. In addition, in some disciplines, such as geotechnical and masonry engineering, the introduction of the Eurocodes signals a move to limit state design from more traditional (e.g. allowable stress) philosophies.

Figure 1.9 summarizes the benefits that the Eurocodes will bring to civil engineering design. They will increase the competitiveness of European companies in their world-wide activities, by providing common design criteria, a common understanding regarding the design of structures and a common basis for research and development in the construction sector; and they will allow the preparation of common design aids and software. Furthermore, they will facilitate the exchange of construction services and the marketing and use of structural components and kits.[19]

1.5   Notes and references

1.   Quotation taken from the 'Introduction to Eurocodes' page on the European Commission's website (http://ec.europa.eu). Follow the links to European Commission > Enterprise > Industry Sectors > Construction > Internal Market > Eurocodes or go directly to: ec.europa.eu/enterprise/construction/internal/essreq/eurocodes/ eurointro_en.htm

2.   EN 1990: 2002, Eurocode − Basis of structural design, European Committee for Standardization, Brussels.

3.   EN 1991, Eurocode 1 − Actions on structures, European Committee for Standardization, Brussels.
     Part 1: General actions
          1-1: Densities, self-weight, imposed loads for buildings
          1-2: Actions on structures exposed to fire
          1-3: Snow loads
          1-4: Wind actions
          1-5: Thermal actions
          1-6: Actions during execution
          1-7: Accidental actions.
     Part 2: Traffic loads on bridges.
     Part 3: Actions induced by cranes and machinery.

Part 4: Silos and tanks.

4. EN 1992, Eurocode 2 — Design of concrete structures, European
   Committee for Standardization, Brussels.
   Part 1: General rules
       1-1: and rules for buildings
       1-2: — Structural fire design.
   Part 2: Concrete bridges — Design and detailing rules.
   Part 3: Liquid retaining and containment structures.

5. EN 1993, Eurocode 3 — Design of steel structures, European
   Committee for Standardization, Brussels.
   Part 1: General rules
       1-1: and rules for buildings
       1-2: — Structural fire design
       1-3: — Supplementary rules for cold-formed members and
             sheeting
       1-4: — Supplementary rules for stainless steels
       1-5: — Plated structural elements
       1-6: — Strength and stability of shell structures
       1-7: — Plated structures subject to out of plane loading
       1-8: — Design of joints
       1-9: — Fatigue
       1-10: — Material toughness and through-thickness properties
       1-11: — Design of structures with tension components
       1-12: — Additional rules for the extension of EN 1993 up to steel
             grades S 700.
   Part 2: Steel bridges.
   Part 3: Towers, masts and chimneys
       3-1: Towers and masts
       3-2: Chimneys.
   Part 4:
       4-1: Silos
       4-2: Tanks
       4-3: Pipelines.
   Part 5: Piling.
   Part 6: Crane supporting structures.

6. EN 1994, Eurocode 4 — Design of composite steel and concrete
   structures, European Committee for Standardization, Brussels.
   Part 1: General rules
       1-1: and rules for buildings
       1-2: — Structural fire design.

Part 2: General rules and rules for bridges.

7. EN 1995, Eurocode 5 – Design of timber structures, European
    Committee for Standardization, Brussels.
    Part 1: General
          1-1: Common rules and rules for buildings
          1-2: Structural fire design.
    Part 2: Bridges.

8. EN 1996, Eurocode 6 – Design of masonry structures, European
    Committee for Standardization, Brussels.
    Part 1: General rules
          1-1: for reinforced and unreinforced masonry structures
          1-2: – Structural fire design.
    Part 2: Design considerations, selection of materials and execution of
    masonry.
    Part 3: Simplified calculation methods for unreinforced masonry
    structures.

9. EN 1999, Eurocode 9 – Design of aluminium structures, European
    Committee for Standardization, Brussels.
    Part 1:
          1-1: General structural rules
          1-2: Structural fire design
          1-3: Structures susceptible to fatigue
          1-4: Cold-formed structural sheeting
          1-5: Shell structures.

10. EN 1997, Eurocode 7 – Geotechnical design, European Committee for
    Standardization, Brussels.
    Part 1: General rules
    Part 2: Ground investigation and testing.

11. EN 1998, Eurocode 8 – Design of structures for earthquake resistance,
    European Committee for Standardization, Brussels.
    Part 1: General rules, seismic actions and rules for buildings.
    Part 2: Bridges.
    Part 3: Assessment and retrofitting of buildings.
    Part 4: Silos, tanks and pipelines.
    Part 5: Foundations, retaining structures and geotechnical aspects.
    Part 6: Towers, masts and chimneys.

12. See EN 1990, ibid., Background to the Eurocode programme.

13. Information taken from the ISO website www.iso.org (see the section entitled 'About ISO').

14. Data taken from 'ISO in figures for the year 2006', available from www.iso.org.

15. Taken from 'ISO in brief' (2006), published by the ISO Central Secretariat, ISBN 92-67-10401-2.

16. Taken from the CEN website www.cen.eu (see section 'About us').

17. See the Foreword to EN 1990, ibid.

18. British Standards Institution (2005) *The BSI guide to standardization – Section 1: Working with British Standards*.

19. See Guidance Paper L, published by the European Commission on its website (http://ec.europa.eu).

Chapter 2

Basis of structural design

'EN 1990 ... is the head document of the Eurocode suite and describes the principles and requirements for safety, serviceability and durability of structures'[1]

2.1   Contents of the Eurocode

Eurocode – Basis of structural design[2] is divided into six sections and four annexes (A-D), as shown in **Figure 2.1**. In this diagram, the size of each segment of the pie is proportional to the number of paragraphs in the relevant section.

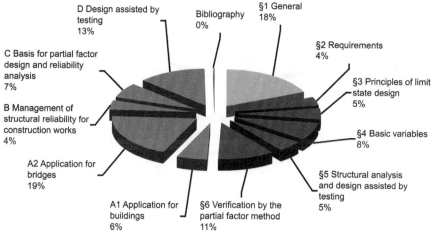

Figure 2.1. Contents of the Eurocode

EN 1990 describes the basis for the design and verification of buildings and civil engineering works, including geotechnical aspects, and gives guidance for assessing their structural reliability (see **Figure 2.2**). It covers the design of repairs and alterations to existing construction and assessing the impact of changes in use. Because of their special nature, some construction works (such as nuclear installations and dams) may need to be designed to provisions other than those given in EN 1990.

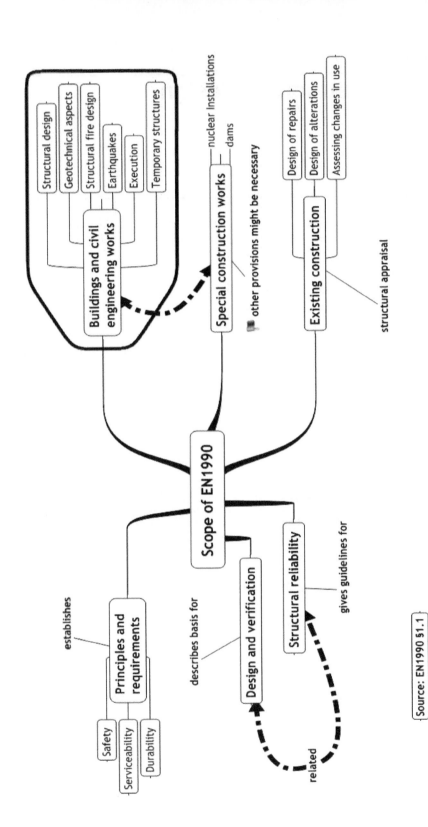

Figure 2.2. Scope of EN 1990, Basic of structural design

Source: EN1990 §1.1

2.2   Requirements

The basic requirements of a structure are to sustain all likely actions and influences, to remain fit for purpose, and to have adequate structural resistance, durability, and serviceability. These requirements must be met for the structure's entire design working life, including construction.

The structure must not suffer disproportionate damage owing to adverse events, such as explosions, impact, or human error. The events to be taken into account are those agreed with client and relevant authorities.

In addition, the design must avoid or limit potential damage by reducing, avoiding, or eliminating hazards. This can be achieved by tying structural members together, avoiding collapse without warning (e.g. by employing structural redundancy and providing ductility), and designing for the accidental removal of a structural member.

The *design working life* is the

> *assumed period for which a structure or part of it is to be used for its intended purpose with anticipated maintenance but without major repair being necessary.* [EN 1990 §1.5.2.8]

Figure 2.3 compares the design working life of various structures according to EN 1990 (dark lines) with the modifications made to these time periods by the UK National Annex to EN 1990 (lighter lines). The most significant change is the extension of Category 5 to 120 years – although this only really affects fatigue calculations.

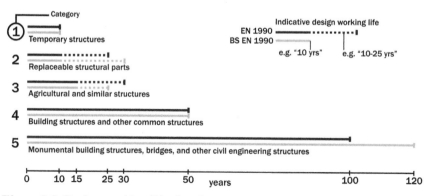

Figure 2.3. Design working life of various structures

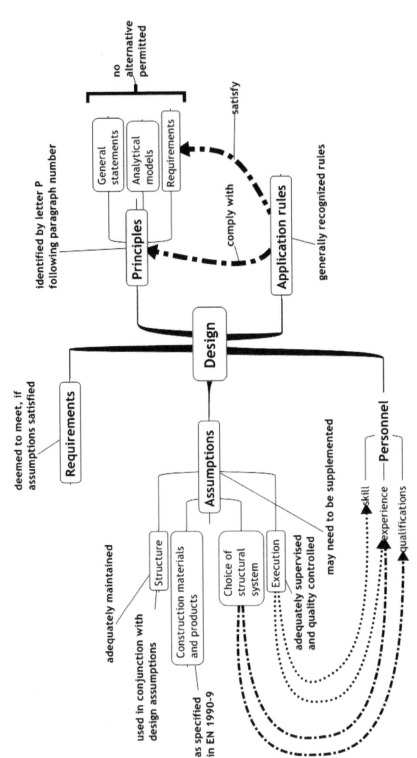

Figure 2.4. Design according to the Eurocode

2.3   Assumptions

EN 1990 makes important assumptions about the way structures are designed and executed (see **Figure 2.4**).

It is assumed that people with appropriate qualifications, skill, and experience will choose the structural system, design the structure, and construct the works. It is also assumed that construction will be adequately supervised and quality controlled; and the structure will be adequately maintained and used in accordance with the design assumptions.

Because they vary from country to country, EN 1990 gives no guidance as to what 'appropriate' qualifications are needed to perform these tasks. Likewise 'adequate' supervision and control is not further defined in the Eurocode.

2.4   Principles and Application Rules

A distinctive feature of the Structural Eurocodes is the separation of paragraphs into Principles and Application Rules (see **Figure 2.4**).

Design which employs the Principles and Application Rules is deemed to meet the requirements provided the assumptions given in EN 1990 to EN 1999 are satisfied.   *[EN 1990 §1.3(1)]*

Principles – identified by the letter 'P' after their paragraph numbers – are general statements and definitions that must be followed, requirements that must be met, and analytical models that must be used. The English verb that appears in Principles is 'shall'.   *[EN 1990 §1.4(2) & (3)]*

Application Rules – identified by the absence of a letter after their paragraph numbers – are generally recognized rules that comply with the Principles and satisfy their requirements. English verbs that appear in Application Rules include 'may', 'should', 'can', etc.   *[EN 1990 §1.4(4)]*

2.5   Principles of limit state design

The Structural Eurocodes are based on limit state principles, in which a distinction is made between ultimate and serviceability limit states.

Ultimate limit states are concerned with the safety of people and the structure. Examples of ultimate limit states include loss of equilibrium, excessive deformation, rupture, loss of stability, transformation of the structure into a mechanism, and fatigue.

Serviceability limit states are concerned with the functioning of the structure under normal use, the comfort of people, and the appearance of the construction works. Serviceability limit states may be reversible (e.g. deflection) or irreversible (e.g. yield).

Limit state design involves verifying that relevant limit states are not exceeded in any specified design situation (see Section 2.6). Verifications are performed using structural and load models, the details of which are established from three basic variables: actions, material properties, and geometrical data. Actions are classified according to their duration and combined in different proportions for each design situation.

Figure 2.5 illustrates the relationship between these various elements of limit state design.

2.6   Design situations

Design situations are conditions in which the structure finds itself at different moments in its working life.

In normal use, the structure is in a *persistent* situation; under temporary conditions, such as when it is being built or repaired, the structure is in a *transient* situation; under exceptional conditions, such as during a fire or explosion, the structure is in an *accidental* situation or (if caused by an earthquake) a *seismic* situation.   *[EN 1990 §3.2(2)P]*

Society is willing to accept that fires and explosions may lead to building damage – necessitating repair – whereas snow and wind should not. None of these events must lead to collapse. The Structural Eurocodes define partial factors for accidental and seismic situations (i.e. exceptional conditions which are unlikely to occur) that are typically 1.0. These factors are considerably lower than those specified for persistent and transient situations (conditions which are more likely to occur), typically 1.2–1.5. The development of different design situations helps to determine what level of reliability the design requires and what actions need to be considered as part of that design situation.

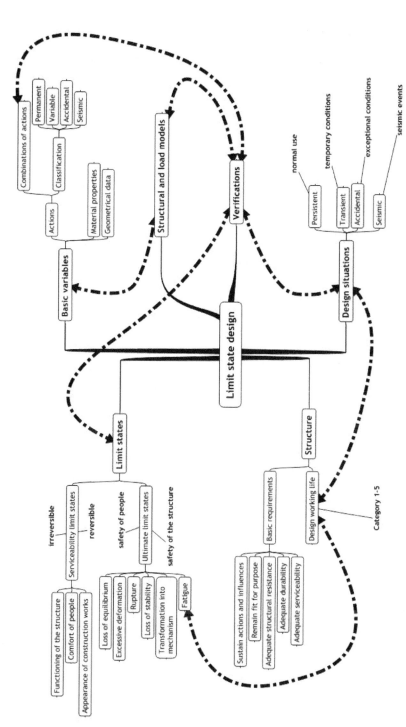

Figure 2.5. Overview of limit state design

2.7   Ultimate limit states

Ultimate limit states (ULSs) are concerned with the safety of people and the structure.   [EN 1990 §3.3(1)P]

EN 1990 identifies three ULSs that must be verified where relevant: loss of equilibrium (EQU); failure by excessive deformation, transformation into a mechanism, rupture, or loss of stability (STR); and failure caused by fatigue or other time-related effects (FAT). (Limit states GEO, UPL, and HYD, which are relevant to geotechnical design, are discussed in Chapters 6 and 7.) These three-letter acronyms are used throughout the Eurocodes as shorthand for the limit states, which, for structures, are defined more fully as follows.

2.7.1   Limit state EQU

Limit state EQU, dealing with static equilibrium, is defined as:

Loss of static equilibrium of the structure ... considered as a rigid body, where minor variations in the [actions or their distribution]... are significant, and the strengths of ... materials ... are generally not governing.
                                                  [EN 1990 §6.4.1(1)P(a)]

Limit state EQU does not occur when the destabilizing design effects of actions $E_{d,dst}$ are less than or equal to the stabilizing design effects $E_{d,stb}$:
$$E_{d,dst} \leq E_{d,stb}$$ [EN 1990 exp (6.7)]
as illustrated in **Figure 2.6**.

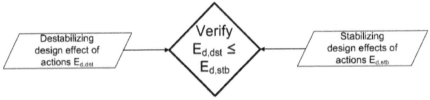

Figure 2.6. Verification of limit state EQU

For example, consider a motorway gantry (see **Figure 2.7**) subjected to a design horizontal wind load $P_d = 250kN$ acting at a height $h = 7.5m$ above the base of the gantry. The design destabilizing (i.e. overturning) moment about the toe of the structure $M_{Ed,dst}$ is given by:
$$M_{Ed,dst} = P_d \times h = 250kN \times 7.5m = 1875kNm$$

If the design self-weight of the gantry is $W_d = 1600kN$ and its base width B = 2.5m, then the design stabilizing (i.e. restoring) moment about the toe $M_{Ed,stb}$ is:

$$M_{Ed,stb} = W_d \times \frac{B}{2} = 1600kN \times \frac{2.5m}{2} = 2000kNm$$

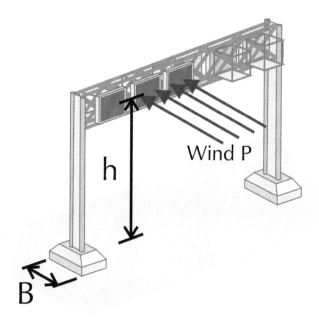

Figure 2.7. Motorway gantry subject to wind load

Limit state EQU is therefore avoided (in the direction of the wind), since:
$$M_{Ed,dst} = 1875kNm \le M_{Ed,stb} = 2000kNm$$

In this book, we define the 'utilization factor' for the EQU limit state as the ratio of the destabilizing and stabilizing effects:
$$\Lambda_{EQU} = \frac{E_{d,dst}}{E_{d,stb}}$$

For a structure to satisfy design requirements, it must have a utilization factor less than or equal to 100%. If Λ exceeds 100%, although it may not necessarily lose equilibrium, the structure is less reliable than required by the Eurocodes. For the motorway gantry of **Figure 2.7**:
$$\Lambda_{EQU} = \frac{M_{Ed,dst}}{M_{Ed,stb}} = \frac{1875kNm}{2000kNm} = 94\%$$

2.7.2   Limit state STR

Limit state STR, dealing with rupture or excessive deformation, is defined as:

> *Internal failure or excessive deformation of the structure ... where the strength of construction materials ... governs.*   *[EN 1990 §6.4.1(1)P(b)]*

To prevent limit state STR from occurring, design effects of actions E_d must be less than or equal to the corresponding design resistance R_d, i.e.:

$$E_d \le R_d$$ *[EN 1990 exp (6.8)]*

as illustrated in **Figure 2.8**.

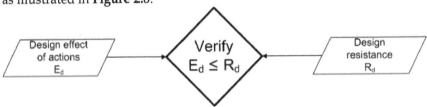

Figure 2.8. Verification of limit state STR

Returning to the example of **Figure 2.7**, the bending moments induced in the motorway gantry under its self-weight are shown in **Figure 2.9**. If the maximum design bending moment in the structure is $M_{Ed} = 500$ kNm and the minimum design bending resistance of the cross-section is $M_{Rd} = 600$ kNm, then limit state STR is avoided, since:

$$M_{Ed} = 500kNm \le M_{Rd} = 600kNm$$

In this book, we define the 'utilization factor' for the STR limit state as the ratio of the effect of actions to its corresponding resistance:

$$\Lambda_{STR} = \frac{E_d}{R_d}$$

For a structure to satisfy design requirements, it must have a utilization factor less than or equal to 100%. If Λ exceeds 100%, although it may not necessarily fail, the structure is less reliable than required by the Eurocodes. For the motorway gantry of **Figure 2.9**:

$$\Lambda_{STR} = \frac{M_{Ed}}{M_{Rd}} = \frac{500kNm}{600kNm} = 83\%$$

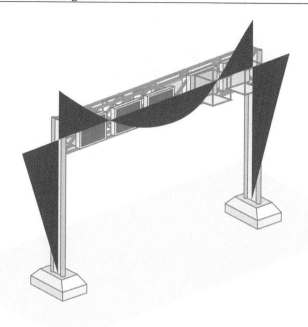

Figure 2.9. Bending moments acting in the motorway gantry of Figure 2.7

2.7.3   Limit state FAT

In materials science, fatigue is the progressive and localised structural damage that occurs when a material is subjected to cyclic loading. Fatigue is mainly relevant to road and rail bridges and tall slender structures subject to wind. The subject receives particular attention in Eurocodes 1 (actions[3]), 3 (steel structures[4]), and 9 (aluminium structures[5]) but is not mentioned in Eurocode 7.

2.8   Serviceability limit states

Serviceability limit states (SLSs) are concerned with the functioning of the structure, the comfort of people, and the appearance of the construction works.   *[EN 1990 §3.4(1)P]*

To prevent serviceability limit states from occurring, design effects of actions E_d – which in this instance are entities such as settlement, distortion, strains, etc. – must be less than or equal to the corresponding limiting value of that effect C_d, i.e.:

$$E_d \leq C_d$$ *[EN 1990 exp (6.13)]*

as illustrated in **Figure 2.10**.

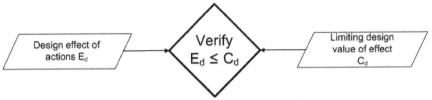

Figure 2.10. Verification of serviceability limit state (SLS)

Returning to the example of **Figure 2.7**, if the maximum settlement that the gantry can tolerate is s_{Cd} = 15 mm and the calculated settlement under the design actions is s_{Ed} = 12 mm, then the serviceability limit state is avoided, since:

$$s_{Ed} = 12mm \leq s_{Cd} = 15mm$$

In this book, we define the 'utilization factor' for the serviceability limit state as the ratio of the effect of actions to its corresponding limiting value:

$$\Lambda_{SLS} = \frac{E_d}{C_d}$$

For a structure to be serviceable, it must have a utilization factor less than or equal to 100%. For the motorway gantry of **Figure 2.7**:

$$\Lambda_{SLS} = \frac{s_{Ed}}{s_{Cd}} = \frac{12mm}{15mm} = 80\%$$

2.9   Actions, combinations, and effects

The use of the word *action* to describe loads (and other entities that act like loads) reminds us of Newton's Third Law of Motion:

'To every action there is always opposed an equal reaction.'[6]

In Eurocode terms, the 'reaction' is known as an *effect*. That is:

$$action \quad = \quad cause$$
$$\downarrow \quad \searrow \quad \downarrow$$
$$reaction \quad = \quad effect$$

The following sub-sections explain the way in which the Structural Eurocodes define actions, combinations of actions, and the effects that arise from them.

2.9.1   Actions

A *direct* action is a set of forces applied to a structure and an *indirect* action a set of imposed deformations or accelerations. A generic action is denoted by the symbol F in the Structural Eurocodes.   *[EN 1990 §1.5.3.1]*

Actions are classified according to their variation in time, as defined in the table below and illustrated in **Figure 2.11**. Permanent ('gravity') actions are denoted G; variable ('live') actions Q; pre-stresses P; and accidental actions A.   *[EN 1990 §1.5.3.3-5 & 4.1.1(1)P]*

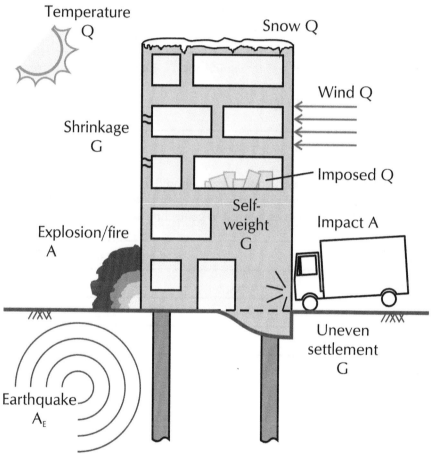

Figure 2.11. *A structure subject to permanent, variable, and accidental actions*

Actions are introduced into design as *characteristic* values (F_k), which may be mean, upper, lower, or nominal values.

Action		Duration	Variation with time	Examples
Permanent	G	Likely to act throughout given reference period	Negligible or monotonic up to a limit value	Self-weight of structures, fixed equipment and road surfacing, water thrust,* shrinkage, uneven settlement
Variable	Q		Neither negligible nor monotonic	Imposed loads on building floors, beams and roofs, wind,† snow,† traffic
Accidental	A	Usually short (unlikely to occur during the design working life)	Significant magnitude	Explosions, impact from vehicles,* seismic* (symbol A_E)

*may be variable, †may be accidental

The self-weight of construction works should be calculated from their nominal dimensions and characteristic weight densities. The table below gives values for materials of interest to geotechnical engineers.

Material		Weight density, γ (kN/m³)
Concrete	normal (plain)	24
	reinforced	25
Steel		77.0–78.5
Dry sand	used as infill for bridges	15.0–16.0
Loose gravel ballast		15.0–16.0
Hardcore		18.5–19.5
Crushed slag		13.5–14.5
Packed stone rubble		20.5–21.5
Puddle clay		18.5–19.5

Values taken from EN 1991-1-1 Annex A

2.9.2 Combinations of actions

Representative actions (F_{rep}) are obtained by assembling suitable combinations of characteristic values (F_k), following the rules given in ENs 1990 and 1991. The representative value of a single generic action is given by:

$$F_{rep} = \psi F_k$$

where ψ is a combination factor, less than or equal to 1.0.

The combination factor ψ is omitted for permanent actions, i.e. a representative permanent action ($G_{rep,j}$) is equal to its characteristic value ($G_{k,j}$). The total design permanent action (G_d) is then obtained from the sum of the representative values multiplied by their appropriate partial factors γ_G (see Section 2.13.1). Hence:

$$G_d = \sum_j \left(\gamma_{G,j} \times G_{k,j} \right) = \sum_j \left(\gamma_{G,sup,j} G_{k,sup,j} \right) + \sum_j \left(\gamma_{G,inf,j} G_{k,inf,j} \right)$$

where the subscripts $_{sup}$ and $_{inf}$ denote unfavourable ('superior') and favourable ('inferior') actions respectively.

In persistent and transient situations, the value of ψ is typically equal to 1.0 for the 'leading' variable action ($Q_{k,1}$), but is less than one ($\psi = \psi_0 < 1.0$) for all 'accompanying' variable actions ($Q_{k,i}$). The total design variable action (Q_d) is then obtained from the sum of the representative values multiplied by their appropriate partial factors γ_Q (see Section 2.13.1). Hence:

$$Q_d = \gamma_{Q,1} \times 1.0 \times Q_{k,1} + \sum_{i>1} \gamma_{Q,1} \times \psi_{0,i} \times Q_{k,i}$$

(Note that only unfavourable variable actions are considered – favourable variable actions are ignored.)

Hence the total design action F_d in persistent and transient design situations is given by:

$$F_d = \sum_{j>1} \gamma_{G,j} G_{k,j} + \gamma_{Q,1} Q_{k,1} + \sum_{i>1} \gamma_{Q,i} \psi_{0,i} Q_{k,i}$$

Alternatively, EN 1990 allows F_d to be calculated as the larger of:

$$F_d = \sum_j \gamma_{G,j} G_{k,j} + \gamma_{Q,1} \psi_{0,1} Q_{k,1} + \sum_{i>1} \gamma_{Q,i} \psi_{0,i} Q_{k,i}$$

and

$$F_d = \xi \sum_j \gamma_{G,sup,j} G_{k,sup,j} + \sum_j \gamma_{G,inf,j} G_{k,inf,j} + \gamma_{Q,1} Q_{k,1} + \sum_{i>1} \gamma_{Q,i} \psi_{0,i} Q_{k,i}$$

where ξ is a reduction factor (a.k.a. 'distribution coefficient') applied to unfavourable permanent actions $G_{k,sup,j}$ only.

An example may help to illustrate the use of these equations in practice. Imagine that the motorway gantry of **Figure 2.7** is subject to imposed load

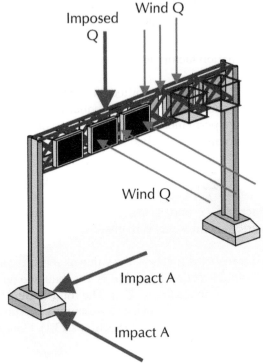

Figure 2.12. Motorway gantry subject to multiple actions

and wind in the vertical and horizontal directions, as shown in **Figure 2.12** and summarized in the table below.

In the table, Combination 1 assumes that the imposed load is the leading variable action (and hence $\psi = 1.0$) and wind is accompanying (with $\psi_0 = 0.6$); Combination 2 assumes that wind is leading ($\psi = 1.0$) and the imposed load is accompanying ($\psi_0 = 0.7$). The design actions that result are given in the row labelled 'Total': Combination 1 gives a (slightly) higher vertical action, but Combination 2 a higher horizontal value. The gantry must be designed to withstand both combinations.

Combination of actions for persistent and transient design situations

Action (type*)	F_k (kN)		γ_F	F_d (kN)					
				Combination 1			Combination 2		
	V†	H‡		ψ	V	H	ψ	V	H
Self-weight (G)	140	0	1.35	-	189	0	-	189	0
Imposed (Q)	100	0	1.5	1.0	150	0	0.7	105	0
Wind (Q)	20	200	1.5	0.6	18	180	1.0	30	300
Total					357	180		324	300

*G = permanent, Q = variable
†V = vertical, ‡H = horizontal (parallel to the motorway)

In accidental situations, the combination factors used in the equations given above are reduced slightly to account for the lower probability of occurrence of that situation. The total design action F_d is given by:

$$F_d = \sum_j \gamma_{G,j} G_{k,j} + \gamma_{A,1} A_{k,1} + \gamma_{Q,1} \psi_{1,1} Q_{k,1} + \sum_{i>1} \gamma_{Q,i} \psi_{2,i} Q_{k,i}$$

where $A_{k,1}$ is the accidental action, ψ_1 is applied (instead of 1.0) to the main variable action, and ψ_2 is applied (instead of ψ_0) to the accompanying variable actions. In accidental situations, the values of the partial factors γ_G, γ_Q, and γ_A are normally 1.0 (see Section 2.13.1).

Extending the example of the motorway gantry to include an accidental action (in this case a traffic impact load shown in **Figure 2.12**) gives revised values for the design actions, as set out in the table below.

In this revised table, Combination 3 assumes that the imposed load is the main variable action (with $\psi_1 = 0.7$) and wind is accompanying ($\psi_2 = 0$); Combination 4 assumes that wind is the main variable action ($\psi_1 = 0$) and the imposed load is accompanying ($\psi_2 = 0.6$). The design actions that result for Combination 3 are more onerous than for Combination 4, but less onerous than Combinations 1 and 2.

Combination of actions for accidental design situations

Action (type*)	F_k (kN)		γ_F	F_d (kN)					
				Combination 3			Combination 4		
	V†	H‡		ψ	V	H	ψ	V	H
Self-weight (G)	140	0	1.0	-	140	0	-	140	0
Imposed (Q)	100	0	1.0	0.7	70	0	0.6	60	0
Wind (Q)	20	200	1.0	0	0	0	0	0	0
Impact (A)	50	80	1.0	1.0	50	80	1.0	50	80
Total					260	80		250	80

*G = permanent, Q = variable, A = accidental
†V = vertical, ‡H = horizontal (parallel to the motorway)

The table below summarizes the values of ψ that must be used for different combinations of actions at ultimate and serviceability limit states.

Combination		Combination factor for characteristic actions				
		$\Sigma G_{k,i}$	P†	$Q_{k,1}$‡	$\Sigma Q_{k,j}$††	A_k or $A_{E,k}$‡‡
Ultimate	Persistent	1	1	1	ψ_0	-
	Transient					
	Accidental	1	1	ψ_1 or ψ_2	ψ_2	1
	Seismic	1	1	-	ψ_0	1
Serviceability	Characteristic	1	1	1	ψ_2	-
	Frequent	1	1	ψ_1	ψ_2	-
	Quasi-permanent	1	1	-	ψ_2	-

Actions: †P = pre-stress; ‡Q_1 = leading variable; ††Q_j = accompanying variable; ‡‡A = accidental; A_E = seismic

Values of ψ are given for imposed loads in EN 1991-1-1, for snow loads in EN 1991-1-3, for wind loads in EN 1990 Annex A1, and for temperature loads in EN 1991-1-5. For buildings, values for ψ_0 typically range between 0.5 and 0.7, for ψ_1 between 0.2 and 0.7, and for ψ_2 between 0.3 and 0.6.

In summary, representative actions arise from considering various combinations of characteristic actions for the particular design situation being verified:

characteristic actions F_k → combinations = representative actions F_{rep}

2.9.3   Effects of actions

In structural engineering, effects of actions are a function of the actions applied to a structure and that structure's dimensions, but not of material strength, i.e.:

$$E_d = E\{F_{d,i}, a_{d,j}\}$$

where the notation E{...} denotes that the design effect E_d depends solely on design actions $F_{d,i}$ and design dimensions $a_{d,j}$. This holds true for linear elastic analysis of structures, but not for plastic analysis. An example may help to illustrate the ideas behind this equation.

Figure 2.13 shows a simply-supported concrete beam subject to an imposed load F at midspan. The presence of the load (an action) results in the beam deflecting and internal stresses occurring in its cross-section (effects of the action). The bending moment at the beam's midspan is given by the equation:

$$M = \frac{FL}{4} + \frac{\rho_c bdL^2}{8}$$

where L, b, and d are the length, breadth, and depth of the beam, respectively, and ρ_c is the weight density of concrete. The second term in this equation represents the bending moment caused by the beam's self-weight.

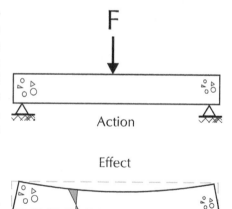

Figure 2.13. Actions and effects for a simply-supported beam

In Eurocode terms: $M_{Ed} = function\{F, \rho_c, b, d, L\} = E\{F_{d,i}, a_{d,j}\}$

2.10   Material properties and resistance

2.10.1   Resistance

The resistance of a structural member is defined as the:

capacity of a member or component, or cross-section of a member or component of a structure, to withstand actions without mechanical failure
[EN 1990 §1.5.2.15]

In structural engineering, resistance is a function of the structure's material strengths and its dimensions, but not of the magnitude of any actions applied to the structure, i.e.:

$$R_d = R\{X_{d,i}, a_{d,j}\}$$

where the notation R{...} denotes that the design resistance (R_d) depends solely on design material strengths $(X_{d,i})$ and design dimensions $(a_{d,j})$. This holds true for non-prestressed beams, but not for pre-stressed beams and columns.

Returning to the example of **Figure 2.13** may help to illustrate the ideas behind this equation. **Figure 2.14** shows a cross-section through the simply-supported beam. The load on the beam causes it to bend, with

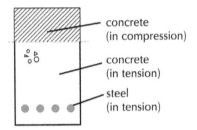

concrete
(in compression)

concrete
(in tension)

steel
(in tension)

Material Properties

Resistance

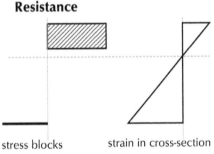

stress blocks   strain in cross-section

Figure 2.14. Material properties and resistance of a concrete beam

the top part going into compression and the bottom part into tension (which is carried by reinforcing steel).

Assuming plane sections remain plane allows the beam's bending resistance to be determined as:

$$M = \frac{A_s f_y d}{4}\left(1 - \frac{f_y A_s}{2 f_c bd}\right)$$

where A_s is the area of steel reinforcement; b and d are the breadth and depth of the beam, respectively; f_y is the yield strength of steel; and f_c is the compressive strength of concrete.

In Eurocode terms: $M_{Rd} = function\{f_y, f_c, A_s, b, d\} = E\{X_{d,i}, a_{d,j}\}$

2.10.2   Material properties

Material properties are introduced into design as *characteristic* values (X_k), with a prescribed probability of not being exceeded in a hypothetically unlimited test series.   *[EN 1990 §1.5.4.1 and 4.2(1)]*

The results of tests on man-made materials, such as concrete and steel, often follow a normal (a.k.a. 'Gaussian') probability density function (PDF), as shown in **Figure 2.15**. The normal distribution arises when a physical property depends on the combination of a large number of individual, random effects.[7] It is encountered frequently in nature and is one of the most important PDFs in the field of statistics. The horizontal axis of **Figure 2.15** measures deviation of the variable X from its mean value and the vertical axis gives the probability density of X.

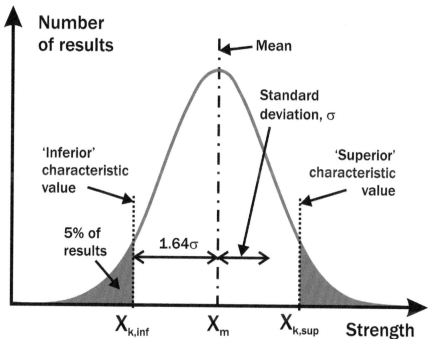

Figure 2.15. Normal strength distribution for man-made materials (e.g. concrete)

The *lower* (or '*inferior*') *characteristic value* $X_{k,inf}$ is defined as the value of X below which 5% of all results are expected to occur. In other words, there is a 95% probability that X will be greater than $X_{k,inf}$. This value is used in situations where overestimating the magnitude of a material property may be unsafe. For example, the lower characteristic value should be used to check that a material is strong enough to carry a particular load. Since strength-checks are a very common design requirement, the qualifier 'lower/inferior' is normally dropped from the description and symbol, leaving X_k as 'the characteristic value'.

Likewise, the *upper* (or *'superior'*) *characteristic value* $X_{k,sup}$ is defined as the value of X above which 5% of all results are expected to occur. There is a 95% probability that X will be lower than $X_{k,sup}$. Although used less frequently than its lower counterpart, the upper characteristic value is important in situations where underestimating the magnitude of a material property may be unsafe. For example, since the force acting on a retaining wall depends on the weight density of the soil behind it, the wall should be designed to withstand an upper estimate of that weight density. Since the upper characteristic value is not used as often as the lower one, it should always be qualified as 'upper/superior' and denoted $X_{k,sup}$.

With prior knowledge of the standard deviation: in situations where the standard deviation σ_X (or variance σ_X^2) of the population is known from prior knowledge (and hence does not need to be determined from the sample), the statistical definitions of $X_{k,inf}$ and $X_{k,sup}$ are:

$$\left.\begin{array}{c} X_{k,inf} \\ X_{k,sup} \end{array}\right\} = \mu_X \mp \kappa_N \sigma_X = \mu_X(1 \mp \kappa_N \delta_X)$$

where μ_X is the mean value of X, the standard deviation of the population, δ_X its coefficient of variation (COV), and κ_N is a statistical coefficient that depends on the size of the population N.

These terms are defined as follows:

$$\mu_X = \frac{\sum\limits_{i=1}^{N} X_i}{N}, \ \sigma_X = \sqrt{\frac{\sum\limits_{i=1}^{N}(X_i - \mu_X)^2}{N}}, \text{ and } \delta_X = \frac{\sigma_X}{\mu_X}$$

The statistical coefficient κ_N is given by:

$$\kappa_N = t_\infty^{95\%}\sqrt{\frac{1}{N}+1} = 1.645 \times \sqrt{\frac{1}{N}+1}$$

where $t_\infty^{95\%}$ is Student's t-value[8] for infinite degrees of freedom at a confidence level of 95% (see **Figure 2.16**).

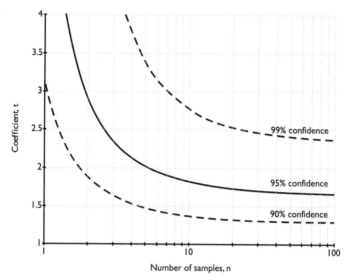

Figure 2.16. *Student's t-values for 90, 95, and 99% confidence levels*

Numerical values of κ_N are given in **Figure 2.17** by the lower line, labelled 'variance known', and vary between ≈ 1.645 for a population size of a hundred and ≈ 2 for a population size of two.

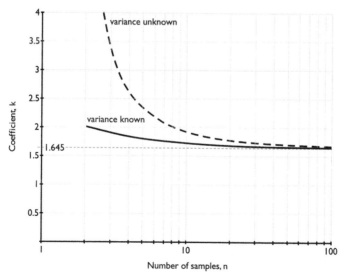

Figure 2.17. *Statistical coefficients for determining the 5% fractile with 95% confidence*

To illustrate the use of this diagram, imagine that a series of forty concrete strength tests measured a mean compressive strength $f_c = 38.6$ MPa. Based on prior experience, the standard deviation of the concrete's strength is assumed to be $\sigma_{fc} = 4.56$ MPa. With $\kappa_{40} = 1.665$ calculated from the equation for κ_N (or taken from **Figure 2.17**), the concrete's lower characteristic strength is then given by:

$$f_{ck} = \overline{f}_c - \kappa_N \sigma_{fc} = 38.6 - 1.665 \times 4.56 = 31.0 \text{ MPa}$$

and its upper characteristic strength by:

$$f_{ck} = \overline{f}_c + \kappa_N \sigma_{fc} = 38.6 + 1.665 \times 4.56 = 46.2 \text{ MPa}$$

With no prior knowledge of the standard deviation: in situations where the variance of the population is unknown ab initio (and hence must be determined from the sample), the statistical definitions of $X_{k,inf}$ and $X_{k,sup}$ change to:

$$\left.\begin{array}{c} X_{k,inf} \\ X_{k,sup} \end{array}\right\} = m_X \mp k_n s_X = m_X (1 \mp k_n V_X)$$

where m_X is the mean value of X, s_X the sample's standard deviation, V_X its coefficient of variation, and k_n is a statistical coefficient that depends on the sample size 'n'. (The use of Latin symbols distinguishes this equation from its variance-known counterpart, which uses the Greek symbols μ, σ, and δ.)

These terms are defined as follows:

$$m_X = \frac{\sum\limits_{i=1}^{n} X_i}{n}, \, s_X = \sqrt{\frac{\sum\limits_{i=1}^{n} (X_i - m_X)^2}{n-1}}, \text{ and } V_X = \frac{s_X}{m_X}$$

Note: the divisor in the expression for standard deviation is (n - 1) not n.

The statistical coefficient k_n is given by:

$$k_n = t_{n-1}^{95\%} \sqrt{\frac{1}{n} + 1}$$

where $t_{n-1}^{95\%}$ is Student's t-value for (n - 1) degrees of freedom at a confidence level of 95% (see **Figure 2.16**).

Numerical values of k_n are given in **Figure 2.17** by the upper line, labelled 'variance unknown', and vary between ≈ 1.645 for a sample size of a hundred and > 3 for a sample size of three. An important feature of this curve is the rapid rise in k_n that occurs as the sample size decreases below about ten. This will have serious implications for the use of statistics for geotechnical problems, as discussed fully in Chapter 5.

Returning to the example of concrete strength tests, in this instance we must calculate the standard deviation s_{fc} from the test results. Imagine that this calculation produced a value of $s_{fc} = 4.56$ MPa identical to the previously assumed value for σ_{fc}. With $k_{40} = 1.706$ calculated from the equation for k_n (or taken from **Figure 2.17**), the concrete's lower characteristic strength is now given by:

$$f_{ck} = \overline{f_c} - k_n s_{fc} = 38.6 - 1.706 \times 4.56 = 30.8 \text{ MPa}$$

and its upper characteristic strength by:

$$f_{ck} = \overline{f_c} + k_n s_{fc} = 38.6 + 1.706 \times 4.56 = 46.4 \text{ MPa}$$

The greater uncertainty in the standard deviation results in slightly more pessimistic values of f_{ck}.

2.11 Geometrical data

Geometrical data are also introduced into design as *characteristic* values (a_k), which may be taken as the nominal values (a_{nom}) given on the design drawings. This has the considerable virtue of not over-complicating design calculations. *[EN 1990 §4.3(1)P]&[EN 1990 §4.3(2)]*

Nominal dimensions → characteristic dimensions

Although geometrical data are random variables, their level of variability is generally small compared with those of actions and material properties. It is therefore usual to treat geometrical data as known values, taking the characteristic value as the nominal value provided on the design drawings. Examples include the size of structural members; the height and inclination of slopes; and footing formation depths.

For structures, nominal dimensions are used for member sizes (and their variation is catered for by partial material factors) but imperfections are considered in the structure's geometry, e.g. the lean of piers. These imperfections are based on tolerances given in the construction specification.

When selecting nominal dimensions, it is usual to choose conservative values relevant to the limit state being analysed taking into account imperfections in setting out, workmanship, and other construction issues. Where a significant variation in dimension is likely, either initially (due to loading, production, setting-out, or erection) or over time (due to loading or various chemical and physical causes),[9] the design dimension should reflect this likely variation (see Section 2.13.3).

2.12   Structural analysis and design by testing

Calculations must be performed using appropriate structural models with relevant variables. The models must be based on established engineering theory and practice and, if necessary, verified experimentally.

[EN 1990 §5.1.1(1)P & (3)]

Design by testing may be used in place of structural analysis, provided it achieves the level of reliability required for the relevant design situation. When a limited number of test results are available, the design must take account of their statistical uncertainty.   *[EN 1990 §5.1.1(2)P]*

2.13   Verification by the partial factor method

2.13.1   Partial factors on actions

Representative actions (F_{rep}) are converted into design values (F_d) by multiplying by an appropriate partial factor (γ_F):

$$F_d = \gamma_F F_{rep}$$

where γ_F takes account of uncertainties in the magnitude of the action, model uncertainties, and dimensional variations. For unfavourable actions, $\gamma_F \geq 1$, whereas for favourable actions $\gamma_F \leq 1$ and the previous equation is qualified as follows (see **Figure 2.18**):

$$F_{d, fav} = \gamma_{F, fav} F_{rep, fav}$$

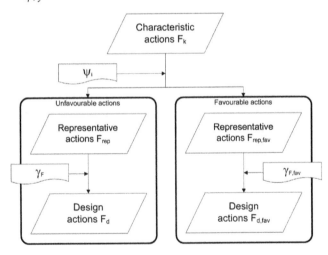

Figure 2.18. *Factors on actions*

Values of γ_F and $\gamma_{F,fav}$ for persistent and transient design situations are given in EN 1990 and vary between 0.9 and 1.5, depending on the duration of the action (see the table below).

For limit state EQU, the factors on permanent actions are fairly small, with $\gamma_G = 1.1$ increasing the adverse effects of unfavourable permanent actions and $\gamma_{G,fav} = 0.9$ reducing the beneficial effects of favourable ones. Unfavourable variable actions are increased by 50% ($\gamma_Q = 1.5$), but favourable variable actions are ignored ($\gamma_{Q,fav} = 0$).

For limit state STR, significant factors are applied to unfavourable permanent and variable actions ($\gamma_G = 1.35$ and $\gamma_Q = 1.5$); favourable permanent and unfavourable accidental actions are taken at their representative values ($\gamma_{G,fav} = \gamma_A = 1.0$); and favourable variable and accidental actions are ignored ($\gamma_{Q,fav} = \gamma_{A,fav} = 0$). In geotechnical design, these partial factors also depend on the Design Approach adopted (see Chapter 6).

EN 1990 also allows limit states EQU and STR to be checked simultaneously by the application of partial factors from the last column of the table.

Action and symbol			Partial factor	EQU	STR	EQU+ STR
Permanent	G	Unfavourable	γ_G	1.1	1.35	1.35*
		Favourable	$\gamma_{G,fav}$	0.9	1.0	1.15*
Variable	Q	Unfavourable	γ_Q	1.5	1.5	1.5
		Favourable	$\gamma_{Q,fav}$	0	0	0

*Provided $\gamma_G = \gamma_{G,fav} = 1.0$ is not more onerous

For accidental design situations, values of γ_G, $\gamma_{G,fav}$, γ_Q, and γ_A are 1.0 and the values of $\gamma_{Q,fav}$ and $\gamma_{A,fav}$ are 0.

2.13.2   Partial factors on material properties

Characteristic material properties (X_k) are converted into design values (X_d) by dividing by an appropriate partial factor (γ_M):

$$X_d = \frac{X_k}{\gamma_M}$$

(see **Figure 2.19**) where γ_M takes account of uncertainties in the magnitude of the material property, model uncertainties, and dimensional variations.

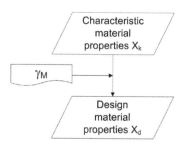

Figure 2.19. Factors on material properties

Values of γ_M for persistent and transient design situations are given in the resistance codes (ENs 1992 to 1999) and vary between 1.0 and 1.5, depending on the material type. In geotechnical design, these partial factors also depend on the Design Approach adopted for limit state STR (see Chapter 6).

For accidental design situations, values of γ_M are 1.0.

2.13.3 Tolerances on geometry

Nominal geometrical dimensions (a_{nom}) are converted into design values (a_d) by adding or subtracting an appropriate safety margin or tolerance (Δa):

$$a_d = a_{nom} \pm \Delta a$$

(see **Figure 2.20**) where Δa takes account of uncertainties in the magnitude of the geometrical dimension.

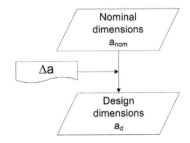

Figure 2.20. Tolerances on geometrical parameters

Values of Δa for persistent and transient design situations are given in the resistance codes (ENs 1992 to 1999) and depend on the sensitivity of the design situation to geometrical imperfections.

Values of Δa for accidental design situations are 0.

2.13.4 Verification of strength for limit state STR

Figure 2.21 combines the previous flow diagrams for limit state STR. Actions (characteristic → representative → design) are shown down the left-hand channel; geometrical parameters (nominal → design) down the central channel; and material properties (characteristic → design) down the right-hand channel. The introduction of combination factors ψ, partial factors γ, and tolerances Δa at appropriate points in this process is shown.

Structural analysis uses design actions and design dimensions to determine design effects of actions; stress analysis employs design material properties and design dimensions to determine design resistance.

Limit state STR is verified when design effects are less than or equal to the design resistance.

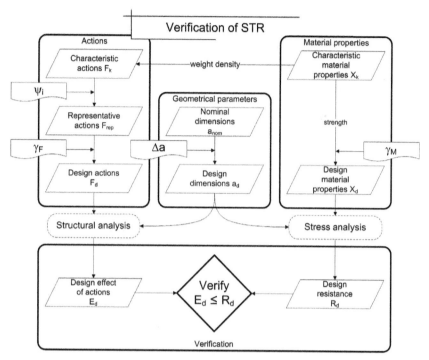

Figure 2.21. Overview of verification of strength

2.13.5   Verification of stability for limit state EQU

Figure 2.22 combines the earlier flow diagrams for limit state EQU. Destabilizing actions (characteristic → representative → design) are on the left; geometrical parameters (nominal → design) in the centre; and stabilizing actions (representative → design) on the right. The introduction of combination factors ψ, partial factors γ, and tolerances Δa at appropriate points in this process is shown.

Structural analysis uses design actions (destabilizing and stabilizing) and design dimensions to determine destabilizing and stabilizing design effects of actions.

Limit state EQU is verified when destabilizing design effects are less than or equal to stabilizing design effects.

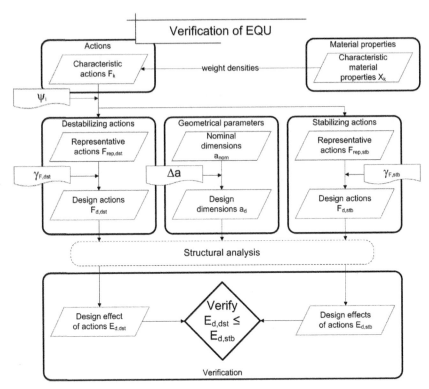

Figure 2.22. *Overview of verification of stability*

2.14   Summary of key points

> *'structural design is an iterative process of applying engineering mechanics and past experience to create a functional, economic, and, most importantly, safe structure for the public to enjoy'[10]*

The Eurocodes – and in particular, EN 1990 – provide a comprehensive and cohesive framework for ensuring the safety of structures. The engineering concepts that are embodied in them have been used in engineering practice for decades and will be familiar to most structural engineers.

The impact of the Structural Eurocodes may be summarized as:

> *'same principles, different rules'[11]*

whereas Eurocode 7 – as we will discover in later chapters – adopts:

> *'same rules, different principles'*

2.15   Worked examples

The worked examples in this chapter look at a shear wall under combined loading (Example 2.1); combination of actions on a pile group supporting an elevated bridge deck (Example 2.2); and the statistical determination of characteristic strength from the results of concrete cylinder tests (Example 2.3).

Specific parts of the calculations are marked ❶, ❷, ❸, etc., where the numbers refer to the notes that accompany each example.

2.15.1   Shear wall under combined loading

Example 2.1 looks at combinations of actions for the foundation shown in **Figure 2.23**.[12] The footing carries imposed loads from the superstructure and a horizontal force and moment from wind.

Notes on Example 2.1

❶ The combination factors for variable actions that are given in EN 1991 depend on the source of loading and the type of structure.

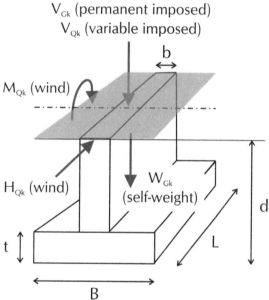

V_{Gk} (permanent imposed)
V_{Qk} (variable imposed)
b
M_{Qk} (wind)
H_{Qk} (wind)
W_{Gk} (self-weight)
d
L
t
B

Figure 2.23. Shear wall subject to vertical and horizontal forces and moment

❷ In Combination 1, the imposed action is leading ($\psi = 1$) and the wind is ignored ($\psi_0 = 0$).

❸ As wind is not included, there is no moment applied and the bearing pressure beneath the base is constant ($\Delta q = 0$).

❹ In Combination 2, the imposed action is leading ($\psi = 1$) and wind is accompanying ($\psi_0 = 0.5$).

❺ As wind is now included, the moment from it causes a variable bearing pressure beneath the base ($q_{av} \pm \Delta q / 2$).

Example 2.1
Shear wall under combined loading
Combination of actions

Design situation

Consider a b = 500mm thick shear wall that is resting on a rectangular footing founded at a depth d = 2m. The footing is B = 2m wide, L = 8m long, and t = 500mm thick. The characteristic weight density of the backfill on top of the footing is $\gamma_k = 16.9\dfrac{kN}{m^3}$ and of unreinforced concrete is

$\gamma_{ck} = 24\dfrac{kN}{m^3}$ (as per EN 1991-1-1). The shear wall is subject to characteristic imposed vertical actions $V_{Gk} = 2000kN$ (permanent) and $V_{Qk} = 1600kN$ (variable) from the superstructure. In addition, wind applies a characteristic variable moment $M_{Qk} = 1200kNm$ and a characteristic horizontal force

Self-weight of foundation (characteristic actions)

Weight of concrete base (permanent) $W_{Gk_1} = \gamma_{ck} \times B \times L \times t = 192\ kN$

Weight of concrete wall (permanent) $W_{Gk_2} = \gamma_{ck} \times b \times L \times (d - t) = 144\ kN$

Weight of backfill (permanent) $W_{Gk_3} = \gamma_k \times (B - b) \times L \times (d - t) = 304.2\ kN$

Total self-weight of foundation $W_{Gk} = \sum W_{Gk} = 640.2\ kN$

Average pressure on foundation due to self-weight alone $\dfrac{W_{Gk}}{B \times L} = 40\ kPa$

Combination factors on variable actions/action effects ❶

Imposed loads in buildings, Category B: office areas: $\psi_{0,i} = 0.7$

Wind loads on buildings, all cases (from BS EN 1990): $\psi_{0,w} = 0.5$

Partial factors on actions/action effects

Unfavourable permanent actions $\gamma_G = 1.35$

Unfavourable variable actions $\gamma_Q = 1.5$

Combination 1 (leading variable = imposed, accompanying = none) ❷

Total permanent vertical action $G_k = W_{Gk} + V_{Gk} = 2640\,kN$

Design vertical action $V_d = \gamma_G \times \left(W_{Gk} + V_{Gk} \right) + \gamma_Q \times 1.0 \times V_{Qk} = 5964\,kN$

Design horizontal action $H_d = \gamma_Q \times \psi_{0,w} \times 0kN = 0\,kN$

Design moment $M_d = \gamma_Q \times \psi_{0,w} \times 0kNm = 0\,kNm$

Average bearing pressure $q_{d,av_1} = \dfrac{V_d}{B \times L} = 372.8\,kPa$ ❸

Pressure change across base $\Delta q_{d_1} = \dfrac{12 M_d}{B \times L^2} = 0\,kPa$

Combination 2 (leading variable = imposed, accompanying = wind) ❹

Permanent vertical action is unchanged

Design vertical action $V_d = \gamma_G \times \left(W_{Gk} + V_{Gk} \right) + \gamma_Q \times 1.0 \times V_{Qk} = 5964\,kN$

Design horizontal action $H_d = \gamma_Q \times \psi_{0,w} \times H_{Qk} = 187.5\,kN$

Design moment $M_d = \gamma_Q \times \psi_{0,w} \times \left(M_{Qk} + H_{Qk} \times d \right) = 1275\,kNm$

Average bearing pressure $q_{d,av_2} = \dfrac{V_d}{B \times L} = 372.8\,kPa$

Pressure change across base $\Delta q_{d_2} = \dfrac{12 M_d}{B \times L^2} = 119.5\,kPa$ ❺

Combination 3 (leading variable = wind, accompanying = none) ❻

Permanent vertical action is unchanged

Design vertical action $V_d = \gamma_G \times \left(W_{Gk} + V_{Gk} \right) + \gamma_Q \times 0kN = 3564\,kN$

Design horizontal action $H_d = \gamma_Q \times 1.0 \times H_{Qk} = 375\,kN$

Design moment $M_d = \gamma_Q \times 1.0 \times \left(M_{Qk} + H_{Qk} \times d \right) = 2550\,kNm$

Average bearing pressure $q_{d,av_3} = \dfrac{V_d}{B \times L} = 222.8\,kPa$

Pressure change across base $\Delta q_{d_3} = \dfrac{12 M_d}{B \times L^2} = 239.1\,kPa$ ❼

<u>*Combination 4 (leading variable = wind, accompanying = imposed)*</u> **⑧**

Permanent vertical action is unchanged

Design vertical action $V_d = \gamma_G \times \left(W_{Gk} + V_{Gk}\right) + \gamma_Q \times \psi_{0,i} \times V_{Qk} = 5244\,kN$

Design horizontal action $H_d = \gamma_Q \times 1.0 \times H_{Qk} = 375\,kN$

Design moment $M_d = \gamma_Q \times 1.0 \times \left(M_{Qk} + H_{Qk} \times d\right) = 2550\,kNm$

Average bearing pressure $q_{d,av_4} = \dfrac{V_d}{B \times L} = 327.8\,kPa$

Pressure change across base $\Delta q_{d_4} = \dfrac{12M_d}{B \times L^2} = 239.1\,kPa$ **⑨**

❻ In Combination 3, wind is leading ($\psi = 1$) and the imposed action is ignored ($\psi_0 = 0$).

❼ As wind is included at its full characteristic value, the moment causes greater variation than in Combination 2.

❽ In Combination 4, wind is leading ($\psi = 1$) and the imposed action is accompanying ($\psi_0 = 0.7$).

❾ The variation in bearing pressure due to wind is the same as in Combination 3.

2.15.2 Elevated bridge deck

Example 2.2 looks at combinations of actions on a pile foundation beneath the elevated bridge deck shown in **Figure 2.24**.[13]

Permanent actions on the foundation include the self-weight of the bridge deck, the pier head, and the pier itself; any superimposed weight; and the

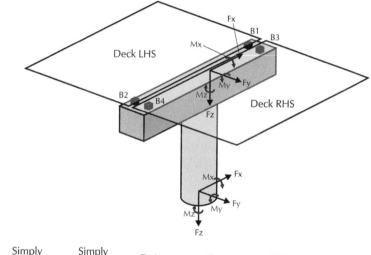

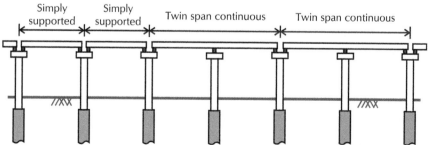

Figure 2.24. Typical viaduct substructure (based on the Dubai Metro)

vertical force due to settlement. The pre-stress in the deck produces an uplift on the foundation. Variable actions include temperature effects, wind on the deck and pier, and traffic actions. Possible accidental actions are from vehicle impact and seismic events.

The tables that follow give the characteristic actions from each of these sources, in the longitudinal (x), transverse (y), and vertical (z) directions defined on **Figure 2.24**. For simplicity, moments have been ignored.

Separate tables are shown for different design situations: persistent and transient at the ultimate limit state (ULS); accidental ULS; seismic ULS; and characteristic at the serviceability limit state (SLS).

The tables give the combination factor ψ and partial factor γ_F for each action, together with their relevant values taken from Eurocode 1.

Finally, the design actions in the x-, y-, and z-directions are calculated from the equation:
$$F_d = F_k \times \psi \times \gamma_F$$

The sum of the F_x, F_y, and F_z components are given at the bottom of the each table.

Notes on Example 2.2

❶ Self-weight is a permanent action and hence ψ is omitted.

❷ The pre-stress attracts a combination factor ψ = 1.0 and (because it produces a favourable effect) a partial factor $\gamma_{P,fav}$ = 1.0 also.

❸ This is the leading variable action in this combination, so ψ = 1.0.

❹ This is an accompanying variable action that, in this combination, uses ψ = ψ_0.

❺ One of the rules for bridges is that wind and temperature actions do not have to be considered together. Hence this action is ignored.

❻ This action does not occur in this particular combination.

❼ For accidental design situations, all partial factors γ_F = 1.0.

Example 2.2 Combination of actions for ULS persistent and transient design situation (leading variable = traffic, accompanying = wind)

Action (G = permanent, Q = variable, A = accidental)	Characteristic F_k (kN)			Combination/partial factors				Design actions F_d (kN)		
	x	y	z	ψ		γ_F		x	y	z
Self-weight (G) ❶ Bridge deck			6764	–	1	γ_G	1.35	0	0	9131
Pier head			1048	–	1	γ_G	1.35	0	0	1415
Pier			852	–	1	γ_G	1.35	0	0	1150
Imposed (G) ❶ Superimposed			2596	–	1	γ_G	1.35	0	0	3505
Settlement			36	–	1	γ_G	1.2	0	0	43
Pre-stress (P) ❷			-136	–	1	$\gamma_{P,fav}$	1	0	0	-136
Variable (Q) Temperature ❺			404	X	X	X	X	0	0	0
Wind ❹	241	486	284	ψ_0	0.6	γ_Q	1.5	217	437	256
Wind ❹	19	19	19	ψ_0	0.6	γ_Q	1.5	17	17	0
Traffic ❾	365	100	954	Lead	1	γ_Q	1.35	493	135	1288
Accidental (A) Vehicle impact ❻	500	1000		X	X	X	X	0	0	0
Seismic ❻	2357	2146		X	X	X	X	0	0	0
Total								727	590	16652

Example 2.2 (cont.) Combination of actions for ULS accidental design situation (leading variable = impact, accompanying = traffic & temperature)

Action (G = permanent, Q = variable, A = accidental)	Characteristic F_k (kN)			Combination/partial factors				Design actions F_d (kN)		
	x	y	z	ψ		γ_F ❼		x	y	z
Self-weight (G) ❶										
Bridge deck			6764	-	1	γ_G	1.0	0	0	6764
Pier head			1048	-	1	γ_G	1.0	0	0	1048
Pier			852	-	1	γ_G	1.0	0	0	852
Imposed (G) ❹										
Superimposed			2596	-	1	γ_G	1.0	0	0	2596
Settlement			36	-	1	γ_G	1.0	0	0	36
Pre-stress (P) ❷			-136	-	1	$\gamma_{P,fav}$	1.0	0	0	-136
Variable (Q)										
Temperature ❽			404	ψ_2	0.5	γ_Q	1.0	0	0	202
Wind ❾	241	486	284	X	X	X	X	0	0	0
Wind ❾	19	19	19	X	X	X	X	0	0	0
Traffic ❽	365	100	954	ψ_2	0.0	γ_Q	1.0	0	0	0
Accidental (A)										
Vehicle impact ❸	500	1000	1000	Lead	1.0	γ_A	1.0	500	1000	0
Seismic ❻			0	X	X	γ_A	X	0	0	0
Total	2357	2146						500	1000	11362

Example 2.2 (cont.) Combination of actions for ULS seismic design situation (leading variable = seismic, accompanying = traffic & temperature)

Action (G = permanent, Q = variable, A = accidental)		Characteristic F_k (kN)			Combination/partial factors				Design actions F_d (kN)		
		x	y	z	Ψ		γ_F		x	y	z
Self-weight (G) ❶	Bridge deck			6764	-	1	γ_G	1	0	0	6764
	Pier head			1048	-	1	γ_G	1	0	0	1048
	Pier			852	-	1	γ_G	1	0	0	852
Imposed (G) ❶	Superimposed			2596	-	1	γ_G	1	0	0	2596
	Settlement			36	-	1	γ_G	1	0	0	36
Pre-stress (P) ❷				-136	-	1	$\gamma_{P,fav}$	1	0	0	-136
Variable (Q)	Temperature ❽			404	Ψ_2	0.5	γ_Q	1	0	0	202
	Wind ❾	241	486	284	X	X	X	X	0	0	0
	Wind ❾	19	19		X	X	X	X	0	0	0
	Traffic ❽	365	100	954	Ψ_2	0	γ_Q	1	0	0	0
Accidental (A)	Vehicle impact ❻	500	1000		X	X	X	X	0	0	0
	Seismic ❸	2357	2146		Lead	1	γ_A	1	2357	2146	0
Total		2357	2146						2357	2146	11362

Example 2.2 (cont.) Combination of actions for SLS characteristic design situation (leading variable = traffic, accompanying = wind)

Action (G = permanent, Q = variable, A = accidental)		Characteristic F_k (kN)			Combination/partial factors				Design actions F_d (kN)		
		x	y	z	ψ	ψ	γ_F ⓾		x	y	z
Self-weight (G) ⓫	Bridge deck			6764	-	1	γ_G	1	0	0	6764
	Pier head			1048	-	1	γ_G	1	0	0	1048
	Pier			852	-	1	γ_G	1	0	0	852
Imposed (G) ❶	Superimposed			2596	-	1	γ_G	1	0	0	2596
	Settlement			36	-	1	γ_G	1	0	0	36
Pre-stress (P) ❷				-136	-	1	$\gamma_{P,fav}$	1	0	0	-136
Variable (Q)	Temperature ❺			404	X	X	X	X	0	0	0
	Wind ❹	241	486	284	ψ_0	0.6	γ_Q	1	145	292	170
	Wind ❹	19	19		ψ_0	0.6	γ_Q	1	11	11	0
	Traffic ❸	365	100	954	Lead	1	γ_Q	1	365	100	954
Accidental (A)	Vehicle impact ❻	500	1000		X	X	X	X	0	0	0
	Seismic ❼	2357	2146		X	X	X	X	0	0	0
Total									521	403	12284

➑ This is an accompanying variable action that, in this combination, uses ψ = ψ₂.

➒ Since, for bridges, wind and temperature actions do not have to be considered together, we have to make a choice of which to include. In this combination, the ψ_2 value for wind action is zero and so it is more onerous to take temperature and ignore wind.

➓ For serviceability limit state design situations, all partial factors $\gamma_F = 1.0$.

2.15.3 Concrete cylinder tests

Example 2.3 considers the results of twenty-five crushing tests on concrete cylinders, a histogram of which is shown in **Figure 2.25**. It is assumed that the probability density function shown is a reasonable representation of the histogram.

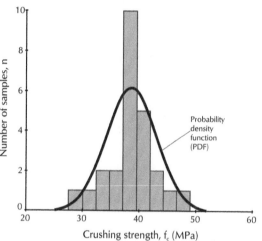

Figure 2.25. Results of compressive strength tests on concrete cylinders

Notes on Example 2.3

➊ The data is taken from *Probability concepts in engineering.*[14]

➋ The denominator in this equation is (n – 1) because we are calculating the standard deviation of a sample of the population, rather than of the population itself (in which case we would divide by n).

➌ This equation is easier to use in hand calculations than the previous one (and gives identical answers).

➍ Physical properties are inherently variable. By applying a minimum value here, we guard against the possibility of using an unrealistically small value of V_X in the subsequent calculations. If the coefficient of variation had been less than 0.1, then we would have needed to assume a higher standard deviation, given by:

$$s_X = 0.1 \times m_X$$

Example 2.3
Concrete cylinder tests
Determination of characteristic cylinder strength

Results
Consider the results of a series of concrete cylinder tests. The measured crushing strengths of the concrete were: 38.6, 36.5, 27.6, 30.3, 37.9, 39.3, 41.4, 38.6, 49.0, 32.4, 37.9, 40.7, 44.1, 40.0, 46.2, 37.2, 34.5, 40.0, 42.7, 38.6, 39.3, 40.7, 37.2, 35.2, and 39.3 MPa. **❶**

Statistical analysis of data
The number of test results is $n = 25$

The sum and mean are $\sum X = 965.2$ MPa and $m_X = \dfrac{\sum X}{n} = 38.6$ MPa

The standard deviation is $s_X = \sqrt{\dfrac{\sum (X - m_X)^2}{n - 1}} = 4.57$ MPa **❷**

which can also be calculated using $s_X = \sqrt{\dfrac{\sum X^2 - n m_X^2}{n - 1}} = 4.57$ MPa **❸**

The coefficient of variation is $V_X = \dfrac{s_X}{m_X} = 0.118$

Minimum coefficient of variation is 0.1, hence:
$$V_X = \max(V_X, 0.1) = 0.118$$ **❹**

With no prior knowledge of concrete strength (variance unknown)
Student's t-value for 95% confidence limit with (n - 1) degrees of freedom is
$$t_{95}(n - 1) = 1.711$$ **❺**

Hence statistical coefficient is $k_n = t_{95}(n - 1) \sqrt{\dfrac{1}{n} + 1} = 1.745$ **❻**

Characteristic concrete strength is then $X_k = m_X - k_n s_X = 30.6$ MPa **❼**

With prior knowledge of concrete strength (variance known)

Assume coeff. of variation is $\delta_X = 0.1$ and mean value $\mu_X = m_X = 38.6$ MPa

Standard deviation is $\sigma_X = \delta_X \times \mu_X = 3.86$ MPa **❽**

Student's t-value for 95% confidence limit with n degrees of freedom is

$t_{95}(n) = 1.708$ **❾**

Hence statistical coefficient is $\kappa_n = t_{95}(n)\sqrt{\dfrac{1}{n} + 1} = 1.742$

Characteristic concrete strength is then $X_k = \mu_X - \kappa_n \sigma_X = 31.9$ MPa **❿**

❺ Student's t-values are given in **Figure 2.16**.

❻ The statistical coefficient k_n is based on $(n - 1)$ degrees of freedom, the 'lost' degree of freedom being the standard deviation of the sample, which we have had to calculate.

❼ This is the 'inferior' characteristic value, which is the one most often needed in calculations of resistance.

❽ The standard deviation is based on an assumed coefficient of variation equal to 0.1.

❾ The statistical coefficient κ_n is based on n degrees of freedom.

❿ The characteristic strength is slightly larger than that calculated at ❼, since both the standard deviation and the statistical coefficient are smaller than before.

2.16   Notes and references

1. Gulvanessian, H., Calgaro, J.-A., and Holický, M. (2002) *Designer's guide to EN 1990, Eurocode: Basis of structural design*, Thomas Telford Publishing.

2. EN 1990: 2002, Eurocode — Basis of structural design, European Committee for Standardization, Brussels.

3. EN 1991, Eurocode 1 — Actions on structures, European Committee for Standardization, Brussels.
    Parts 1–4: General actions — Wind actions.
    Part 2: Traffic loads on bridges.
    Part 3: Actions induced by cranes and machinery.

4. EN 1993, Eurocode 3 — Design of steel structures, European Committee for Standardization, Brussels.
    Parts 1–9: General rules — Fatigue.

5. EN 1999, Eurocode 9 — Design of aluminium structures, European Committee for Standardization, Brussels.
    Parts 1–3: Structures susceptible to fatigue.

6. Isaac Newton (1687) *Principia Mathematica*: Laws of Motion 3 (translated from the Latin *'Actioni contrarium semper et aequalem esse reactionem'* by Andrew Motte, 1729).

7. This follows from the central limit theorem, 'one of the most important theorems in probability theory'. See p. 168 of Ang, A. H-S., and Tang, W. H. (2006) *Probability concepts in engineering: emphasis on applications in civil and environmental engineering*, John Wiley and Sons Ltd.

8. BS 2846-4: 1976 Guide to statistical interpretation of data - Part 4: Techniques of estimation and tests relating to means and variances, British Standards Institution.

9. Gulvanessian, ibid., p59.

10. From the Wikipedia article on 'Structural design' downloaded January 2008 (see en.wikipedia.org/wiki/Structural_design).

11. Chris Hendy, pers. comm. (2007).

12. The original version of this example appeared in: Curtin, W.G., Shaw, G., Parkinson, G.I., and Golding, J.M. (1994) *Structural foundation designer's manual*, Blackwell Science.

13. Information for this example was kindly provided by Chris Hendy and Claire Seward of Atkins, Epsom (pers. comm, 2007).

14. Ang and Tang, ibid., Example 6.1, pp. 250–251. Values converted from ksi to MPa.

Chapter 3

General rules for geotechnical design

'... within the UK, the extent to which geotechnical design has been codified [is] much less than in other sectors. ...the introduction of EN 1997 (Geotechnical design) will represent a marked change in UK practice. ...the needs of geotechnical designers ... to adapt ... will be significant.'[1]

3.1 Scope of Eurocode 7 Part 1

Eurocode 7 – Geotechnical design, Part 1 – general rules[2] is divided into twelve sections and nine annexes, as shown in **Figure 3.1** and **Plate 5** of this book's

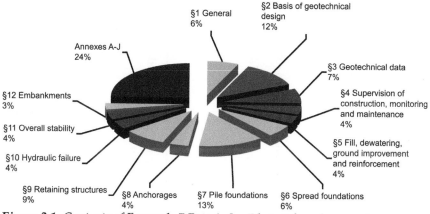

Figure 3.1. Contents of Eurocode 7 Part 1. See Plate 5 for colour version.

colour section. In this diagram, the size of each segment of the pie is proportional to the number of paragraphs in the relevant section. Part 1 provides a general framework for geotechnical design; definition of ground parameters; characteristic and design values; general rules for site investigation; rules for the design of the main types of geotechnical structures; and some assumptions on execution procedures.

The table below shows both the common (white background) and different (grey) sub-section headings in Sections 6–9 and 11–12 of EN 1997-1.

Section and sub-section headings employed in EN 1997-1

	Common† and §11–12	§6 Spread foundations	§8 Anchorages	§7 Pile foundations	§9 Retaining structures
§x.1	General				
§x.2	Limit states				
§x.3	Actions, geometrical data, and design situations (or variations on this title)				
§x.4	Design methods and construction considerations (or variations on this title)				
§x.5	Ultimate limit state design			Pile load tests	Determination of earth pressures
§x.6	Serviceability limit state design			Axially loaded piles	Water pressures
§x.7	Supervision and monitoring	Foundations on rock; additional design considerations	Suitability tests	Transversely loaded piles	Ultimate limit state design
§x.8	-	Structural design of spread foundations	Acceptance tests	Structural design of piles	Serviceability limit state design
§x.9	-	Supervision and monitoring (or variations on this title)			(missing)

†Common headings apply to §6–9, except where replaced by specific headings shown

Sections 6–9 and 11–12 of EN 1997-1, dealing with the rules for geotechnical structures, each present the information in a different manner reflecting the authorship of these sections. This difference in authorship has led to inconsistencies in the sub-section headings for each geotechnical structure and to the associated level of detail. For example, §6.5 for spread foundations provides a relatively short section on ultimate limit state design with no illustrations, whereas this topic is covered in detail in §9.7 for retaining structures with a large number of diagrams.

The order of the sections is not ideal. The topics covered by Sections 10 and 11 can apply to any site, regardless of the type of foundation that is going to be built. Section 10 deals with problems relating to the flow of water through soils and rocks (i.e. particulars of limit states HYD and UPL). Section 11 deals primarily with slope stability, an issue relevant to other geotechnical structures, such as retaining walls and footings. In our opinion, Sections 10 and 11 would have been better placed before Section 6, 'Spread foundations'.

3.2   Design requirements

3.2.1   Commitment to limit state design

Perhaps the most significant requirement of Eurocode 7 is the following commitment to limit state design:

> *For each geotechnical design situation it shall be verified that no relevant limit state ... is exceeded.*   [EN 1997-1 §2.1(1)P]

For many geotechnical engineers across Europe, this represents a major change in design philosophy, away from the traditional allowable (a.k.a. permissible) stress design involving a single, lumped factor of safety.

Traditional geotechnical design, using lumped factors of safety, has proved satisfactory over many decades and much experience has been built on such methods. However, the use of a single factor to account for all uncertainties in the analysis – although convenient – does not provide a proper control of different levels of uncertainty in various parts of the calculation. A limit state approach forces designers to think more rigorously about possible modes of failure and those parts of the calculation process where there is most uncertainty. This should lead to more rational levels of reliability for the whole structure. The partial factors in Eurocode 7 have been chosen to give similar designs to those obtained using lumped factors – thereby ensuring that the wealth of previous experience is not lost by the introduction of a radically different design methodology.

Limit state philosophy has been used for many years in the design of structures made of steel, concrete, and timber. Where these structures met the ground was, in the past, a source of analytical difficulties. The Eurocodes present a unified approach to all structural materials and should lead to less confusion and fewer errors when considering soil-structure interaction.

Limit states should be verified by calculation, prescriptive measures, experimental models and load tests, an observational method, or a combination of these approaches. These are discussed later in this chapter. Not every limit state needs to be checked explicitly: when one clearly governs, the others may be verified by a control check.

3.2.2   Complexity of design

A welcome requirement of Eurocode 7 is the mandatory assessment of risk for all design situations:

> ... the complexity of each geotechnical design shall be identified together with the associated risks ... a distinction shall be made between light and simple structures and small earthworks ... with negligible risk [and] other geotechnical structures.   [EN 1997-1 §2.1(8)P]

The idea here is that when negligible risk is involved, the design may be based on past experience and *qualitative* geotechnical investigations. In all other cases, *quantitative* investigations are required.

There are many schemes for assessing risk that may be used in conjunction with a Eurocode 7 design. For example, the approach outlined in the UK Highways Agency's document HD22/02[3] requires designers to identify possible hazards for a project or operation within that project. The cause of each hazard is detailed and its probability and impact both assessed on scale of 1 to 4. These two numbers are multiplied together to provide a 'risk rating', which helps the designer decide whether measures are required to mitigate the risk. The exercise is repeated when the mitigating measures are in place to show that the risk has been reduced below acceptable levels. Should the risk still be too high, further measures would be needed or the project may need to be abandoned or redesigned. Such risk assessments clearly satisfy the requirements of Eurocode 7 to identify the complexity and risks of geotechnical design.

3.2.3   Geotechnical categories

To assist geotechnical engineers in classifying risk, Eurocode 7 introduces three Geotechnical Categories, their design requirements, and the design procedure they imply — as summarized below. The Geotechnical Categories are defined in a series of Application Rules, not Principles, and hence alternative methods of assessing geotechnical risk could be used.

GC	Includes ...	Design requirements	Design procedure
1	Small and relatively simple structures ... with negligible risk	Negligible risk of instability or ground movements; ground conditions are 'straightforward'; no excavation below water table (or such excavation is 'straightforward')	Routine design and construction (i.e. execution) methods
No examples given in EN 1997-1			
2	Conventional types of structure and foundation with no exceptional risk or difficult soil or loading conditions	Quantitative geotechnical data and analysis to ensure fundamental requirements are satisfied	Routine field and lab testing Routine design and execution
Examples: spread, raft, and pile foundations; walls and other structures retaining or supporting soil or water; excavations; bridge piers and abutments; embankments and earthworks; ground anchors and other tie-back systems; tunnels in hard, non-fractured rock, not subject to special water-tightness or other requirements			
3	Structures or parts of structures not covered above	Include alternative provisions and rules to those in Eurocode 7	
Examples: very large or unusual structures; structures involving abnormal risks or unusual or exceptionally difficult ground or loading conditions; structures in highly seismic areas; structures in areas of probable site instability or persistent ground movements that require separate investigation or special measures			

The design requirements and procedure for GC3 warrant comment, since it is not immediately clear what 'include alternative provisions and rules' means. Eurocode 7 does not say 'use alternative Principles and rules' (even though the word 'provisions' could be taken to include the Principles). Our understanding of this is that designs should follow the Principles of Eurocode 7, but the Application Rules provided in the standard may not be sufficient on their own to satisfy those Principles. Hence alternative (and/or additional) rules may be required.

It is not necessary to classify all parts of a project in one Geotechnical Category. Indeed, many projects will comprise a mixture of GC1 and GC2 (and in some cases GC3) elements, as illustrated in **Figure 3.2**.

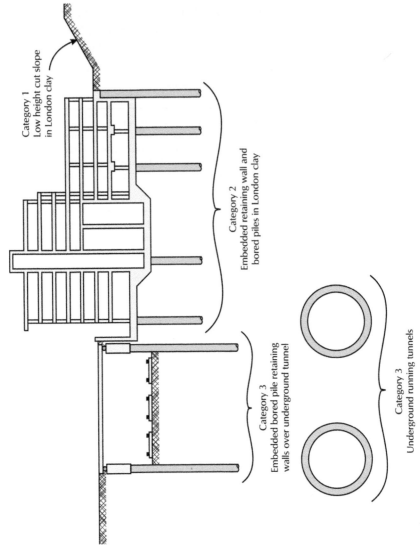

Figure 3.2. Example project involving different Geotechnical Categories

Eurocode 7 gives examples of structures in the three Geotechnical Categories, as shown in the table above. A structure's Geotechnical Category influences the level of supervision and monitoring called for in Section 4 of EN 1997-1 (discussed later in this chapter).

Structures in GC1 can safely be designed and built by habit, because they involve no unusual features or circumstances.

The examples in GC2 form the bread and butter work of many geotechnical design offices (what we term the 'usual suspects'), requiring thoughtful but not unusual design and construction.

The design of structures in GC3 requires careful thought because of their exceptional nature; the keywords are 'large or unusual', 'abnormal or exceptionally difficult', 'highly seismic', 'probable instability', etc.

The magnitude and scope of geotechnical investigations must reflect the structure's Geotechnical Category, as summarized in the table below. Since the ground conditions may influence the category chosen for the structure or parts of it, they should be established early on in the investigation (perhaps through a desk study or preliminary field work).

[EN 1997-1 §3.2.1(2)P and 3.2.1(4)]

GC	Risk	Geotechnical investigation requirements
1	Negligible	Normally limited (verifications often based on local experience)
2	Not exceptional	Provisions of EN 1997-2 apply
3	Exceptional	Amount of investigations at least the same as for Category 2 projects. The circumstances that place a project in Category 3 may necessitate additional investigations and more advanced tests.

3.3   Limit states

When designing a geotechnical structure, the engineer needs to identify the possible ultimate and serviceability limit states that are likely to affect the structure. Ultimate limit states are those that will lead to failure of the ground or the structure; serviceability limit states are those that result in unacceptable levels of deformation, vibration, noise, or flow of water or contaminants (for example).

Eurocode 7 identifies five ultimate limit states for which different sets of partial factors are provided: failure or excessive deformation in the ground (GEO) and internal failure or excessive deformation of the structure (STR) are discussed at length in Chapter 6. Loss of static equilibrium (EQU), loss of equilibrium or excessive deformation due to uplift (UPL), and hydraulic heave, piping, and erosion (HYD) are discussed in Chapter 7.

The following ultimate limit states should be checked for all geotechnical structures: loss of overall stability (of the ground and/or associated structures); combined failure in the ground and structure; and structural failure due to excessive ground movement.

The following serviceability limit states should be checked for all geotechnical structures: excessive settlement; excessive heave; and unacceptable vibrations.

3.4   Actions and design situations

3.4.1   Design situations

Design situations have a key role to play in the selection of actions to include in design calculations and in the choice of partial factors to apply to both actions and material properties. Design situations are:

> *sets of physical conditions representing the real conditions occurring during a certain time interval for which the design will demonstrate that relevant limit states are not exceeded.*
>
> *[EN 1990 §1.5.2.2]*

The table below summarizes the design situations defined in EN 1990.

Design situation	Real conditions	Time interval*	Probability	Example
Persistent	Normal	~ DWL	Certain	Everyday use
Transient	Temporary	<< DWL	High	Construction or repair
Accidental	Exceptional	Very short	Low	Fire, explosion, impact, local failure
Seismic				Earthquake

*DWL = Design Working Life

Eurocode 7 requires consideration of both short- and long-term design situations, to reflect the sometimes vastly different resistances obtained from drained and undrained soils. At first sight, the requirements of EN 1997-1 appear to cut across those of EN 1990. However, it is not difficult to combine these ideas to cater for common geotechnical problems:

Design situation	Real conditions	Term	Example
Persistent	Normal	Long	Buildings and bridges founded on coarse soils and fully-drained fine soils
		Short	Partially-drained slope in fine soils (with DWL less than 25 years)
Transient	Temporary	Long	Temporary works in coarse soils
		Short	Temporary works in fine soils
Accidental	Exceptional	Long	Buildings and bridges founded on coarse soils and quick-draining fine soils
Seismic		Short	Buildings and bridges founded on slow-draining fine soils

3.4.2 Geotechnical actions

EN 1997-1 lists twenty different types of action that should be included in geotechnical design. These include obvious things, such as the weight of soil, rock, and water; earth pressures and ground-water pressures; and removal of load or excavation of ground – and less obvious things, such as movements caused by caving; swelling and shrinkage caused by climate change; and temperature effects, including frost action.   [EN 1997-1 §2.4.2(4)]

Retaining structures are often subject to a large set of actions, including the weight of backfill, surcharges, the weight of water, wave and ice forces, seepage forces, collision forces, and temperature effects.   [EN 1997-1 §9.3.1]

Likewise, many slopes are subject to a large set of actions, including previous or continuing movements from vibrations, climatic variations, removal of vegetation, and wave action.   [EN 1997-1 §11.3.2(P)]

Embankments suffer the erosion effects of overtopping, ice, waves, and rain on their slopes and crests.   [EN 1997-1 §12.3(4)P]

In situations where structural stiffness has a significant influence on the distribution of actions, for example in the design of raft foundations or pile groups, that distribution should be determined by soil-structure interaction analysis.   *[EN 1997-1 §6.3(3) and 7.3(4)]*

Consolidation, swelling, creep, landslides, and earthquakes can all impose significant additional actions on piles and other deep foundations. When considering these effects, the worst situation may involve upper values of ground strength and/or stiffness.   *[EN 1997-1 §7.3.2.1(2)]*

For example, assessment of the upper values of ground strength may be important when determining the potential loading on a pile due to negative skin friction. In this situation, the stronger the ground, the greater the load that acts on the pile. The design, therefore, should consider upper values of strength (the 'superior' values discussed in Chapter 2) when assessing design actions and lower values for resistance ('inferior' values).

3.4.3   Distinction between favourable and unfavourable actions

The Eurocodes make an important distinction between favourable (or stabilizing) and unfavourable (destabilizing) actions, which is reflected in the values of the partial factors γ_F applied to each type of action. As discussed in Chapters 6 and 7, unfavourable/destabilizing actions are typically increased by the partial factor (i.e. $\gamma_F > 1$) and stabilizing actions are decreased or left unchanged (i.e. $\gamma_F \leq 1$).

Consider the design of the T-shaped gravity retaining wall shown in **Figure 3.3**. To provide sufficient reliability against bearing capacity failure, we should treat the self-weight of the wall and the soil on top of its heel (W) as unfavourable (since it increases the effective stress beneath the wall base) — but as favourable for sliding and overturning (since it reduces the effective stress beneath the wall base and increases the clockwise restoring moment about 'O').

The imposed surcharge q is unfavourable for bearing, sliding, and overturning where it acts to the right of the virtual plane shown in **Figure 3.3**. But where it extends to the left of the virtual plane, it has the same effect as the wall's self-weight W.

Unfortunately, the distinction between favourable and unfavourable actions is not always as straightforward as this. Consider next the vertical and horizontal thrusts U_v and U_h owing to ground water pressure acting on the wall's boundaries. The horizontal thrust U_h is unfavourable for bearing, sliding, and overturning; whereas the vertical thrust U_v is favourable for bearing (since it helps to counteract W), but unfavourable for sliding (reducing the effective stress beneath the wall base) and overturning (since it increases the anticlockwise overturning moment about 'O'). But it is illogical to treat an action as both favourable and unfavourable in the same calculation — how can the horizontal component of the ground water pressure be treated differently to its vertical component?

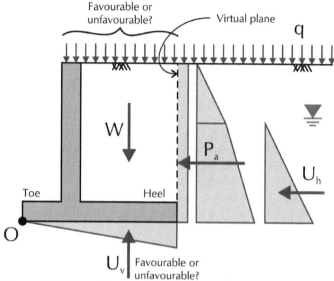

Figure 3.3. Examples of favourable and unfavourable actions

Eurocode 7 deals with this issue in what has become known as the 'Single-Source Principle' (although in fact it is merely a note to an Application Rule):

> *Unfavourable (or destabilising) and favourable (or stabilising) permanent actions may in some situations be considered as coming from a single source. If ... so, a single partial factor may be applied to the sum of these actions or to the sum of their effects.*   [EN 1997-1 §2.4.2(9)P NOTE]

This note allows the thrusts U_h and U_v to be treated in the same way — either both unfavourable or both favourable, whichever gives the more onerous design condition.

The Single-Source Principle has a profound effect on the outcome of some very common design situations, as explained more fully in Chapters 9 to 14. It also precludes the use of submerged weights in design calculations: by replacing the gross weight W and water thrust U_v in **Figure 3.3** by the submerged weight $W' = W - U_v$, the choice is implicitly made to treat both the self-weight and the water thrust as favourable or unfavourable — which does not tally with our discussion of their 'favourableness' given above.

3.4.4   Should water pressures be factored?

According to Eurocode 7, for ultimate limit states:

> *design values [of groundwater pressures] shall represent the most*
> *unfavourable values that could occur during the design lifetime of the*
> *structure.*   *[EN 1997-1 §2.4.6.1(6)P]*

whereas for serviceability limit states:

> *design values shall be the most unfavourable values which could occur in*
> *normal circumstances.*   *[EN 1997-1 §2.4.6.1(6)P]*

To many engineers, the first definition conjures up the idea of 'worst credible' groundwater pressures, i.e. the most adverse water pressures that are physically possible during the structure's working life. The second definition suggests less severe conditions, e.g. the most adverse water pressures that are likely to occur without something exceptional happening. Extreme water pressures 'may be treated as accidental actions'.

[EN 1997-1 §2.4.6.1(6)P]

In many cases, water pressures are calculated from an assumed water level, which should therefore be the most unfavourable water level that could occur. Eurocode 7 goes on to say:

> *Design values of ground-water pressures may be derived either by applying*
> *partial factors to characteristic water pressures or by applying a safety*
> *margin to the characteristic water level*   *[EN 1997-1 §2.4.6.1(8)]*

The interpretation of this Application Rule is an issue that provokes strong opinions from the engineers with whom we have discussed the matter. **Figure 3.4** shows some of the possible interpretations of §2.4.6.1(8).

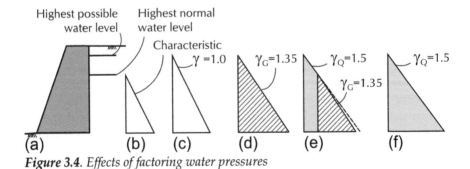

Figure 3.4. *Effects of factoring water pressures*

Consider a gravity wall retaining water as shown in **Figure 3.4**(a). Based on a knowledge of the ground's hydro-geology, we might identify the highest water level expected behind the wall 'in normal circumstances' and the highest possible water level 'during the design lifetime of the structure'. On **Figure 3.4**, the characteristic water pressures are shown by triangle (b) and possible design pressures — depending on your interpretation of §2.4.6.1(8) — by 'triangles' (c)-(f).

In **Figure 3.4**(c), the water pressure is regarded as being at its design value when its level is at its highest possible and hence no factor is applied to it (i.e. $\gamma = 1.0$). In **Figure 3.4**(d), the water pressure is treated as a permanent action and factored by $\gamma_G = 1.35$.[†] In **Figure 3.4**(e), the *additional* pressure due to water rising from its highest normal to its highest possible level is treated as variable and factored by $\gamma_Q = 1.5$, while the remaining pressure (hatched) is treated as permanent and factored by $\gamma_G = 1.35$. In **Figure 3.4**(f), all the water pressure is treated as variable and factored by $\gamma_Q = 1.5$.

The differences between (d), (e), and (f) are relatively minor and of less importance than selecting suitable water levels in the first place. However, the choice between (c) and one of (d) to (f) depends on how you answer the question *should water pressures be factored?* This question provokes strong debate. For many geotechnical engineers, it is illogical to apply a partial factor to a quantity whose ultimate value is relatively well known (particularly if the highest possible water level is placed at ground surface). For others, it is illogical to treat water pressures any differently to other actions, especially effective earth pressures, which are normally factored by γ_G. From a practical point of view, applying factors to effective earth pressures but not to water pressures would make numerical analysis extremely difficult (if not impossible).

Two arguments favour the approach of applying factors to water pressures. First, structural engineers have traditionally applied partial factors between 1.2 and 1.4 to retained liquid loads[4] and ground water pressures[5] (1.2 where the maximum credible water level can be clearly defined; otherwise 1.4). This practice is continued in Eurocode 1 Part 4[6] for liquid induced loads during tank operation (where $\gamma_F = 1.2$) and in the head Eurocode[7], which gives $\gamma_G = 1.35$ for permanent and $\gamma_Q = 1.5$ for variable ground- and free-water pressures. If the design omits these factors, then it must make compensating adjustments to the structure's partial material factors in order to attain the same level of reliability.

[†]Chapter 6 discusses the Design Approaches in which these factors apply.

Second, it is common for geotechnical engineers to perform numerical analyses using unfactored parameters and then to apply a factor of safety to the resultant structural effects, such as bending moments and shear forces. In doing so, the analyst has implicitly applied the same partial factor to water pressures as to effective earth pressures. If you perform a calculation with the partial factor $\gamma_G = 1.35$ applied to effective earth pressures only (and not to water pressures), you will get different bending moments and shear forces from those obtained from the numerical analysis.

The main argument against applying factors to water pressures is that it results in physically unreasonable values, a feature that is exacerbated by the Eurocodes' use of factors between 1.35 and 1.5, rather than the traditional 1.2–1.4.

One approach that provides a balance between providing reliability and maintaining realism in the design is shown in **Figure 3.5**.

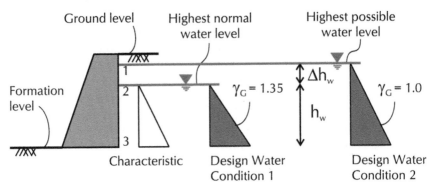

Figure 3.5. Recommended treatment of water pressures for design

When partial factors $\gamma_G > 1.0$ are applied to effective earth pressures, then pore water pressures should also be multiplied by $\gamma_G > 1.0$ but calculated from highest normal (i.e. serviceability) water levels — i.e. no safety margin is applied. On **Figure 3.5**, we have called this 'Design Water Condition 1'.

Alternatively, when partial factors $\gamma_G = 1.0$ are applied to effective earth pressures, then pore water pressures should also be multiplied by $\gamma_G = 1.0$ but calculated from highest possible (i.e. ultimate) water levels — after an appropriate safety margin has been applied. **Figure 3.5**, this is 'Design Water Condition 2'.

The table below summarizes this approach and indicates the relative magnitudes of the water thrusts acting against the wall.

Limit state	DWC*	Partial factor γ_G	Safety margin Δh_w	Water thrust
Characteristic	-	1.0	0	$0.5 \times \gamma_w h_w^2$
Ultimate	1	1.35	0	$0.675 \times \gamma_w h_w^2$
	2	1.0	> 0	$0.5 \times \gamma_w \left(h_w + \Delta h_w \right)^2$

*Design Water Condition
γ_w = weight density of water; other symbols are defined in **Figure 3.5**

The question 'Should water pressures be factored?' warrants further study before definitive rules can be proposed. In the meantime, engineers should judge for themselves which approach to take, depending on how they have selected characteristic water levels for 'normal' and 'possible' circumstances.

3.5 Design and construction considerations

3.5.1 Durability

The significance of environmental conditions must be assessed so that materials are suitably protected. *[EN 1997-1 §2.3(1)P]*

3.5.2 Design considerations relating to construction

Account shall be taken of the possible differences between the ground properties and the geotechnical parameters obtained from test results and those governing the behaviour of the geotechnical structure.
[EN 1997-1 §2.4.3(3)P]

Parameters selected for design should reflect the effects of construction on their operating values. *[EN 1997-1 §2.4.3 (4)]*

3.5.3 Execution

Sections 6–9 and 11–12 of EN 1997-1 give little practical guidance on execution, but instead refer the reader to separate Euronorms that provide more detailed information.

Frost damage of building foundations will be avoided if the soil is not frost-susceptible, the foundation is below the frost-free depth, or frost is eliminated by insulation. §6 refers to EN ISO 13793[8] for guidance on frost protection measures. *[EN 1997-1 §6.4(2 and 3)]*

§7 refers to ENs 1536, 12063, 12699, and 14199 for guidance on execution of piles. These standards are discussed at length in Chapter 15.

[EN 1997-1 §7.1(3)P]

§8 refers to EN 1537 for guidance on execution of anchorages. This standard is discussed at length in Chapter 15.   *[EN 1997-1 §8.4-8.5 and 8.7-8.9]*

§9 makes no reference to execution standards for retaining structures, although ENs 1536, 1537, 1538, and 12063 are relevant. These standards are discussed at length in Chapter 15.

§11 makes no reference to execution standards for slopes and cuttings, although ENs 1537, 14475, 14490, and 15237 may be relevant. These standards are discussed at length in Chapter 15.

§12 makes no reference to execution standards for embankments.

3.6   Geotechnical design

'The Eurocodes adopt, for all civil and building engineering materials and structures, a common design philosophy based on the use of separate limit states and partial factors, rather than 'global' factors (of safety); this is a substantial departure from much traditional geotechnical design practice ... an advantage of BS EN 1997-1 is that its design methodology is largely identical with that for all of the structural Eurocodes, making the integration of geotechnical design with structural design more rational.'[9]

As noted earlier, limit states should be verified by calculation; prescriptive measures; experimental models and load tests; an observational method; or a combination of these approaches. These are discussed in the following sub-sections.   *[EN 1997-1 §2.1(4)]*

3.6.1   Design by calculation

Design by calculation involves a number of elements, which are summarized in **Figure 3.6**. These include three basic variables: actions (e.g. the weight of soil and rock, earth and water pressures, traffic loads, etc.), material properties (e.g. the density and strength of soils, rocks, and other materials), and geometrical data (e.g. foundation dimensions, excavation depths, eccentricity of loading, etc.).

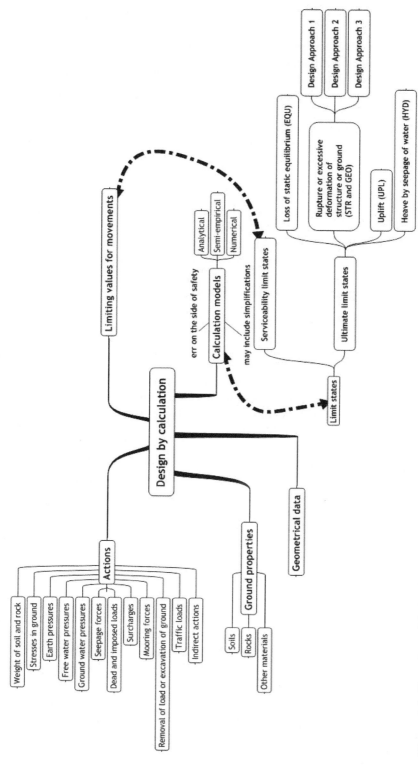

Figure 3.6. Overview of design by calculation

The basic variables are entered into calculation models, which may include simplifications but are required to 'err on the side of safety'. These models may be analytical (e.g. bearing capacity theory), semi-empirical (e.g. the alpha method of pile design), or numerical (e.g. finite element analysis).

The calculation models are used to verify that limit states are not exceeded. For serviceability limit states, these models must demonstrate that predicted displacements do not exceed limiting values of movement, which are usually project-specific. For ultimate limit states, they must demonstrate that effects of actions do not exceed the available resistance. Ultimate limit states include rupture or excessive deformation of the structure or ground (limit states STR and GEO — discussed further in Chapter 7) and loss of equilibrium, uplift, and hydraulic failure (EQU, UPL, and HYD — discussed further in Chapter 8). Design Approaches provide choice in the way STR and GEO are verified.

3.6.2   Design by prescriptive measures

'Prescriptive measures' are a combination of conservative design rules and strict control of execution that, if adopted, avoid the occurrence of limit states.

The design rules, which often follow local convention, are commonly set by local or national authorities, via building regulations, government design manuals, and other such documents. These design rules may be given in a country's national annex to EN 1997-1.

Design by prescriptive measures may be more appropriate than design by calculation, especially when there is 'comparable experience', i.e. documented (or other clearly established information) in similar ground conditions, involving similar structures – suggesting similar geotechnical behaviour.   *[EN 1997-1 §1.5.2.2 and 2.5(2)]*

Annex G of EN 1997-1 provides a sample method for deriving presumed bearing resistance for spread foundations on rock, which originally appeared in BS 8004.[10]

3.6.3   Design by testing

Eurocode 7 acknowledges the role of large- and small-scale model tests in justifying the design of geotechnical structures by calculation, prescriptive measures, or observation. However, apart from requiring time and scale effects to be considered and differences between the test and real construction to be allowed for, EN 1997-1 provides very little guidance on design by testing.

3.6.4 Design by observation

In a similar way to design by testing, Eurocode 7 acknowledges the role of the Observational Method in the design and construction of geotechnical structures, but provides little guidance on how to implement it.

Certain actions must be taken: establish limits of behaviour, assess the range of possible behaviour, devise a plan of monitoring, devise a contingency plan, and adopt it if behaviour goes outside acceptable limits. More detailed guidance on implementing the Observational Method may be found in documents such as CIRIA report R185.[11]

3.7 Supervision, monitoring, and maintenance

Eurocode 7 has specific requirements to ensure the quality and safety of a structure:

> ... the construction processes and workmanship shall be supervised; the performance of the structure shall be monitored during and after construction; [and] the structure shall be adequately maintained.
>
> [EN 1997-1 §4.1(1)P]

EN 1997-1 qualifies this requirement by saying these tasks should be undertaken 'as appropriate'. Thus, if construction does not need supervising, or the structure does not need monitoring or maintaining, then the design could explicitly rule out the need for them.

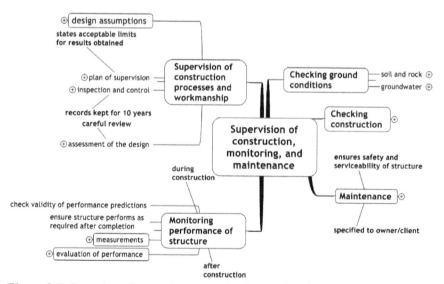

Figure 3.7. *Overview of supervision, monitoring, and maintenance*

Figure 3.7 gives an overview of the activities that are specified by Eurocode 7 as part of its supervision, monitoring, and maintenance requirements. They include supervision of the construction process and workmanship, monitoring the performance of the structure, checking ground conditions, checking construction, and maintenance.

3.7.1   Supervision

Figure 3.8 overleaf shows a detail from **Figure 3.7**, dealing with the supervision and checking requirements only. Certain activities are required for structures in particular Geotechnical Categories only – these are indicated by the numbers on the diagram. (On this diagram, GDR = Geotechnical Design Report, as discussed in Chapter 16.)

Supervision involves checking design assumptions are valid, identifying differences between actual and assumed ground conditions, and checking construction is carried out according to the design. A plan of supervision should be prepared, stating acceptable limits for any results obtained, and indicating the type, quality, and frequency of supervision.

The amount and degree of inspection and control and the checking of ground conditions and construction all depend on the Geotechnical Category (GC) in which the structure is placed (see Section 3.2.3). If an alternative method of assessing geotechnical risk is used, then the amount of inspection and control should be selected appropriately for the risk involved. The fact that EN 1997-1 relates its requirements for supervision to the Geotechnical Categories may make their adoption difficult to avoid.

3.7.2   Monitoring

Figure 3.9 shows another detail from **Figure 3.7**, covering the monitoring of the structure's performance during and after construction.

Monitoring involves measurements of ground deformations, actions, contact pressures, and such like and an evaluation of the performance of the structure. For structures placed in Geotechnical Category 1, such evaluation is simple, qualitative and based largely on inspection. For Category 2 structures, the movement of selected points on the structure should be measured. Finally, for Category 3 structures, monitoring also involves analysis of the construction sequence. If an alternative method of assessing geotechnical risk is used, then the degree of monitoring should be selected appropriately for the risk involved.

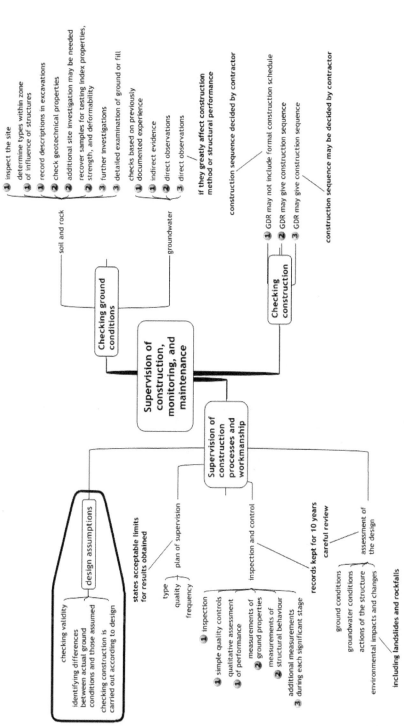

Figure 3.8. Supervision of construction process and workmanship

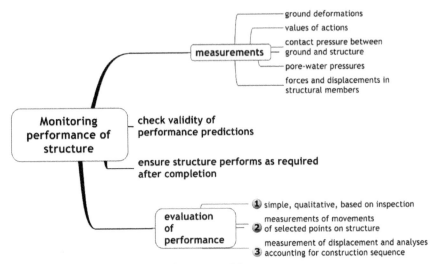

Figure 3.9. Monitoring the performance of the structure

3.7.3   Maintenance

Maintenance of a structure ensures its safety and serviceability and the requirements for it should be specified to the owner or client. These requirements include identifying those parts of the structure that require regular inspection, warning about work that should not be undertaken without prior design review, and indicating the frequency of inspection.

3.7.4   Practical recommendations

Although EN 1997-1 imposes specific requirements for supervision, monitoring, and maintenance, it gives few practical recommendations as to the actions that need to be taken to meet those requirements.

This is demonstrated by the small number of paragraphs on this subject in the whole of Part 1 (thirty-three), broken down as follows: §5 Fill, dewatering, etc., eight paragraphs; §6 Spread foundations, two; §7 Pile foundations, eight; §8 Anchorages, eight; §9 Retaining structures, none; §11 Overall stability, two; and §12 Embankments, five. The table below summarizes the information given in §4 of EN 1997-1 with regards to supervision, monitoring, and maintenance.

3.8   The Geotechnical Design Report

The Geotechnical Design Report and its companion, the Ground Investigation Report, are discussed at length in Chapter 16.

	Geotechnical Category		
	1	2	3
Inspection and control	Limited to inspection, simple quality controls, & qualitative assessment of structure's performance	Often requires measurements of ground properties or behaviour of structures	Additional measures required during each significant construction stage
Checking ground conditions (soil and rock)	Site inspection, determining soil/rock types within zone of influence, recording descriptions of soil/rock exposed in excavations	Check geotechnical properties; additional SI may be needed; recover representative samples to determine index properties, strength, and deformability	Further investigations and examination of details of ground or fill conditions important to the design
Checking ground-water conditions	Based on previously documented experience or indirect evidence	Direct observations if construction method or structure's performance greatly affected	
Checking construction	Formal construction schedule not normally included in GDR (construction sequence normally decided by contractor)	GDR may give construction sequence envisaged in the design	
Evaluation of performance (monitoring)	Simple, qualitative, and inspection-based	Measurements of movements of selected points on structure	Measurement of displacements and analyses, accounting for construction sequence

3.9 Summary of key points

'Reaching agreement between the EU Member States on [a] unified geotechnical methodology ... led to the introduction of concepts and terminology that may not be familiar to many designers of foundations and other geotechnical structures.'[12]

Reading it for the first time, many engineers find Eurocode 7 Part 1 difficult to understand.[13] Contributing to this is an abundance of 'Eurospeak' (i.e. English written in a way that makes it easier for non-English-speaking Europeans to understand).[14] However, Eurocode 7 is worthy of prolonged study. It brings together many excellent principles in a comprehensive and rational manner, consistent with structural design, while still acknowledging the difficulties and peculiarities of geotechnical engineering.

3.10   Notes and references

1. Nethercott H. et al. (2004) *National Strategy for Implementation of the Structural Eurocodes*, Institution of Structural Engineers.

2. BS EN 1997-1: 2004, Eurocode 7 – Geotechnical design, Part 1 – general rules, British Standards Institution, London, 168pp.

3. Highways Agency (2002) *Managing geotechnical risk*, HD22/02.

4. See §2.2.2 of BS 8007: 1987, Code of practice for design of concrete structures for retaining aqueous liquids, British Standards Institution.

5. See Table 2.1 of BS 8110-1: 1997, Structural use of concrete – Part 1: Code of practice for design and construction, British Standards Institution.

6. See Annex B.2.1(1) of EN 1991, Eurocode 1 – Actions on structures, Part 4: Silos and tanks, European Committee for Standardization, Brussels.

7. See Table A2.4(B) of EN 1990: 2002, Eurocode – Basis of structural design, European Committee for Standardization, Brussels.

8. BS EN ISO 13793: 2001, Thermal performance of buildings – Thermal design of foundations to avoid frost heave, British Standards Institution.

9. Driscoll, R., Powell, J., and Scott, P. (2007) *A designers' [sic] simple guide to BS EN 1997*, London: Dept for Communities and Local Government.

10. BS 8004: 1986, Code of practice for foundations, British Standards Institution.

11. Nicholson, D., Tse, C-M., and Penny, C. (1999) *The Observational Method in ground engineering: principles and applications*, CIRIA, p. 214.

12. Driscoll et al., ibid.

13. DiMaggio, J. et al. (1998) *Report of the Geotechnical Engineering Study Tour (GEST)*, FHWA International Technology Scanning Program.

14. See the slightly tongue-in-cheek article by Stuart Alexander in *The Structural Engineer*, 15th November 2005.

Chapter 4

Ground investigation and testing

'[Eurocode 7 Part 2] ... will bring only small changes in the requirements of the site investigation industry. ... the mandatory reporting of ground investigation will become more prescribed and ... appropriate communication of information will be obligatory. ... the greater general emphasis on the assessment of deformation [will] lead to a greater need for SI providers to consider ground deformation parameters.'[1]

This chapter discusses the essential tasks of investigating and testing the ground. It highlights EN 1997-2's key role in providing a consistent suite of procedures for use across the whole of Europe. EN 1997-2 makes extensive reference to supporting documents, in particular ISO standards, which provide detailed information not present in Eurocode 7 itself.

4.1 Standards for geotechnical investigation and testing

4.1.1 Eurocode 7 Part 2

Eurocode 7 - Geotechnical design, Part 2 – ground investigation and testing[2] is divided into six sections and twenty-four annexes, as illustrated in **Figure 4.1** and **Plate 6** (in the book's colour section).

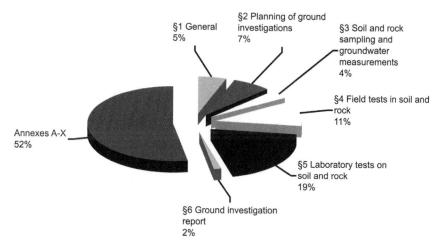

§1 General 5%

§2 Planning of ground investigations 7%

§3 Soil and rock sampling and groundwater measurements 4%

§4 Field tests in soil and rock 11%

§5 Laboratory tests on soil and rock 19%

§6 Ground investigation report 2%

Annexes A-X 52%

*Figure 4.1. Contents of Eurocode 7 Part 2. See **Plate 6** for a colour version.*

EN 1997-2 provides detailed rules for site investigations, general test specifications, derivations of ground properties and the geotechnical model of the site, and examples of calculation methods based on field and laboratory testing.

4.1.2   Complementary standards

EN 1997-2 refers extensively to a new suite of international and European standards, prepared jointly by ISO technical committee TC 182 and CEN TC 341, as summarized in **Figure 4.2**.

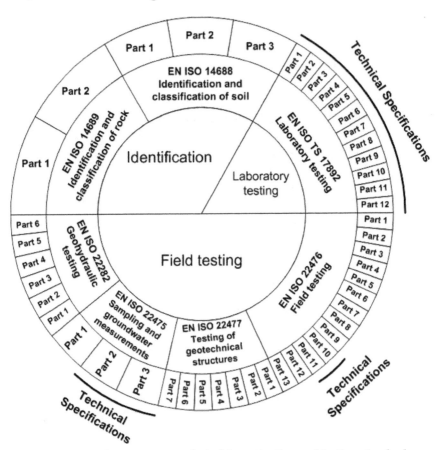

Figure 4.2. Complementary geotechnical investigation and testing standards

Two of these groups of standards (EN ISOs 14688 and 14689) are concerned with the identification and classification of soil and rock and are discussed in Sections 4.3 and 4.4 below.

Four of the groups of standards (EN ISOs 22282, 22475, 22476, and 22477) cover field testing and are discussed in Section 4.7.

Finally, one group of standards (EN ISO TS 17892[3]) deals with laboratory testing and is discussed in Section 4.7.6.

Each of the standards within each group is divided into a number of parts, as shown in the outer ring of **Figure 4.2**. The entire suite comprises nearly fifty standards or specifications.

The relationship between EN 1997-2 and these complementary standards is shown in **Figure 4.3**. The diagram shows which sections of EN 1997-2 refer to which EN ISO standard.

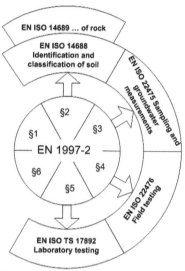

Figure 4.3. Links between EN 1997-2 and its complementary standards

4.2 Planning ground investigations

4.2.1 Aims of a geotechnical investigation

The aims of a geotechnical investigation are to establish soil, rock, and groundwater conditions; determine the properties of soils and rocks; and gather additional relevant knowledge about the site.

According to Eurocode 7 Part 1:

> *Geotechnical investigations shall provide sufficient data concerning the ground and the ground-water conditions ... for a proper description of the essential ground properties and a reliable assessment of the characteristic ... ground parameters to be used in design calculations.*
>
> *[EN 1997-1 §3.2.1(1)P]*

The scope of geotechnical investigations is illustrated in **Figure 4.4**.

According to EN 1997-2, *preliminary* investigations must allow the engineer to assess the general suitability of the site, compare alternative sites, estimate changes caused by the proposed works, identify any borrow areas, and plan the subsequent design and control investigations. These tasks (which are Principles) are boxed in **Figure 4.4**.

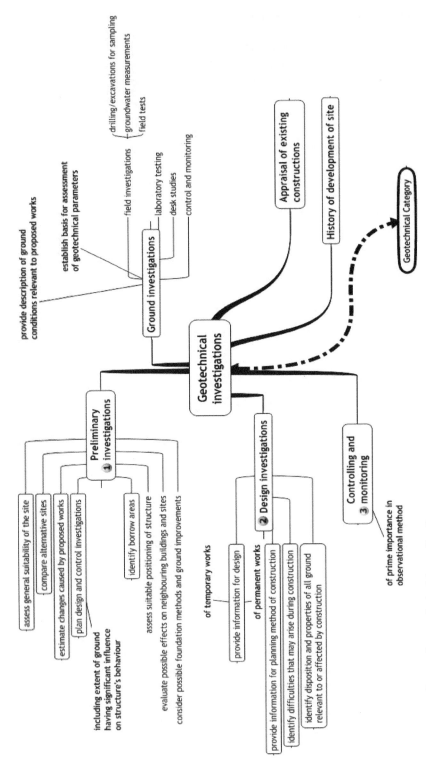

Figure 4.4. Scope of geotechnical investigations

Preliminary investigations should allow the engineer to assess a suitable position for the structure; evaluate its possible effects on adjacent buildings; and consider possible foundations and ground improvements.

Design investigations must provide information for designing temporary and permanent works and for planning the construction method; identify difficulties that may arise during construction; and identify the disposition and properties of all ground relevant to or affected by the construction.

Control investigations are used to monitor and control construction and are of prime importance in the Observational Method[4] of design.

As **Figure 4.4** indicates, geotechnical investigations may involve ground investigations (i.e. desk studies, field investigations, and laboratory testing), appraisal of existing constructions, and research into the site's historical development. The amount of investigation needed is related to the Geotechnical Category to which the project has been assigned.

In many situations only one investigation will be performed as the budget available for investigation is small or the project size does not warrant staged investigation. In these circumstances the aims of the preliminary and design investigations must be fulfilled by one investigation.

4.2.2   Spacing of investigation points

Annex B.3 of EN 1997-2 provides guidance on the spacing of investigation points for geotechnical investigations, as summarized in the table below.

Structure		Spacing	Arrangement
High-rise and industrial		15–40m	Grid
Large area		≤ 60m	Grid
Linear	roads, railways, channels, pipelines, dikes, tunnels, retaining walls	20–200 m	-
Dams and weirs		25–75m	vertical sections
Special	bridges, stacks, machinery foundations	2–6 per foundation	

Guidance on spacing is available from other sources.[5] However, all such guidance is provided as a starting point for assessing the scope of investigations and will need modification to account for site specific requirements. In particular the spacing of exploration points will need to reflect the expected variation in the underlying geology of the site as well as the type and size of structure.

4.2.3   Depth of investigation points

Annex B.3 of EN 1997-2 provides recommendations for the minimum depth of investigation below the lowest point of high-rise structures and civil engineering projects; rafts; embankments and cuttings; linear structures such as roads, airfields, trenches, pipelines; tunnels and caverns; excavations; cut-off walls; and piles. **Figure 4.5** illustrates some of these recommendations.

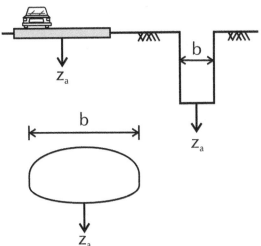

The minimum depth of excavation z_a for roads (and airfields) is:

$$z_a \geq 2m$$

for trenches (and pipelines):

$$z_a \geq 2m \text{ and } z_a \geq 1.5b$$

and for small tunnels and caverns:

$$b < z_a < 2b$$

where b is the width of the structure as defined in **Figure 4.5**.

Figure 4.5. *Minimum depth of investigation below (top-left) roads, (top-right) trenches, and (bottom) tunnels and caverns*

Recommendations for other geotechnical structures are presented in Chapters 10 to 14.

The selection of investigation depth needs to consider factors other than geometrical criteria, as does the spacing of exploration positions. It is not generally necessary to penetrate rock strata by more than 5m unless the anticipated loading is exceptionally high or the geology suggests that significantly weaker strata may be underneath. Further, it is normal to limit investigations to depths at which the expected increase in stress due to foundation loading will be less than 10% of the existing overburden pressure.

4.3   Identification and classification of soil

Identification and classification of soil is covered by International Standard EN ISO 14688, which is divided into three parts covering description (Part 1), classification (Part 2), and data transfer (Part 3).[6] EN ISO 14688 is referenced extensively in EN 1997-2.

4.3.1   Soil description

Figure 4.6 illustrates the logic for identifying soils according to EN ISO 14688-1. The main soil types are divided into made ground, organic soil, volcanic soil, and very coarse, coarse, and fine soils. Very coarse soils are sub-divided into large boulders, boulders, and cobbles; coarse soils into gravels and sands; and fine soils into silts and clays. Gravel, sand, and silt are further qualified as coarse, medium, or fine.

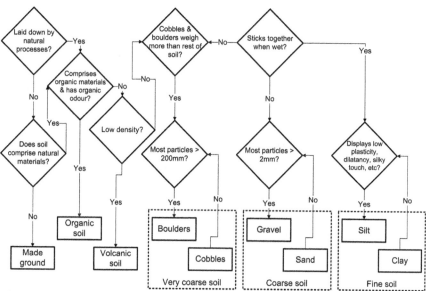

Figure 4.6. Flow-chart for identification of soils

Figure 4.7 indicates the boundaries between these descriptions, based on particle-size, and summarizes their two-letter abbreviations (e.g. Gr for gravel, Sa for sand, Cl for clay). The boundary between the particle size of fine and coarse soils is now 0.063 mm, which is the same as the corresponding EN ISO sieve size and slightly larger than the 0.06 mm boundary specified in BS 5930.[7]

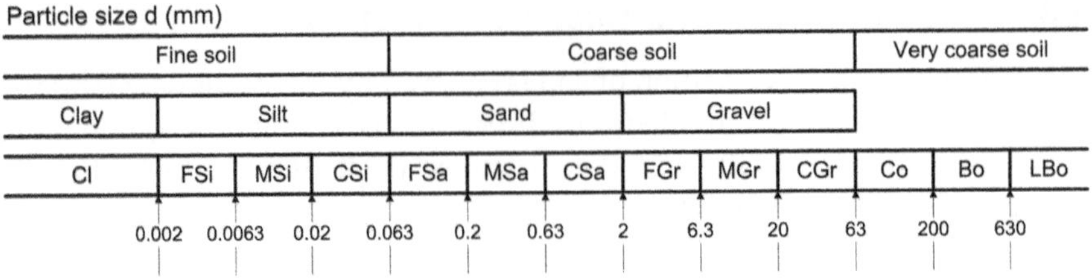

Particle size d (mm)

	Fine soil					Coarse soil					Very coarse soil	

	Clay		Silt			Sand			Gravel					

Cl		FSi	MSi	CSi	FSa	MSa	CSa	FGr	MGr	CGr	Co	Bo	LBo

0.002 0.0063   0.02   0.063   0.2   0.63   2   6.3   20   63   200   630

Figure 4.7. Particle-size fractions

The principal fraction of a composite soil should be indicated by a capital letter (e.g. Sa for sand or SAND) and the secondary fraction by lower-case letters (e.g. gr for gravelly). Some examples of these abbreviations are:

saGr = *sandy gravel* or (in the UK) *sandy GRAVEL*
msaCl = *medium sandy clay* or (in the UK) *medium sandy CLAY*

The shape of a soil's particle size distribution (or grading curve) is described by the terms *multi-graded, medium-graded, even-graded,* and *gap-graded* and quantified by the uniformity coefficient C_U and coefficient of curvature C_C, defined as:

$$C_U = \frac{d_{60}}{d_{10}} \text{ and } C_C = \frac{(d_{30})^2}{d_{10} \times d_{60}}$$

where d_n is the particle size where n% of the soil is smaller than this size.

	Shape of grading curve			
	Multi-graded	Medium-graded	Even-graded	Gap-graded
C_U	>15	6-15	< 6	Usually high
C_C	1-3	< 1	< 1	Any (usually < 0.5)

4.3.2   Density of coarse soils

The density of coarse soils is classified in EN ISO 14688-2 by the terms *very loose, loose, medium dense, dense* and *very dense*, according to the value of their density index I_D, defined as:

$$I_D = \frac{e_{max} - e}{e_{max} - e_{min}}$$

where e is the soil's void ratio, e_{max} its maximum voids ratio, and e_{min} its minimum. See **Figure 4.8**.

Density index I_D

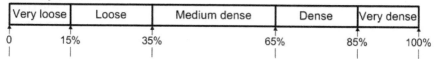

Very loose	Loose	Medium dense	Dense	Very dense

0   15%   35%   65%   85%   100%

Figure 4.8. Density of coarse soils

4.3.3   Consistency and strength of fine soils

The consistency of fine soils is described by the terms *very soft, soft, firm, stiff,* and *very stiff* and classified by the value of their consistency index I_C, defined as:

$$I_C = \frac{w_L - w}{w_L - w_P}$$

where w is the soil's water content, w_L its liquid limit, and w_P its plastic limit. See **Figure 4.9**.

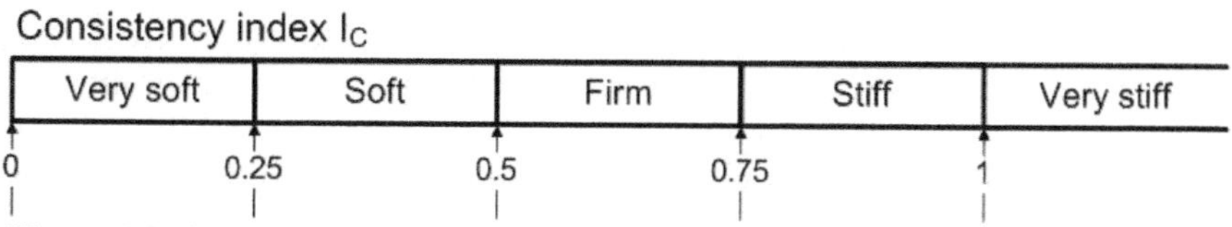

Consistency index I_C

Very soft	Soft	Firm	Stiff	Very stiff

0   0.25   0.5   0.75   1

Figure 4.9. Consistency of fine soils

The related terms, plasticity index I_P and liquidity index I_L are defined as:

$$I_P = w_L - w_P \text{ and } I_L = \frac{w - w_P}{w_L - w_P} = 1 - I_C$$

The strength of fine soils is described by the terms *extremely low, very low, low, medium, high, very high,* and *extremely high* and classified by the undrained strength c_u measured in a field or laboratory strength test. See **Figure 4.10**.

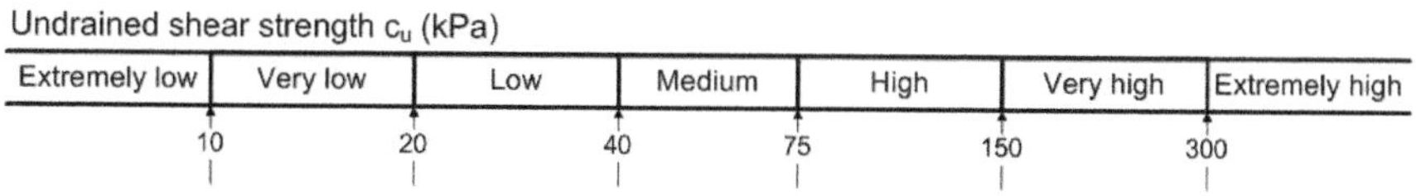

Undrained shear strength c_u (kPa)

Extremely low	Very low	Low	Medium	High	Very high	Extremely high

10   20   40   75   150   300

Figure 4.10. Strength of fine soils

This is a change to traditional UK practice, which employed the terms 'very soft' to 'hard' to describe both the consistency and strength of cohesive soils (very soft c_u < 20kPa; soft 20–40kPa; firm 40–75kPa; stiff 75–150kPa; very stiff 150–300kPa; and hard > 300kPa). The category 'extremely low' is new to UK practice, subdividing the previous 'very soft' category into two: extremely and very low.

When both the consistency and strength are measured, the soil might be described as, for example, a 'stiff fissured high strength CLAY'. The soil's consistency would be based on the field log and its strength on the results of subsequent laboratory tests.

4.4   Identification and classification of rock

Identification and classification of rock is covered by International Standard EN ISO 14689, which is divided into two parts covering description and classification (Part 1) and data transfer (Part 2).[8] EN ISO 14689 is referenced extensively in EN 1997-2.

4.4.1   Rock description

Rock description according to EN ISO 14689-1 is based on the terms published by the International Society of Rock Mechanics (ISRM),[9] in preference to those in BS 5930.

Two stages are identified in the description of rocks: rock material and rock mass. The description of the rock material shall include colour, grain size, matrix, weathering and alteration effects, carbonate content, stability of rock material and unconfined compressive strength. For the rock mass terms include type of rocks, structure, discontinuities, weathering and groundwater.

4.4.2   Rock material strength

The strength of rock is described by the terms *extremely weak, very weak, weak, medium strong, strong, very strong,* and *extremely strong* and classified by the unconfined compressive strength q_u measured in a laboratory strength test. See **Figure 4.11**.

Unconfined compressive strength q_u (MPa)

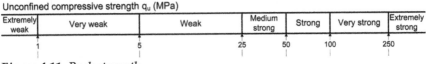

Figure 4.11. Rock strength

This differs from BS 5930,[10] which specified the following terms and associated strengths: very weak q_u < 1.25 MPa; weak 1.25–5 MPa; moderately weak 5–12.5 MPa; moderately strong 12.5–50 kPa; strong 50–100 MPa; very strong 100–200 MPa; and extremely strong > 200 MPa.

4.4.3   Discontinuities

Discontinuities are key to the engineering performance of rock masses. When describing discontinuities a number of features need to be identified including: dip and dip direction, spacing and block shape, persistence, roughness, aperture, infilling and seepage. **Figure 4.12** provides details of the block size, discontinuity spacing and bedding thickness of rock.

Average length of rock block sides (mm)

Very small		Small	Medium	Large	Very large

Discontinuity spacing (mm)

Extremely close	Very close	Close	Medium	Wide	Very wide

Bedding thickness (mm)

Thinly	Thickly laminated	Very thin	Thin	Medium	Thick	Very thick

6   20   60   200   600   2000

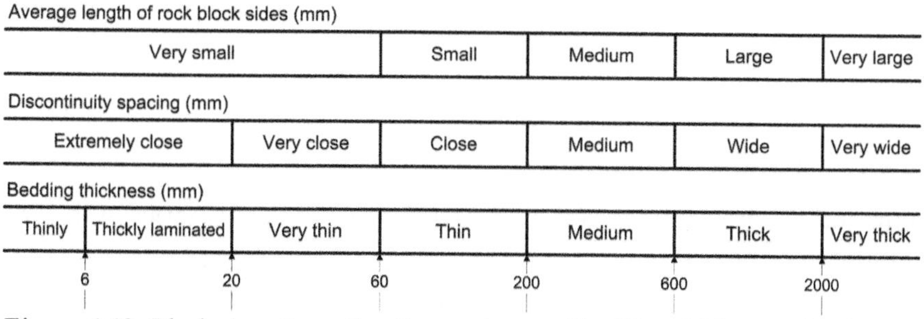

Figure 4.12. Block size, discontinuity spacing, and bedding thickness of rock

The descriptive terms used and the dimensions associated with them are the same as specified in BS 5930.

Figure 4.13 gives the EN ISO 14689-1 definition of aperture sizes.

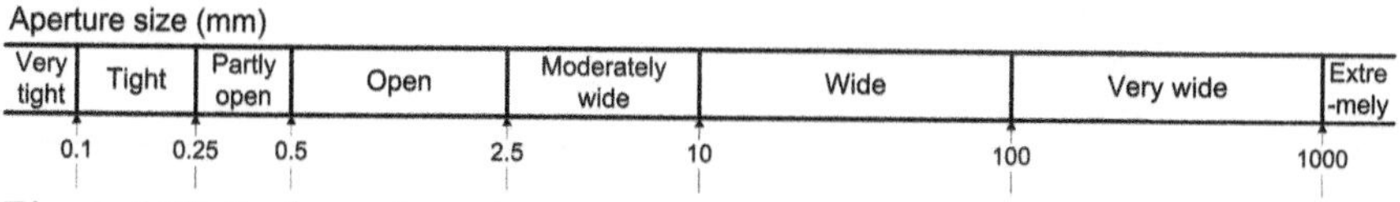

Aperture size (mm)

Very tight	Tight	Partly open	Open	Moderately wide	Wide	Very wide	Extre-mely

0.1   0.25   0.5   2.5   10   100   1000

Figure 4.13. Rock aperture sizes

The above provides greater sub-division of aperture sizes compared to BS 5930. The changes are:
- three sub-divisions for >10mm, compared with only one in BS 5930;
- one more sub-division in the range up to 10mm;
- in providing additional sub-divisions, some of the descriptive terms now have different definitions: for example, 'tight' now refers to 0.1–0.25 mm not 0.1–0.5 mm and 'open' refers to 0.5–2.5 mm not 2.5–10 mm, which becomes 'moderately wide'.

4.4.4   Weathering

When describing the rock material a four basic terms are to be used: fresh, discoloured, disintegrated, and decomposed. These basic terms may be amplified by use of qualifying terms, e.g. 'wholly' or 'partially' discoloured.

For a rock mass, a six stage weathering scale (0 to 5) is provided: fresh, slightly weathered, moderately weathered, highly weathered, completely weathered, and residual soil. More appropriate weathering classification systems have been developed for specific rock types and EN ISO 14689 allows these to be used provided they comply with the general guidance given in the standard.

The weathering terms adopted by EN ISO 14689 are a significant departure from BS 5930: 1999, reverting to the outdated terms given in BS 5930: 1981.

4.5   Soil and rock sampling

4.5.1   Sampling methods

Eurocode 7 Part 2 divides sampling methods into three categories, A to C, as summarized below.

Category	Sampling method	Comment	Sample quality class
A	Tube samplers, rotary coring, block sampling	U100 samplers are not appropriate for soft clays	1–5
B	SPT split-spoon, Mostap samples from window sampling		3–5
C	Bulk bag samples	Fabric of soil totally destroyed	5

4.5.2   Quality classes

Quality classes are provided to define the quality of samples required to allow the satisfactory measurement of soil and rock properties. The measurement of deformation characteristics is limited to Class 1 samples with strength predominantly from Class 1 samples but, for some soils, Class 2 samples may be used. Class 5 samples may only be used for identifying soil type and sequence of layers.

4.5.3   Applicability of sampling to obtain parameters

EN 1997-2 provides guidance on the applicability of Category A to C sampling methods to obtain geotechnical parameters, as summarized in the table below. When considering the sampling method it is also important to understand that the same sampling method may not be capable of producing samples of sufficient quality in all soils. Thus a Category A sampler may only produce Class 2 samples in some soils.

Applicability of sampling category to obtain parameters

Sampling method and sample quality classes		†	Soil or rock type		
			Rock	Coarse soil	Fine soil
Category A	1–5	H	Rock type Extension of layers Particle size Water content Density Shear strength Compressibility Permeability Chemical tests	Soil type Extension of layers Particle size Water content Chemical tests	Soil type Extension of layers Particle size Water content Density Shear strength Compressibility Permeability Chemical tests Atterberg limits
		M	-	Density Shear strength Compressibility Permeability	-
Category B	3–5	H	Rock type Extension of layers Particle size Water content Density	Soil type Extension of layers Particle size Chemical tests	Soil type Extension of layers Particle size Chemical tests Atterberg limits
		M	-	Water content	Water content
		L	-	Density	Density
Category C	5	H	-	-	-
		M	Rock type Extension of layers Particle size	Soil type	Soil type
		L	-	Extension of layers Water content	Extension of layers Water content

†Applicability: H = high, M = medium, L = low

4.5.4   Minimum testing

The volume of testing to be carried out depends on the quality of data already available for the site and the extent of prior knowledge of the properties of materials on the site. Where there is extensive experience, minimal amounts of testing are recommended. The amount is also a function of the potential variability in the parameters to be assessed.

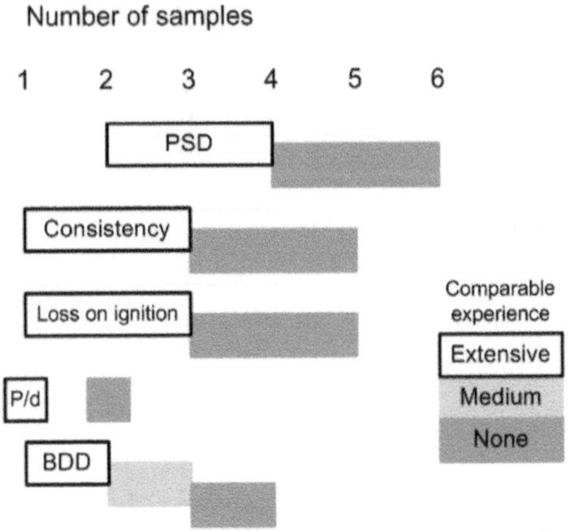

Figure 4.14 summarizes Eurocode 7's recommendations for the minimum number of samples that should be tested in each stratum. Here PSD = particle size distribution; Consistency = Atterberg tests; P/d = particle density determination; and BDD = bulk density determination.

Figure 4.15 summarizes Eurocode 7's recommendations for the minimum number of tests that should be carried out on samples from each stratum. Here TXL φ and TXL c_u = triaxial tests to determine the effective angle of shearing resistance and undrained shear strength, respectively; DSS = direct shear tests; E_{oed} = incremental oedometer tests to determine soil modulus; k = permeability tests; and UCS = unconfined compressive strength tests on rock.

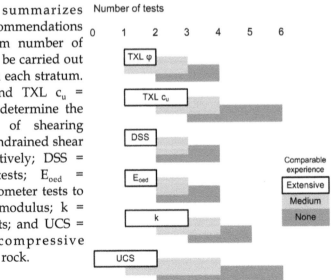

Figure 4.14. Number of samples to be tested in each stratum

Figure 4.15. Number of tests to be carried out in each stratum

A difficulty with these recommendations is the lack of a definition for 'stratum' — simple rules cannot substitute for sound engineering judgment.

4.6   Groundwater measurements

Groundwater measurements are covered by International Standard EN ISO 22475, which is divided into three parts covering technical principles for execution (Part 1), qualification criteria for enterprises and personnel (Part 2), and conformity assessment of enterprises and personnel by third party (Part 3).[11] At the time of writing (early 2008), only Part 1 has been published and the other parts are in draft awaiting approval.

Geohydraulic testing is covered by International Standard EN ISO 22282, which is divided into six parts covering general rules (Part 1), water permeability tests in a borehole using open systems (Part 2), water pressure tests in rock (Part 3), pumping tests (Part 4), infiltrometer tests (Part 5), and water permeability tests in a borehole using closed systems (Part 6).[12] At the time of writing (early 2008), all of these standards are in draft awaiting approval before publication.

4.6.1   Groundwater measuring systems covered

§3.6 of EN 1997-2 gives information about open and closed groundwater measuring systems. In open systems, the piezometric groundwater head is measured by an observation well, usually provided with an open pipe. In closed systems, the groundwater pressure is measured by a transducer at the chosen location.

4.6.2   Planning and execution

§3.6.2 of EN 1977-2 provides guidance on the factors that need to be considered when planning and executing a groundwater investigation.

Particular attention should be paid to the selection of the type and location of the pore pressure/groundwater monitoring device to ensure that it is fit for purpose. This is a key factor as there are large numbers of potential devices/installations and not all will provide the information required. For example, if rapid changes in groundwater/pore pressure are likely, a simple standpipe piezometer is inadequate — a device that takes measurements at regular time intervals, such as a diver, should be specified.

4.6.3   Applicability of groundwater measurement systems

EN 1997-2 provides guidance on the applicability of groundwater measurement systems for assessing groundwater related geotechnical parameters, is summarized in the table below.

Applicability of groundwater measurement systems for obtaining parameters

System	†	Soil or rock type		
		Rock	Coarse soil	Fine soil
Open	H		Groundwater level Pore water pressure	
	M	Groundwater level Pore water pressure	Permeability	Groundwater level Pore water pressure
	L			Permeability
Closed	H	Groundwater level Pore water pressure	Groundwater level Pore water pressure	Groundwater level Pore water pressure
	M		Permeability	Permeability

†Applicability: H = high, M = medium, L = low

The quality and reliability of the information obtained from closed systems is greater than that from open systems. However, no in situ system is effective at assessing the permeability of soils or rocks. This is largely due to the difficulties in interpretation of the results of in situ permeability tests.

4.7   Field tests in soil and rock

Field testing is covered by International Standard EN ISO 22476, which is divided into thirteen parts.[13]

4.7.1   Field tests covered

Section 4 of EN 1997-2 gives information about nine different categories of field test, as listed below.

Field tests covered by Eurocode 7 Part 2

Test			Section	Annex	EN ISO
Standard penetration		SPT	4.6	F	22476-3
Cone penetra-tion	Electrical	CPT	4.3	D	22476-1
	Electrical w/pwp	CPTU			
	Mechanical	CPTM			22476-12

Test			Section	Annex	EN ISO
Field vane		FVT	4.9	I	22476-9
Plate loading		PLT	4.11	K	22476-13
Pressure-meter (PMT)	Pre-bored	PBP	4.4	E	22476-5
	Ménard	MPM			22476-4
	Self-boring	SBP			22476-6
	Full displacement	FDP			22476-8
Flexible dilatometer		FDT	4.5	-	22476-5
Dynamic probing		DP	4.7	G	22476-2
Weight sounding		WST	4.8	H	22476-10
Flat dilatometer		DMT	4.1	J	22476-11

4.7.2 Objectives

The objectives of each test are given in §4.x.1 of EN 1997-2 (where the x can be determined from the table above). For example, the objectives of CPTs and CPTUs are stated as:

to determine the resistance of soils and soft rock to the penetration of a cone and the local friction on a sleeve

[EN 1997-2 §4.3.1(1)]

4.7.3 Specific requirements

The specific requirements of each test are given in §4.x.2 of EN 1997-2 by reference to the appropriate part of EN ISO 22476 (see table above).

4.7.4 Evaluation of tests results

The way in which the results of each test should be evaluated is specified in §4.x.3 of EN 1997-2 by reference to the appropriate part of EN ISO 22476 (see table above).

4.7.5   Obtaining derived values from the test results

EN 1997-2 provides a number of example methods that can be used to obtain derived values from the field tests covered.

Annex F provides methods for deriving density index I_D from SPT blow count N_{60} using the following formula:

$$\frac{N_{60}}{I_D^2} = a + b\sigma'_{v0}$$

where σ'_{v0} = the initial vertical effective stress in the ground and 'a' and 'b' are dimension-less constants. The Annex gives values of I_D related to N_{60} for normally consolidated sands but does not provide guidance on relevant values for a or b. It also provides correlations between I_D and angle of shearing resistance for silica sands.

Methods are given in Annex D for the derivation of the angle of shearing resistance φ' and drained Young's modulus E' from cone resistance q_c, as measured in CPT, CPTM, and CPTU tests. Two methods are provided: one is in the form of a table giving values of φ' and E' for different ranges of q_c (for quartz and feldspar sands, silty soils, and gravels); the other is the equation below (for poorly graded sands situated above the water table):

$$\varphi' = 13.5° \times \log_{10}\left(\frac{q_c}{MPa}\right) + 23°$$

Annex K provides a method for deriving undrained strength c_u from the ultimate load p_{ult} measured in plate load tests, based on simple bearing capacity theory:

$$c_u = \frac{p_u - \gamma z}{N_c}$$

where γ = the ground's weight density, z = the depth of the test, and N_c is a bearing capacity factor for a circular plate ($N_c = 6$ for tests conducted near the ground surface; and $N_c = 9$ for tests conducted down a borehole, at depths greater than four times the plate size).

The methods provided for deriving soil parameters from in situ tests are only examples and the designer should not be limited to these correlations. Where there are accepted and locally substantiated correlations or derivations based on empiricism or theoretical analysis these may be used within the context of the code.

4.7.6 Applicability of field tests to obtain parameters

EN 1997-2 provides guidance on the applicability of field investigation methods for determining geotechnical parameters, as summarized below.

Field test and abbreviation		†	Soil or rock type		
			Rock	Coarse soil	Fine soil
Standard penetration test	SPT	H	-	-	Soil type Particle size
		M	-	Soil type Extension of layers Particle size Water content Density Shear strength Compressibility Chemical tests	Extension of layers Water content Density Compressibility Chemical tests Atterberg limits
		L	-	-	Shear strength
Cone penetration test with or without pore pressure measure- ment	CPT CPTM CPTU	H	-	Extension of layers Compressibility	Extension of layers Shear strength
		M	-	Soil type Groundwater level Pore water pr. Density Shear strength	Soil type Pore water pr. Density Compressibility Permeability
		L	Rock type	Permeability	-
Plate loading test	PLT	H	-	Shear strength Compressibility	Shear strength Compressibility
		M	Shear strength	-	-
		L	-	-	-
Dynamic probing, light, medium, heavy, super heavy	DPL DPM DPH DPSH	H	-	Extension of layers	-
		M	-	Density Shear strength Compressibility	Extension of layers Compressibility
		L	-	Soil type	Soil type Shear strength

†Applicability: H = high, M = medium, L = low

Although the table is useful in identifying where it is likely that a particular in situ test will be useful, there are some situations where the categorization could be questioned. For example, the table indicates that standard penetration tests (SPTs) have low applicability for assessing undrained shear strength c_u in fine soils, but experience suggests that established correlations between c_u and SPT blow count are as reliable – if not more reliable – than laboratory tests for determining c_u.

4.8 Laboratory tests in soil and rock

4.8.1 Laboratory tests covered

Section 5 of EN 1997-2 gives information about seven different categories of laboratory test on soil, plus a further three categories on rock, as listed below.

Test		Section	Annex	EN ISO
Classification, identification, and description of soil		5.5	M	14688-1 14688-2
Chemical testing of soil and groundwater		5.6	N	-
Strength index testing		5.7	O	TS 17892-6
Strength testing of soil	UC UUTX CTX DSS	5.8	P	TS 17892-7 TS 17892-8 TS 17892-9 TS 17892-10
Compressibility and deformation testing (OED)		5.9	Q	TS 17892-5
Compaction testing (Proctor, CBR)		5.10	R	-
Permeability testing		5.11	S	TS 17892-11
Classification of rock	ident. and description w/c, density, porosity	5.12	U	14689-1 TS 17892-3
Swelling testing of rock (various)		5.13	V	-
Strength testing of rock material (UC, point load, DSS, Brazil, TX)		5.14	W	-

In the final column of this table, it is important to note the use of the letters 'TS' to signify EN ISO documents that are 'technical specifications' rather than standards. These TSs have a limited lifespan, after which they are either withdrawn or 'promoted' to the status of a full standard. In the UK, the EN ISO TSs for laboratory testing are considered to be inferior documents to BS 1377 and hence have not been published by British Standards Institution. A decision on the final status of the TSs is not likely to be made before 2010.

4.8.2   Objectives

The objective and scope of each test are given in §5.x.1 of EN 1997-2 (where the x can be determined from the table above). For example, the purpose of strength testing soil is stated as:

> *to establish the drained and/or undrained shear strength parameters*
> *[EN 1997-2 §5.8.1(1)]*

4.8.3   Requirements

The requirements of each test are given in §5.x.2 of EN 1997-2 by reference to the appropriate part of EN ISO 14689 or EN ISO-TS 17892 (see table above). The existing UK standard BS 1377 is being updated to remove any conflicts with the relevant EN ISO standards (in particular, Section 9 of BS 1377 which deals with in situ testing). However, at the time of writing (early 2008), there are no plans to incorporate the requirements of EN ISO-TS 17892 (which are considered to be technically inferior) into BS 1377.

4.8.4   Evaluation and use of tests results

The way in which the results of each test should be evaluated is specified in §5.x.3 of EN 1997-2 by reference to the appropriate part of EN ISO 14689 or EN ISO-TS 17892 (see table above). However, as noted above BS 1377 will continue in the UK.

4.8.5   Applicability of laboratory tests to obtain parameters

EN 1997-2 provides guidance on the applicability of laboratory tests for determining common geotechnical parameters, as summarized in the table below.

Laboratory test (and abbreviation)	† Soil type		
	Coarse (Gr/Sa)	Fine (Si)	Fine (Cl)
Bulk density determination (BDD)	ρ	ρ	ρ
Oedometer (OED)	-	$E_{oed}\ C_c\ c_v$	$E_{oed}\ C_c\ c_v$
	P $E_{oed}\ C_c$	-	k
Particle size analysis (PSD)	k	-	-
Direct simple shear (DSS)	-	c_u	c_u
Ring shear (RS)	c'_R/φ_R	c'_R/φ_R	c'_R/φ_R
Translational shear box (SB)	c'/φ	c'/φ	c'/φ
	P c'_R/φ_R	$c'_R/\varphi_R\ c_u$	$c'_R/\varphi_R\ c_u$
Strength index (SIT)	-	c_u	c_u
Triaxial (TX)	$E'\ G\ c'/\varphi$	$E'\ G\ c'/\varphi\ c_u\ c_v$	$E'\ G\ c'/\varphi\ c_u\ c_v$
	P $E_{oed}\ C_c$		

†Applicability: P = partial
ρ = bulk density
E_{oed} = oedometer modulus
C_c = compression index
c_v = coefficient of consolidation
k = permeability
c_u = undrained shear strength
c'_R/φ_R = residual shear strength, c'/φ = drained effective shear strength
E' = Young's modulus, G = shear modulus

4.9 Testing of geotechnical structures

Testing of geotechnical structures is covered by International Standard EN ISO 22477, which is divided into seven parts covering static pile load tests (Parts 1–3), dynamic pile load tests (Part 4), and testing of anchorages (Part 5), nailing (Part 6), and reinforced fill (Part 7).[14] At the time of writing (early 2008), only Part 1 has been approved for publication while the other parts are either in development or in abeyance.

4.10 Summary of key points

'There are ... a substantial number of standards and technical specifications emanating from ... the European standards organisation. These will significantly affect the [geotechnical and engineering geology] industry and cover everything from qualifications to field and laboratory tests ...'[15]

Furthermore:

'There are some significant changes to the way in which we describe soil and rock contained within the new European standards.'[16]

The standardization of ground investigation and testing has undergone a massive leap forward in the past few years, reflected by the number (over forty) and scope of the International Standards published on the subject.

4.11 Worked examples

The worked examples in this chapter use the guidance given in Eurocode 7 Part 2 on the scope of ground investigations to illustrate where there are significant changes from current UK practice. Example 4.1 considers the specification of field work for a hotel site in North West England; Example 4.2 looks at how changes in soil description will affect the borehole logs; and Example 4.3 discusses the specification of laboratory tests.

Specific parts of the calculations are marked ❶, ❷, ❸, etc., where the numbers refer to the notes that accompany each example.

4.11.1   Specification of investigation points

Example 4.1 considers a site in the North West of England where a new hotel complex is proposed (see **Figure 4.16**), comprising a five storey hotel building and associated car parking plus amenity areas. The section of the site to be occupied by the hotel building will be approximately 80m x 60m.

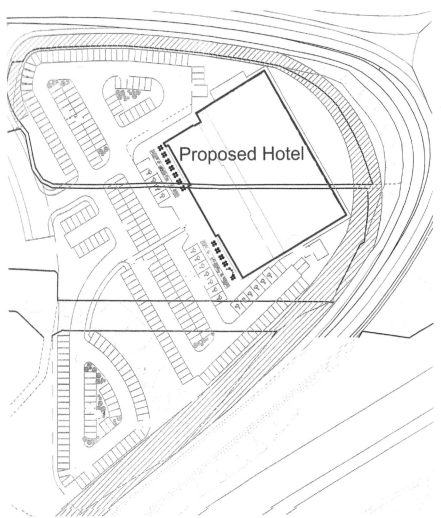

Figure 4.16. Five storey hotel requiring a suitable ground investigation

The general geology of the area indicates drift deposits comprising glacial till or sand and gravel overlying Westphalian coal measures. The drift deposits may be up to 15m deep and vary laterally.

Notes on Example 4.1

❶ Recommended spacing of investigation points for high rise and industrial structures is given in Annex B of EN 1997-2.

❷ The function 'ceil' rounds up to the nearest integer, hence ceil (80/40 + 1) = 3 and ceil (60/40 + 1) = 3. And 3 x 3 = 9.

❸ The function 'floor' rounds down to the nearest integer, hence floor (80/15 + 1) = 6 and floor (60/15 + 1) = 5. And 6 x 5 = 30.

❹ How many investigation points to provide between 9 and 30 is a matter of engineering judgement. Since the hotel building is relatively low rise, we have chosen a value (12) towards the lower end of this range. Eurocode 7 does not require that all these points be boreholes, but sufficient investigation to depth is required to determine any vertical or horizontal variation in ground properties.

❺ The minimum depth of investigation for individual footings is 6m.

❻ Unless a competent stratum is encountered at shallower depth, the minimum depth of investigation for a raft is 90m! It is likely at this site that coal measures will be encountered within 15m of ground surface. Thus, the investigation may be terminated 2–5m into the competent stratum (i.e. 17–20m depth). At least one of the early investigation points should be taken deeper into the underlying stratum to check for weaker layers. The rest of the investigation may then be adjusted to account for the information gathered.

❼ The minimum depth of investigation for a piled foundation is 60m. Similar reasoning to ❻ above may allow that depth to be shortened.

❽ The scope of investigation recommended by EN 1977-2 is much greater than that which has traditionally been carried out in the UK. For this project, eight boreholes were drilled to a maximum depth of 13.2m supplemented by 20 trial pits. Only three boreholes and seven trial pits were within the hotel's footprint; none of the boreholes penetrated the underlying coal measures, but some terminated in a dense to very dense sand or gravel.

Example 4.1
Specification of investigation points

Design situation

Consider the design of a ground investigation for a new hotel complex, comprising a five storey hotel building and associated car park and amenity facilities. The building will occupy an area $B = 80m \times L = 60m$ of the site.

Number of investigation points

For high-rise and industrial structures, investigation points should be on a grid between $s_{min} = 15m$ and $s_{max} = 40m$ spacing. **①**

Minimum number of investigation points is:

$$n_{min} = ceil\left(\frac{B}{s_{max}} + 1\right) \times ceil\left(\frac{L}{s_{max}} + 1\right) = 9 \quad ②$$

Maximum number of investigation points is:

$$n_{min} = floor\left(\frac{B}{s_{min}} + 1\right) \times floor\left(\frac{L}{s_{min}} + 1\right) = 30 \quad ③$$

Since the building is relatively low-rise, select a low number between these limits, say $n = 12$ **④**

Depth of investigation points for individual shallow foundations

Assume individual shallow foundations will be $b_f = 2m$ wide

Minimum depth of investigation is $z_{spread} = max\left(6.0m, 3 \times b_f\right) = 6m$ if using spread footings **⑤**

Depth of investigation points for raft

Assume raft will be $b_{raft} = 60m$ wide

Minimum depth of investigation is $z_{raft} = max\left(6.0m, 1.5 \times b_{raft}\right) = 90m$ **⑥**

Depth of investigation points for pile foundations

Assume individual piles will have base diameter $D_F = 600mm$ and the minimum overall width of the pile group is $b_g = L = 60m$

Minimum depth of investigation is $z_{pile} = max\left(5.0m, 3 \times D_F, b_g\right) = 60m$ **⑦**

Summary

Allowing for competent stratum at $t = 15m$, proposed maximum depth of investigation is $z_{max} = t + 5m = 20m$ **⑧**

4.11.2   Borehole log

Example 4.2 presents a typical borehole log from the hotel site, prepared according to BS 5930. The notes that follow discuss the changes necessary to bring the log into line with EN 1997-2 and EN ISO 14688.

Notes on Example 4.2

❶ This description remains the same after the introduction of Eurocode 7 and its associated standards, in particular EN ISO 14688.

❷ According to BS 5930, the term 'stiff' implies the clay has an undrained strength c_u between 75 and 150 kPa. Under EN ISO 14688, the term 'stiff' refers solely to the soil's consistency as assessed by field inspection of the clay samples. Where measurements of undrained strength are made, an additional term may be added to the description, e.g. 'medium strength' for c_u = 40–75 kPa. Two triaxial compression tests gave undrained strengths of 59 and 72 kPa. Hence a Eurocode 7 compatible soil description in this instance would be 'Stiff brown medium strength sandy becoming gravelly CLAY', where the CLAY may be abbreviated to 'Cl'. The terms 'stiff' and 'medium strength' are not inconsistent.

❸ This description remains the same under EN ISO 14688, but could be abbreviated to 'Dense brown clSa and Gr'.

❹ This description remain the same under EN ISO 14688.

❺ According to BS 5930, this sand is described as 'dense' because the measured standard penetration test (SPT) blow count is in the range 30–50. However, Annex F of Eurocode 7 Part 2 correlates blow counts corrected for rod energy and depth, the $(N_1)_{60}$ value, with the relative density descriptive terms. The range of corrected blow counts are different from BS 5930. However, it has been recommended[17] that the descriptive terms for relative density continue to be based on the uncorrected N-value and the ranges given in BS 5930.

❻ The description remains the same under EN ISO 14688. See ❺ above.

Example 4.2
Borehole log

Decoding Eurocode 7 by Andrew Bond and Andrew Harris Chapter 4 Ground investigation and testing		Borehole NO **BH 01** Sheet 1 of 1

Project Name WITLEY HOTEL	Project No. GEO	Co-ords: 627366E · 127654N	Hole Type Cable
Location: Warrington		Level: 15.36mAOD	Scale 1:75
Client: Euronorm Ltd		Dates: 03/05/2007	Logged By AJB

Well	Water Strikes	Samples & In Situ Testing Depth (m)	Type	Results	Depth (m)	Level (m AOD)	Legend	Stratum Description	
		0.00-0.80	B					Brown friable gravelly TOPSOIL ❶	
		1.00	D					Stiff brown sandy becoming gravelly CLAY	-1
		1.30	D					❷	
		1.50-1.95	U						-2
		2.00	D						
		2.50-2.95	SPT B	14					-3
		3.50-3.95	D						-4
		4.00	D						
		4.50-4.95	SPT B	17					-5
		5.50	D						-6
		6.00-6.45	U						
		6.50	D						
		7.00	D		7.10	8.26			-7
								Dense brown clayey SAND and GRAVEL ❸	
		7.50-7.95	SPT B	35					-8
		8.50	D		8.30	7.06		Stiff brown boulder CLAY	
					8.70	6.66		Dense brown fine to coarse gravelly SAND	-9
		9.00-9.45	SPT B	44				❹	
		10.00	D		10.10	5.26		Dense dark brown fine to coarse SAND ❺	-10
		10.50-10.95	SPT B	39					-11
		11.50-11.95	SPT B	46					
					42.00	3.36			-12
								End of borehole 12.00m ❻	
									-13

Remarks:

4.11.3 Specification of laboratory tests

Example 4.3 considers the amount of laboratory testing required for the hotel site according to Eurocode 7 Part 2. The guidance given in Annexes M to S allows the number of tests to be carried out on each stratum to be determined. This number depends on how much prior knowledge of the site exists and the expected variation in ground properties.

Assuming that the borehole log given in Section 4.11.2 is typical, but we have little previous experience of the site, we have indicated the amount of testing needed to allow the design of shallow or piled foundations in the box below. In practice, not all these tests would be carried out on samples from one borehole. Instead an assessment would be made of the lateral extent of each stratum and samples from different boreholes (but from the same stratum) would be used. More testing would be required of parameters whose value varies with depth in order to identify this variation.

Notes on Example 4.3

❶ We assume that effective stress testing of clays will not be necessary for this project.

❷ It may be better to establish effective stress parameters for granular soils from in situ tests rather than from laboratory tests. Eurocode 7 allows relevant in situ tests to be used in place of laboratory tests.

❸ The boulder clay is only 0.4m thick in the borehole log of Section 4.11.2 and it may therefore be impractical to carry out all the testing specified owing to a lack of material to test.

❹ The symbol E_{oed} is used here to mean all associated and equivalent consolidation parameters, such as the coefficient of consolidation c_v and coefficient of compressibility m_v.

Example 4.3
Specification of laboratory tests

Stratum	PSD	Att†	De‡	c_u	c', φ'	E_{oed} ❶
Brown friable gravelly TOPSOIL (Made Ground)	-*	-	-	-	-	-
Stiff brown sandy becoming gravelly CLAY	4	3	4	5	- ❶	4
Dense brown clayey SAND & GRAVEL	4	3w	-	-	3 ❷	-
Stiff brown boulder CLAY ❸	4	3	4	5	- ❶	4
Dense brown gravelly fine to coarse SAND	4	3w	-	-	3 ❷	-
Dense dark brown fine to coarse SAND	4	3w	-	-	3 ❷	-

*- indicates not applicable; †Atterberg limits (w_L and w_P) and water content (w); w after number indicates water content only; ‡Density

4.12   Notes and references

1. Driscoll, R.M.C., Powell, J.J.M., and Scott, P.D. (2008) *EC7 – implications for UK practice*, CIRIA RP701.

2. BS EN 1997-2: 2007, Eurocode 7 – Geotechnical design, Part 2 – Ground investigation and testing, British Standards Institution, London, 196pp.

3. BS EN ISO 17892, Geotechnical investigation and testing — Laboratory testing of soil, British Standards Institution.
    Part 1: Determination of water content.
    Part 2: Determination of density of fine grained soil.
    Part 3: Determination of density of solid particles – Pycnometer method.
    Part 4: Determination of particle size distribution.
    Part 5: Incremental loading oedometer test.
    Part 6: Fall cone test.
    Part 7: Unconfined compression test.
    Part 8: Unconsolidated undrained triaxial test.
    Part 9: Consolidated triaxial compression tests.
    Part 10: Direct shear tests.
    Part 11: Permeability tests.
    Part 12: Determination of Atterberg Limits.

4. Nicholson, D., Tse, C-M, and Penny, C. (1999) *The Observational Method in ground engineering: principles and applications*, CIRIA R185.

5. See, for example: Clayton, C.R.I., Simons, N.E., and Matthews, M.C. (1984) *Site investigation: A handbook for engineers*, London, Granada Publishing; or Coduto, D. P. (2001) *Foundation design: Principles and practices*, Prentice Hall.

6. BS EN ISO 14688, Geotechnical investigation and testing — Identification and classification of soil, British Standards Institution.
    Part 1: Identification and description.
    Part 2: Principles for a classification.
    Part 3: Electronic exchange of data on identification and description of soil.

7. BS 5930: 1999, Code of practice for site investigations, British Standards Institution.

8. BS EN ISO 14689, Geotechnical investigation and testing — Identification and classification of rock, British Standards Institution.
    Part 1: Identification and description.

Part 2: Electronic exchange of data on identification and description of rock.

9. International Society of Rock Mechanics (1980) 'Basic geotechnical description of rock masses', *Int. J. Rock Mech. Min. Sci. & Geomech. Abstr.*, 18, pp 85–110.

10. BS 5930, ibid., see §44.

11. BS EN ISO 22475, Geotechnical investigation and testing — Sampling methods and groundwater measurements, British Standards Institution.
    Part 1: Technical principles for execution.
    Part 2: Qualification criteria for enterprises and personnel.
    Part 3: Conformity assessment of enterprises and personnel by third party.

12. BS EN ISO 22282, Geotechnical investigation and testing — Geohydraulic testing, British Standards Institution.
    Part 1: General rules.
    Part 2: Water permeability tests in a borehole using open systems.
    Part 3: Water pressure tests in rock.
    Part 4: Pumping tests.
    Part 5: Infiltrometer tests.
    Part 6: Water permeability tests in a borehole using closed systems.

13. BS EN ISO 22476, Geotechnical investigation and testing — Field testing, British Standards Institution.
    Part 1: Electrical cone and piezocone penetration tests.
    Part 2: Dynamic probing.
    Part 3: Standard penetration test.
    Part 4: Ménard pressuremeter test.
    Part 5: Flexible dilatometer test.
    Part 6: Self-boring pressuremeter test.
    Part 7: Borehole jack test.
    Part 8: Full displacement pressuremeter test.
    Part 9: Field vane test.
    Part 10: Weight sounding test.
    Part 11: Flat dilatometer test.
    Part 12: Mechanical cone penetration test.
    Part 13: Plate loading test.

14. BS EN ISO 22477, Geotechnical investigation and testing — Testing of geotechnical structures, British Standards Institution.

Part 1: Pile load test by static axially loaded compression.
Part 2: Pile load test by static axially loaded tension.
Part 3: Pile load test by static transversely loaded tension.
Part 4: Pile load test by dynamic axially loaded compression test.
Part 5: Testing of anchorages.
Part 6: Testing of nailing.
Part 7: Testing of reinforced fill.

15. Powell, J. and Norbury, D. (2007) 'Prepare for EC7', *Ground Engineering*, 40(6), pp 14–17.

16. Baldwin, M., Gosling, D., and Brownlie, N. (2007) 'Soil and rock descriptions', *Ground Engineering*, 40(7), pp 14–24.

17. Baldwin et al., ibid.

Chapter 5

Ground characterization

'In dealing with real world problems, uncertainties are unavoidable'[1]

5.1   From test results to design

Ground characterization is the process of deducing suitable values for geotechnical parameters from the results of field or laboratory tests. Ultimately, these values will be used in design calculations, after the application of appropriate partial factors to cater for uncertainties in the available data.

There are three distinct steps in this process, as shown in **Figure 5.1**. Put simply, derivation involves converting test results into derived values (X); characterization involves choosing a suitable characteristic value (X_k) from those derived values; and, finally, factorization is the process of applying a partial factor to the characteristic value to make it more reliable for design purposes (X_d).

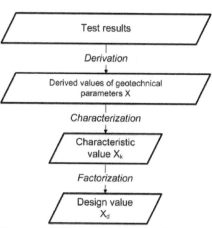

Figure 5.1. Overview of ground characterization

The remainder of this chapter considers the first two of these steps: derivation (Section 5.2) and characterization (Section 5.3).

5.2   Deriving geotechnical parameters

5.2.1   Overview

The derived value of a geotechnical parameter is defined in Eurocode 7 as:

> *[the] value ... obtained by theory, correlation or empiricism from test results*
> *[EN 1997-1 §1.5.2.5]&[EN 1997-2 §1.6(3)]*

As the flow-chart of **Figure 5.2** illustrates, test results may be converted into derived values X by use of correlations (such as that between cone penetration resistance and angle of shearing resistance in sand), theoretical considerations (such as conversion of triaxial compression into plane strain strengths for clays), or through empirical rules (such as those between standard penetration test blow count and undrained strength for clays).

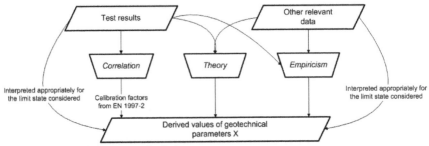

Figure 5.2. *Deriving geotechnical parameters*

When available, test results may be supplemented by other relevant data, such as that from nearby sites (i.e. comparable experience) or from research studies of the materials encountered.

Derived values may also be assessed directly, provided the engineer knows for which limit state the derived value is required.

Figure 5.3 gives an example of a geotechnical parameter (the undrained strength of two clays) derived from two different tests: standard penetration tests, with blow counts suitably converted into undrained strength (black symbols, SPT); and undrained triaxial compression tests (white symbols, TX).[2] This data is discussed further in Section 5.3.5 to illustrate the selection of characteristic values.

5.2.2   Correlations

Eurocode 7 Part 2[3] gives correlations for common field investigation methods and widely-used geotechnical parameters:

Annex D, cone penetration test: φ, E', E_{oed}
Annex F, standard penetration test: I_D, φ
Annex G, dynamic probing test: I_D, φ, E_{oed}
Annex H, weight sounding test: φ, E'
Annex J, flat dilatometer test: E_{oed}
Annex K, plate loading penetration test: c_u, E_{PLT}, k

where φ is the soil's angle of shearing resistance, E' its drained Young's modulus, E_{oed} the one-dimensional (oedometer) modulus, I_D relative density,

c_u undrained strength, E_{PLT} the plate-loading modulus, and k the coefficient of subgrade reaction.

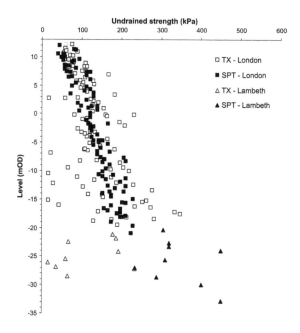

Figure 5.3. Derived values of undrained strength in London clay (squares) and Lambeth clay (triangles)

Correlations may be derived from a theoretical relationship between one parameter and another (e.g. drained Young's modulus E' and shear modulus G) or from some form of empirically derived relationship. Although theories exist that relate field test results to geotechnical parameters, these theories do not fully model the soil behaviour and engineers therefore rely more commonly on empirical rules.

Most correlations are limited to particular ground strata and/or geotechnical situations. Care must be exercised when extrapolating these correlations to strata and/or situations which differ from those on which they are based.

Figure 5.4 shows the correlation between SPT blow count and relative density of a coarse soil that is included in Annex F to EN 1997-2.

5.2.3   Theory

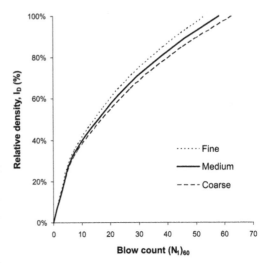

A theory is 'a plausible or scientifically acceptable general principle or body of principles offered to explain phenomena'.[4]

A theory allows the derivation of a parameter from a measured value or it describes how one geotechnical parameter may be related to one or more other parameters.

Figure 5.4. Correlation between SPT blow count and relative density of coarse soil

The field vane test is widely used to determine the undrained shear strength of low strength fine soils (e.g. soft clays). The test involves measuring the torque required to continuously turn the blades of the vane through the soil. The soil's undrained strength c_u is obtained from:

$$c_u = \frac{T}{K}, \text{ with } K = \pi \frac{D^2 H}{2}\left(1 + \frac{D}{3H}\right)$$

where T is the measured torque and K is a constant that depends on the vane's dimensions and shape (D and H being the diameter, i.e. the width of two blades, and height). The equation given above for K assumes a uniform distribution of shear strength across the ends and perimeter of the cylinder. For many vanes, H/D = 2 and so K = 3.66D^3. The theory on which this equation is based clearly does *not* depend on any factor of safety and hence it can be used equally well within the framework of Eurocode 7 as it was in traditional practice.

Theories may have universal application but are often limited to particular situations and their users should understand and work within those limitations. Normally, theories have wider application than empirically-derived formulae, which only apply to the data sets on which they are based.

There are reasonably robust theories to explain observations made under laboratory conditions, but theories become less reliable as the number of unknowns increases.

5.2.4 Empiricism

Following the introduction of Eurocode 7, existing empirical rules appear to have one of three fates: some may continue to be used 'as-is', some may need to be re-cast, and others will fall into disuse.

Many existing empirical rules were developed before the widespread use of partial factors in geotechnical design. Some of these rules implicitly include a factor-of-safety, usually to avoid excessive movement of the structure or to prevent the mobilization of high soil stresses. An example of this type of empirically derived correlation is the relationship provided by Terzaghi and Peck[5] for allowable bearing pressure and SPT blow count in sands, shown in **Figure 5.5**. Because the Structural Eurocodes ensure reliability through the application of partial factors, rules such as this cannot be used within the framework of Eurocode 7, unless they are suitably modified.

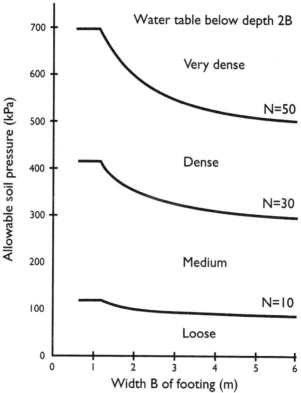

Figure 5.5. Terzaghi and Peck's correlation between allowable bearing pressure and SPT blow count in sand

Figure 5.6 shows Stroud and Butler's correlation – between standard penetration test blow count (N), undrained shear strength (c_u), and plasticity index (I_p) of clays – which is widely used in the UK. The relationship shown is typical of correlations

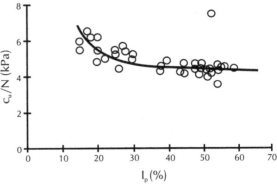

that are likely to survive **Figure 5.6**. *Stroud and Butler's correlation between* the introduction of *SPT blow count and undrained strength of fine soil* Eurocode 7, since it involves no use of safety factors and so can be used unmodified.

5.2.5   Direct assessment

Direct assessment of soil parameters is typically only possible on the basis of laboratory tests. For example, derived values of strength, permeability, and stiffness (drained or undrained) may be determined directly from appropriate triaxial tests. Such direct assessment must, however, take appropriate account of any sample disturbance that could influence the derived value.

5.2.6   Symbols

Symbols used in Eurocode 7 Part 1 for geotechnical parameters include c' for cohesion (in terms of effective stress), c_u for undrained shear strength, K_0 for the coefficient of earth pressure at rest, γ for weight density, and φ' for the angle of shearing resistance (in terms of effective stress) – with X being used as a general symbol for an unspecified material property.   *[EN 1997-1 §1.6(1)]*

A longer list of symbols for geotechnical parameters is given in Eurocode 7 Part 2.   *[EN 1997-2 §1.8(1)]*

In our opinion, the failure of Eurocode 7 to provide a more comprehensive list of 'standard' symbols is a lost opportunity to remove confusion between engineers who have to check each other's calculations or work together on the same calculations. In addition, the opportunity to standardize symbols common to both geotechnical and structural engineering has not been grasped, except at the highest level of abstraction, where, for example, E is used for effects of actions and R for resistance, etc.

Throughout this book, we have adopted a set of symbols that follows the notation used in many of the non-geotechnical Eurocodes. In particular, we have used simple capital letters to denote major variables, for example:

F for a general force
V for a vertical force
H for a horizontal force
M for a bending moment
q for surcharge
u for pore pressure

To each of these symbols we then attach subscripts that provide the necessary additional information that the Eurocodes require in order for the appropriate combination and partial factors to be applied. For example:

V_{Gk} = characteristic (k) permanent (G) vertical force (V)
$M_{Q,rep}$ = representative (rep) variable (Q) moment (M)
u_{Ad} = design (d) accidental (A) pore pressure (u)

Furthermore, we identify design effects of actions and design resistances by using the subscripts 'Ed' and 'Rd' attached to the main symbol, rather than use the symbols E_d and R_d alone; e.g. M_{Ed} rather than plain E_d; M_{Rd} rather than plain R_d.

5.3   Obtaining the characteristic value

EN 1990 defines a characteristic material property as follows:

> *where a low value ... is unfavourable, the characteristic value should be defined as the 5% fractile value; where a high value ... is unfavourable, ... as the 95% fractile value*   *[EN 1990 §4.2(3)]*

This definition works well for man-made materials, such as steel and concrete, but – as discussed below – fails to account for the remarkable variability of geomaterials and for the practical difficulty in obtaining measurements of relevant material properties in soil and rock.

5.3.1   Problems applying statistics to geotechnical parameters

Geotechnical engineering poses many challenges, not least of which is the need to determine ground conditions at each and every construction site. The adequacy of many site investigations leaves much to be desired, resulting in designs being prepared on the basis of insufficient information. Furthermore, many geomaterials are highly variable in nature, which makes it inherently difficult to establish not only their location but also their mechanical and chemical properties.

'The sources of uncertainty may be classified into two main types: (1) those that are associated with natural randomness [aleatory]; and (2) those that are associated with inaccuracies in our prediction and estimation of reality [epistemic].'[6]

It is not atypical for important design parameters to vary across a wide range of values, as illustrated in **Figure 5.7**. In addition, studies[7] of the coefficient of variance of various soil properties have shown it is typically much higher than for man-made materials (see table below). This is an example of aleatory uncertainty.[8]

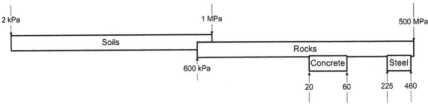

Figure 5.7. *Range of soil and rock strengths encountered in nature, compared with strengths of man-made materials*

Coefficient of variation (COV) of geotechnical and man-made materials[9]

Material	Parameter		COV
Soil	coefficient of shearing resistance	$\tan \varphi$	5–15%
	effective cohesion	c'	30–50%
	undrained strength	c_u	20–40%
	coefficient of compression	m_v	20–70%
	weight density	γ	1–10%
Concrete	resistance of beams and columns		8–21%
Steel			11–15%
Aluminium			8–14%

Epistemic uncertainty is highly relevant to geotechnical engineers, who rarely have sufficient test data with which to justify a statistical approach to parameter selection.

In 1991, the Ground Board of the UK's Institution of Civil Engineers published a report[10] on site investigation, which recommended:

> 'National guidelines on the extent, intensity and quality of ground investigation should be produced for the benefit of clients, planners and engineers. These guidelines should follow the philosophy of Eurocode 7 on geotechnical categories.'

We hope that the introduction of Eurocode 7 will provide greater encouragement for all parties to procure adequate ground investigation.

5.3.2 Cautious estimate

Because of the inherent difficulties in selecting characteristic geotechnical parameters on the basis of EN 1990's statistical definition, Eurocode 7 're-defines' the characteristic value as:

> a cautious estimate of the value affecting the occurrence of the limit state
> [EN 1997-1 §2.4.5.2(2)P]

But what is a cautious estimate? EN 1997-1, unfortunately, gives little guidance, so we must revert to commonly accepted definitions:

> cautious [adj.] careful to avoid potential problems or dangers
> estimate [n.] an approximate calculation or judgement [11]

Combining these definitions gives:

> cautious estimate = approximate calculation/judgement that is careful to avoid problems or dangers

The words 'affecting ... the limit state' are an important rider to the definition of a characteristic geotechnical parameter. It means that there is no such thing as *the* characteristic value – rather there are potentially several characteristic values, one for each limit state being considered.

For example, when selecting a characteristic angle of shearing resistance for use in the design of a pile foundation, we might select a lower value for calculating end bearing resistance (which involves primarily compression) than for calculating shaft friction (which involves direct shear). Thus in this example: $\varphi_{k,base} < \varphi_{k,shaft}$.

A consequence of EN 1997-1's definition is that the characteristic value can only be selected during the design of the structure, not before (for example as part of a ground investigation), since determination of limit states is a design activity.

5.3.3   Representative values

Prior to the publication of Eurocode 7, the design of retaining walls in the UK was based on 'representative' soil parameters, defined in BS 8002[12] as:

> *conservative estimates ... of the properties of the soil as it exists in situ ... properly applicable to the part of the design for which it is intended*

For soil parameters such as weight density, which show little variation in value, the representative value 'should be the mean value of the test results'. Where greater variations occur (or where values cannot be fixed with confidence), the representative value 'should be a cautious assessment of the lower limit ... of the acceptable data'.

In practice, the difference between BS 8002's representative value and Eurocode 7's cautious estimate is merely one of semantics.

5.3.4   Moderately conservative and worst credible values

The terms 'moderately conservative' and 'worst credible' describe two commonly-used UK design approaches, which were originally set out in a seminal report covering the design of embedded retaining walls in stiff clays (CIRIA 104).[13]

According to CIRIA 104, the safety factors applied to soil parameters should depend on how those parameters, together with loads and geometry, have been selected. The table opposite summarizes the main aspects of the two design approaches and indicates the equivalent level of reliability in Eurocode terms.

At the time CIRIA 104 was written, the moderately conservative design approach was the method most often used by experienced engineers in the UK. Regrettably, CIRIA 104 does not define precisely what is meant by 'moderately conservative'. Referring once again to the dictionary:

> *moderately [adv.] average in amount, intensity, or degree*
> *conservative [adj.] (of an estimate) purposely low for the sake of caution*[14]

Combining these definitions gives:

> moderately conservative = a low, cautious average

Parameter/factor		Design approach	
		Moderately conservative	Worst credible
Soil	select ...	'conservative best estimate' values	'worst realistic' values
Loads			values that are 'very unlikely to be exceeded' (but not worst physically possible)
Geometry			
Safety factors applied to soil strength (F_s)	apply ...	generous values	less conservative values
	total stress analysis	1.5 (temporary works)	not recommended
	effective stress analysis	1.1–1.2 (temporary works) 1.2–1.5 (permanent works)	1.0 (temporary works) 1.2 (permanent works)
Equivalent safety level		5% fractile?	0.1% fractile

According to CIRIA 104's successor, CIRIA C580,[15] soil parameters selected by this method are equivalent to representative values defined in BS 8002 (see Section 5.3.3) and to characteristic values defined in Eurocode 7 (Section 5.3.2). The difference between CIRIA 104's moderately conservative value and Eurocode 7's cautious estimate is again merely one of semantics.

5.3.5   How much ground is involved?

An important factor in selecting the characteristic value of a ground parameter X is an assessment of how much ground is relevant to the occurrence of the limit state.

Failure of part of the ground may not lead to an ultimate limit state being exceeded by the structure – for example, forces are often redistributed from highly-stressed regions to adjacent lower-stressed regions. Because of this, it is the *average* value of the material strength (or other relevant material property) that governs the occurrence of the limit state. The characteristic value X_k should be selected as a cautious estimate of the *spatially averaged* value of X representing the relevant volume of soil or rock. In statistical terms, Eurocode 7 requires a 95% confident assessment of the mean value of X.

When a small volume of ground is involved, the characteristic value X_k should be selected as a cautious estimate of the spatially averaged value of X within that small volume (otherwise known as the 'local' value of X). This value may be significantly lower than the one selected for a larger volume of ground, since there is less averaging being done. Statistically, the characteristic value X_k in this instance will be closer to the 5% fractile discussed in Chapter 2.

For example, consider the settlement of a footing resting on a layered soil profile, in which one particular layer (e.g. peat) is much more compressible than the others. In this situation, the coefficient of compressibility (m_v) of the peat dominates the calculation of total settlement and hence we are particularly interested in the *local* value of m_v in this layer.

Figure 5.8 shows our evaluation of the characteristic undrained strengths of the London and Lambeth Clays first shown in **Figure 5.3** and chosen for pile design.

The black symbols give values of undrained strength c_u derived from 100 standard penetration tests (SPTs), assuming the following ratio between c_u and SPT blow count N:

$$c_u = 4.5 \times N$$

which was chosen from Stroud and Butler's correlation with plasticity index shown in **Figure 5.6**.

The white symbols give values of c_u measured in 91 triaxial compression tests (TXs) on U100 samples. We have ignored the circled data points on **Figure 5.8** since they give unrealistically low strengths, believed to be the result of sample disturbance.

In the London Clay (above -20m OD), we have chosen a cautious estimate of the spatially averaged strength, suitable for calculating the average skin friction over the pile shaft with 95% confidence. Because the points are not particularly scattered, the selected line is close to the middle of the data. If we had been worried about locally low values controlling shaft friction, we would have chosen the dashed line, which is a lower bound to the SPT results. We have attached greater importance to the SPT results since they provide – albeit indirectly – measurements of the clay strength in situ, whereas direct measurement of strength in laboratory tests on small samples may not be representative of field conditions.

In the Lambeth Clay (below -20m OD), we have chosen a cautious estimate of the spatially-averaged strength, for a smaller volume of soil (down to -33m OD), suitable for calculating the pile's end-bearing resistance. Because the data points are highly scattered in the Lambeth Clay and end-bearing involves a relatively small volume of soil, we also gave serious consideration to choosing the cautious estimate of the local strength given by the lower dashed line. There is some disparity between the strengths obtained from the SPTs and those from the triaxial tests, the dashed line being a better fit to the latter.

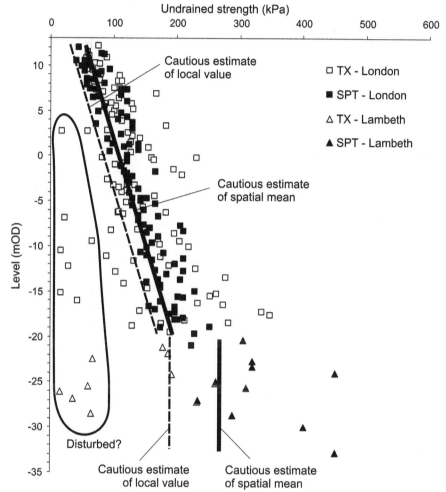

Figure 5.8. *Difference between local and spatially-averaged characteristic values*

5.3.6   Well-established experience

Eurocode 7 requires the selection of characteristic values to be 'complemented by well-established experience', i.e. practical contact with and observation of facts or events and knowledge or skill acquired over time.[16]

Well-established experience also includes simple rules-of-thumb for estimating soil parameters, such as the following[17] for determining the peak (φ) and constant-volume (φ_{cv}) angles of shearing resistance of siliceous sands and gravels:

$$\varphi = 30° + \begin{Bmatrix} 0, \text{ rounded} \\ 2, \text{ sub-angular} \\ 4, \text{ angular} \end{Bmatrix} + \begin{Bmatrix} 0, \text{ uniform} \\ 2, \text{ moderate grading} \\ 4, \text{ well-graded} \end{Bmatrix} + \begin{Bmatrix} 0, N < 10 \\ 2, N = 20 \\ 6, N = 40 \\ 9, N = 60 \end{Bmatrix}$$

$$\varphi_{cv} = 30° + \begin{Bmatrix} 0, \text{ rounded} \\ 2, \text{ sub-angular} \\ 4, \text{ angular} \end{Bmatrix} + \begin{Bmatrix} 0, \text{ uniform} \\ 2, \text{ moderate grading} \\ 4, \text{ well-graded} \end{Bmatrix}$$

5.3.7   Standard tables of characteristic values

Geotechnical engineers are sometimes required to estimate ground parameters for a particular stratum without the benefit of test results in that stratum. This is often the case for made ground or other near-surface soils, for occasional layers of coarse soil (i.e. gravel or sand), and for thin layers of fine soil (i.e. silt or clay). In the absence of test results, ground parameters may be selected from standard tables of values, such as those that appear in many geotechnical text books.[18]

The 'standard table' below shows how to determine a clay's constant-volume angle of shearing resistance (φ_{cv}) from its plasticity index (I_p):[19]

Plasticity index, I_p	15%	30%	50%	80%
Angle of shearing resistance, φ_{cv}	30°	25°	20°	15°

When choosing a characteristic value from tables such as these, Eurocode 7 requires a 'very cautious value' to be selected. It is important, therefore, to understand the engineering basis of the table, so that the degree of caution embodied in the published values can be taken into account when selecting the characteristic value.

5.3.8   Summary of geotechnical characterization

Figure 5.9 summarizes the ways in which a characteristic value X_k may be chosen according to Eurocode 7. In most situations, a cautious estimate of X_k is made from derived values of the geotechnical parameter X; when there is sufficient data, statistical methods may be used to select a 95% confident value (upper or lower, as appropriate); in the absence of data, reference made to standard tables of characteristic values may be used, with increased caution. In all cases, the resultant value should be checked against well-established experience.

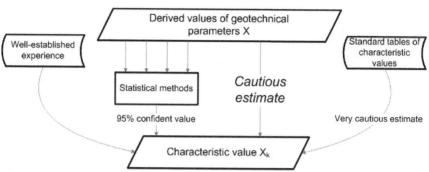

Figure 5.9. Characterizing geotechnical parameters

5.4   Case studies selecting characteristic values

This section of the book presents a series of case studies that investigate the process of selecting a characteristic value (or values) from field and laboratory tests. The first study (see Section 5.4.1) considers two stiff over-consolidated clays found in London; the second (Section 5.4.2) looks at the soft marine clay found in Singapore; and the third (Section 5.4.3) considers dense, well-graded gravels at Gravesend in Kent.

5.4.1   London and Lambeth clays at Holborn

Figure 5.10 shows the results of more than a hundred standard penetration tests performed at a site in Holborn, London, where over six hundred piles were to be installed by conventional rotary bored techniques.[20] The (uncorrected) blow counts increase steadily with depth in the London Clay (solid triangles) and then jump up in value in the Lambeth Clay (open triangles). This data is typical of many sites in London.

As part of our research for this book, we asked more than one hundred engineers and engineering geologists to select a characteristic line (or lines) through this data, on the basis of Eurocode 7's definition of the characteristic value as a 'cautious estimate of the value affecting the occurrence of the limit state'. The lines on **Figure 5.10** show the outcome of this study.

In the London Clay (above about −21m OD), most engineers selected a line that comes within a distinct and narrow band some seven blows wide; the blow count defined by this band increases with depth at a rate just over one blow per metre. Outside of this band are several extreme interpretations that appear difficult if not impossible to justify on the basis of available data.

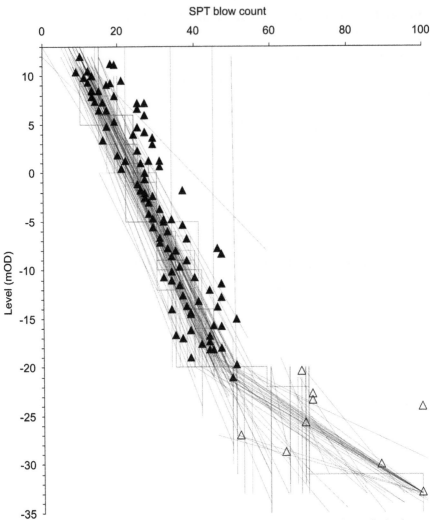

Figure 5.10. *Results of standard penetration tests in London and Lambeth clays, with engineers' interpretation of the characteristic value*

In the Lambeth Clay (below −21m OD), a variety of interpretations were submitted. Some involve a linear increase in blow count with depth (at a rate approaching 3 blows/m); others assume a constant value between 50 and 70. **Figure 5.10** shows that engineers' interpretation of data becomes more scattered where that data is less certain (owing to its greater variability and paucity).

Figure 5.11 shows the results of more than a hundred undrained triaxial compression tests performed in the laboratory on U100 samples taken from nine different boreholes from the site in Holborn discussed above. The undrained shear strength measured in both clays is highly scattered, caused in large part by sample disturbance prior to laboratory testing.

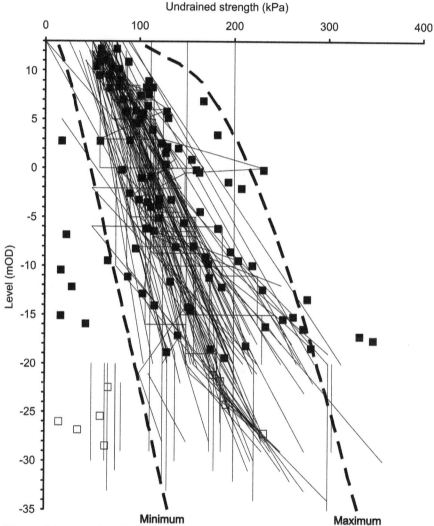

Figure 5.11. Results of triaxial compression tests in London and Lambeth clays, with engineers' interpretation of the characteristic value

The dashed lines on **Figure 5.11** indicate theoretical limits to the undrained strength, assuming the following simple relationship between c_u, effective overburden pressure σ_v', and over-consolidation ratio (OCR):

$$\left(\frac{c_u}{\sigma_v'}\right) = \left(\frac{c_u}{\sigma_v'}\right)_{nc} \times OCR^{0.8}$$

where $(c_u/\sigma_v')_{nc}$ has been assumed to be 0.23. This relationship assumes the removal of overburden pressure due to geological processes: for the line labelled 'minimum', 5m of overburden is removed and, for the one labelled 'maximum', 70m. There are seven results in London Clay (solid squares) and five results in Lambeth Clay (open squares) that are below the minimum — these may be considered physically impossible values for the in situ clay. There are also four results in London Clay that are above the maximum — these too may be physically unreasonable values (although they might be believable if more than 70m of overburden was removed).

We asked the same one hundred engineers as before to select a characteristic line (or lines) through this data based on a 'cautious estimate'. The thin solid lines on **Figure 5.11** show the outcome.

The bandwidth of the interpretations in the London Clay increases from perhaps 50kPa at +10m OD to about 100kPa at –20m OD. Rather alarmingly, the ratio of the most optimistic to the most pessimistic assessment of c_u is as much as three-fold. The seven results below the minimum line and the four above the maximum were (almost always) ignored.

In the Lambeth Clay, estimates of c_u ranged from 50kPa to 300kPa — the lower value is clearly influenced by the five physically impossible values that fall below the minimum line.

5.4.2   Singapore marine clay

Figure 5.12 shows the results of about forty field vane tests performed at a site in Singapore, where bored piles and barrettes were installed for a residential development.[21] Ground conditions typically comprise up to 35m of soft to firm marine CLAY, overlying silty SAND/sandy SILT, overlying SANDSTONE.

The marine clay has low to medium sensitivity (S_t = 2–7). Laboratory tests suggest it is under-consolidated (over-consolidation ratio, OCR < 1) but this may be in error owing to sample disturbance – independent studies suggests that Singapore marine clay is lightly over-consolidated (OCR = 1.5–2.5) and this is supported by the soil's liquid limit exceeding its natural water content.

The data shown on **Figure 5.12** indicates that the marine clay's undrained strength c_u increases with depth at a steady rate and lies slightly above the following relationship with vertical effective stress σ'_v:

$$c_u = 0.22\sigma'_v$$

(shown by the dashed line), which is commonly used to estimate the undrained strength of normally consolidated clays.

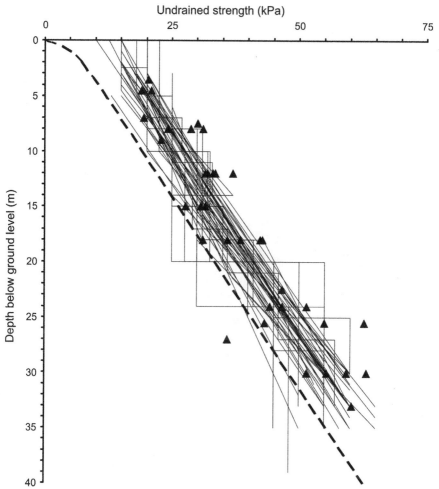

Figure 5.12. *Results of field vane tests in Singapore marine clay, with engineers' interpretation of the clay's characteristic undrained strength*

As for the Holborn case study, we asked more than one hundred engineers and engineering geologists to select a characteristic line (or lines) through this data, on the basis of Eurocode 7's definition of the characteristic value as a cautious estimate. The lines on **Figure 5.12** show the outcome of this study.

A majority of engineers drew a straight line through the data points; others preferred a stepped relationship. The bandwidth of the interpretations is about 10kPa and remains fairly constant with depth.

Figure 5.13 shows the results of more than eighty undrained triaxial compression tests performed in the laboratory on U100 samples taken from eight different boreholes from the same site in Singapore. The measured shear strength is highly scattered, caused in large part by sample disturbance prior to laboratory testing.

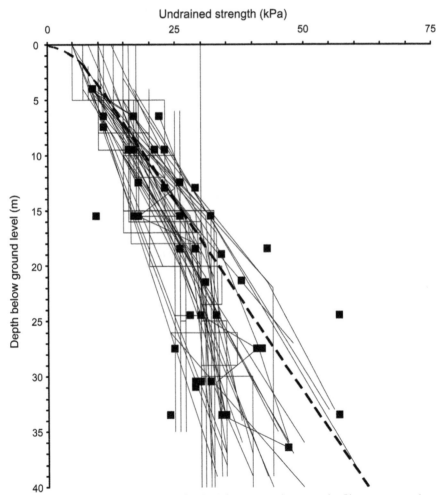

Figure 5.13. Results of undrained triaxial compression tests in Singapore marine clay, with engineers' interpretation of the clay's characteristic undrained strength

Also shown on **Figure 5.13** is the relationship between undrained strength and vertical effective stress given above for normally consolidated clays. The results are generally below the line, confirming the assessment based on the vane tests that sample disturbance occurred during sampling.

We asked the same one hundred engineers to select a characteristic line (or lines) through this data based on a 'cautious estimate'. **Figure 5.13** shows the outcome.

The interpretations are far more variable than for the field vane results, which reflects the greater variability in the laboratory measurements. The engineers were particularly conservative in their estimates of c_u deeper than 20m, where a value of 35kPa was popular.

It is interesting to note that the dashed line — which represents a typical c_u/σ'_v relationship for normally consolidated clay — has a better agreement with the triaxial than the vane test results. However, the laboratory data has considerable scatter, indicating that sample disturbance (both physical and stress-related) was significant. Disturbance is less likely for the in situ vane tests. If a correction factor[22] of 0.8 is applied to the vane test results (based on the soil's plasticity index of 40–60%), better agreement would be obtained with the c_u/σ'_v relationship.

5.4.3 Thames Gravels at Gravesend

Figure 5.14 shows the results of about forty standard penetration tests (SPTs) performed at a site near Gravesend, in Kent, where about one thousand driven-cast-in-situ piles were to be installed.[23] Two site investigations ('A' and 'B') were undertaken at the site and gave markedly different results.

In Investigation A, the SPT blow counts obtained in the Thames Gravels between 14m and 22m below ground level show very little variation with depth, as can been seen from the open white symbols on **Figure 5.14**. By contrast, in Investigation B, the blow counts obtained are highly variable and appear to increase slightly with depth — witness the closed black symbols.

As for the Holborn and Singapore studies, we asked more than forty civil and geotechnical engineers to select a characteristic line (or lines) through this data, based on a 'cautious estimate'. The engineers were given the data for Investigations A and B separately.

Many of the engineers questioned whether the data from Investigation A is credible, since it shows very little scatter in a stratum that is expected to be highly variable. We have therefore ignored Investigation A in this case study.

The interpretations obtained for Investigation B are shown by the lines on **Figure 5.14**.

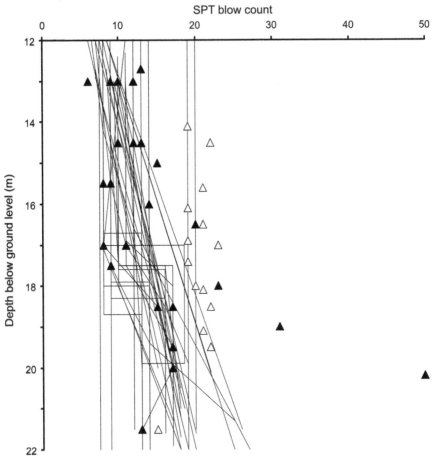

Figure 5.14. *Results of standard penetration tests in Thames Gravels, with engineers' interpretation of the characteristic value for Investigation B (black symbols). Interpretation for Investigation A (white symbols) is not shown*

It is difficult to understand how, when asked for a cautious estimate of the characteristic value of blow count N, the engineers could produce such a wide spread of interpretations. The data presented is fairly typical of the quality and variability of data obtained in most site investigations. This exercise therefore poses serious questions about how the industry selects design parameters.

5.4.4   Conclusions from the case studies

Our conclusion from these studies is that engineers are not particularly good at selecting a cautious estimate of the characteristic value, particularly when the available data is scattered. Statistical treatment of large data sets (such as these) may help to guide engineers in this task (see Section 5.5).

5.5   Statistical methods for ground characterization

> '[Use of statistics] demands a high order of statistical technique, available from very few designers who have committed their time to training and experience in geotechnical engineering.'[24]

The characteristic value of a material property X_k is defined in Chapter 2 as follows:

$$X_k = m_X \mp k_n s_X$$

where m_X is the mean of X, s_X its standard deviation, and k_n a statistical coefficient that depends on the number of samples n.

This definition may be stated more simply as:

Characteristic value = mean value $\mp$ epistemic x aleatory uncertainties

Statistical methods for determining the epistemic and aleatory uncertainties of man-made materials are presented in Chapter 2. The following sub-sections discuss the way natural materials such as the ground depart from the assumptions made there (and embodied in EN 1990).

5.5.1   Normal or log-normal distribution?

A normal (a.k.a. Gaussian) distribution arises when a physical property depends on the combination of a large number of individual, random effects.[25] The normal distribution is encountered frequently in nature and is one of the most important *probability density functions* in the field of statistics. Many man-made materials have properties that follow a normal distribution, e.g. the compressive strength of concrete and the yield strength of steel.

A log-normal distribution arises when a physical property depends on the *product* of a large number of individual, random effects. The log-normal distribution is especially useful in cases where the data cannot take on negative values. Since this applies to many geotechnical parameters (e.g. the angle of shearing resistance of soils φ, the undrained shear strength of fine soils c_u, and other parameters with a coefficient of variation greater than about 30%), the log-normal distribution is of particular interest to geotechnical engineers.

Consider the test results shown in **Figure 5.15**, which were obtained from cone penetration tests in clay tills.[26] If we assume that the results follow a normal distribution (shown by the dashed bell-shaped curve on **Figure 5.15**), then the mean cone resistance is 1.99 MPa and its standard deviation 0.73 MPa. Using the equations given in Chapter 2, the lower ('inferior') and upper ('superior') characteristic values of q_c are then:

$$\left.\begin{matrix}q_{ck,inf}\\q_{ck,sup}\end{matrix}\right\} = m_{qc} \mp k_n s_{qc} = 1.99 \mp 1.647 \times 0.73 = \begin{cases}0.78 MPa\\3.20 MPa\end{cases}$$

However, the test results more closely follow a log-normal distribution, shown by the *skewed* bell-shaped curve of **Figure 5.15** and defined by the expression:

$$P(X,\lambda,\varsigma) = \frac{1}{\varsigma X\sqrt{2\pi}}e^{\left(\frac{-(\ln X-\lambda)^2}{2\varsigma^2}\right)}, X \geq 0$$

where X in this case is the cone resistance; $P(X, \lambda, \varsigma)$ is the probability density function for X; λ is the mean value of $\ln(X)$; and ς the standard deviation of $\ln(X)$.

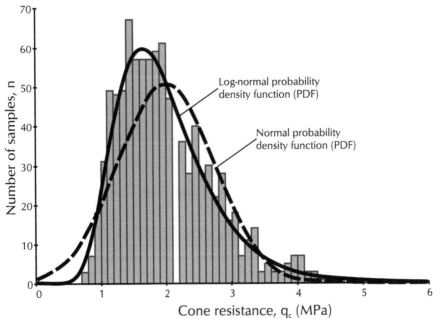

Figure 5.15. Results of cone penetration tests in clay tills

The mean λ_X and standard deviation ζ_X of the log-normal distribution are related to the corresponding values for the normal distribution, as follows:[27]

$$\lambda_X = \ln(\mu_X) - \frac{\zeta_X^2}{2}$$

$$\zeta_X = \sqrt{\ln\left[1+\left(\frac{\sigma_X}{\mu_X}\right)^2\right]} = \sqrt{\ln\left[1+\delta_X^2\right]}$$

where, in this instance, ζ_X = 0.357 and λ_X = 0.624. The inferior and superior characteristic values are then given by:

$$\left.\begin{array}{l} q_{ck,\text{inf}} \\ q_{ck,\text{sup}} \end{array}\right\} = e^{\lambda_{qc} \mp k_n \zeta_{qc}} = e^{0.624 \mp 1.647 \times 0.357} = \begin{cases} 1.04 MPa \\ 3.36 MPa \end{cases}$$

The log-normal distribution gives an inferior characteristic value that is 32% higher than obtained with the normal distribution and a superior value some 5% higher.

Many geotechnical parameters more closely follow a log-normal distribution and statistics based on the (more complicated) equations given in this chapter are preferred to statistics based on the equations given in Chapter 2.

5.5.2   Calculating the 95% confident mean value

Chapter 2 presents the statistical basis for selecting the upper and lower characteristic values of a material property, based on the 5% fractiles of the normal distribution — i.e. values that have a 5% probability of being exceeded.

As discussed in Section 5.3.5, many geotechnical designs involve a large volume of ground for which the spatial average of the material property is of greater relevance than a 5% fractile. In statistical terms, what we must calculate are the 95% confidence limits to the 50% fractile.

For this situation, if the variance of the population is known from prior knowledge (and hence does not need to be determined from the sample), the statistical definitions of $X_{k,\text{inf}}$ and $X_{k,\text{sup}}$ are:

$$\left.\begin{array}{l} X_{k,\text{inf}} \\ X_{k,\text{sup}} \end{array}\right\} = \mu_X \mp \kappa_N \sigma_X = \mu_X(1 \mp \kappa_N \delta_X)$$

where μ_X is the mean value of X, σ_X the standard deviation of the population, and δ_X its coefficient of variation (COV) — terms defined in Chapter 2. Here, however, the statistical coefficient κ_N is given by:

$$\kappa_N = t_\infty^{95\%} \sqrt{\frac{1}{N}} = 1.645 \times \sqrt{\frac{1}{N}}$$

where $t_\infty^{95\%}$ is Student's t-value and N is the size of the population. The key difference between this definition of κ_N and the one given in Chapter 2 is the term inside the square root: $(1/N)$ here, but $(1/N + 1)$ for the 5% fractile.

Alternatively, if the variance of the population is unknown ab initio (and hence must be determined from the sample), the statistical definitions of $X_{k,inf}$ and $X_{k,sup}$ change to:

$$\left.\begin{array}{l} X_{k,inf} \\ X_{k,sup} \end{array}\right\} = m_X \mp k_n s_X = m_X(1 \mp k_n V_X)$$

where m_X is the mean value of X, s_X the sample's standard deviation, V_X its coefficient of variation — terms defined in Chapter 2. The statistical coefficient k_n is given by:

$$k_n = t_{n-1}^{95\%} \sqrt{\frac{1}{n}}$$

where $t_{n-1}^{95\%}$ is Student's t-value for $(n - 1)$ degrees of freedom at a confidence level of 95% and n is the sample size.

Numerical values of κ_n are given in **Figure 5.16** by the lower solid line, labelled 'variance known', and vary between 0.164 for n = 100 and 1.163 for n = 2. Values of k_n are given by the upper solid line, labelled 'variance unknown', and vary between 0.166 for n = 100 and 4.464 for n = 2. For comparison, the dashed lines on **Figure 5.16** show the equivalent values of k_n (for estimating the 5% fractile), which were discussed in Chapter 2.

Returning to the cone penetration test results of **Figure 5.15** (which have mean 1.99 MPa and standard deviation 0.73 MPa), the lower and upper characteristic values of q_c — based on 95% confident mean values — are:

$$\left.\begin{array}{l} q_{ck,inf} \\ q_{ck,sup} \end{array}\right\} = m_{qc} \mp k_n s_{qc} = 1.99 \mp 0.054 \times 0.73 = \begin{cases} 1.95 MPa \\ 2.03 MPa \end{cases}$$

(assuming the data follows a normal distribution), whereas the 5% fractiles were 0.78 and 3.20 MPa.

An important feature of **Figure 5.16** is the rapid divergence between the curves for n < 5. The 'variance unknown' curve rises rapidly as the number of samples decreases.

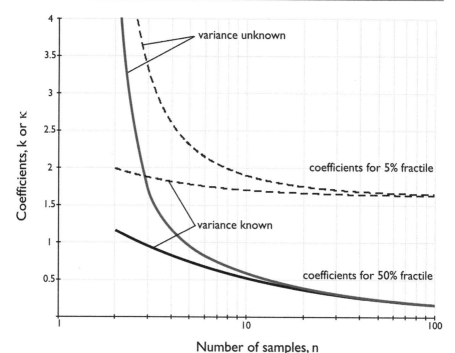

Figure 5.16. *Statistical coefficients for determining the 50% fractile with 95%
confidence (coefficient for 5% fractile from Chapter 2)*

It is common in many ground investigations to gather just a handful of
measurements in each stratum and so the uncertainty in any deductions
made from those measurements (on the basis of statistics) is inherently large.
This uncertainty may be reduced if the variance of the parameter being
measured is known from previous observation — witness the gap between
the curves labelled 'variance known' and 'unknown' in **Figure 5.16**.
However, it is rare that we have sufficient prior knowledge of the strata
encountered in ground investigations for this to be useful to us.

5.5.3   Statistics for parameters that vary with depth

Many geotechnical parameters vary with confining pressure and hence show
a distinct correlation with depth below the ground surface. In such cases, the
statistical methods described above – which apply to a single variable – must
be replaced by multi-variate statistics.

The characteristic value of a material property X that varies with depth z is
given by:

$$X_k = m_X + \left[\frac{\sum_{i=1}^{n}(X_i - m_X)(z_i - m_z)}{\sum_{i=1}^{n}(z_i - m_z)^2} \right] \times (z - m_z) \mp \varepsilon_n$$

where m_X is the mean of X, m_z is the mean of z, and ε_n the error at depth z.

For the 5% fractiles, the error ε_n is given by:

$$\varepsilon_n = t_{n-2}^{95\%} \times s_e \times \sqrt{\left(1 + \frac{1}{n}\right) + \frac{(z - m_z)^2}{\sum_{i=1}^{n}\{(z_i - m_z)^2\}}}$$

and for the 50% fractiles, by:

$$\varepsilon_n = t_{n-2}^{95\%} \times s_e \times \sqrt{\left(\frac{1}{n}\right) + \frac{(z - m_z)^2}{\sum_{i=1}^{n}\{(z_i - m_z)^2\}}}$$

where the standard error s_e is given by:

$$s_e = \sqrt{\frac{\sum_{i=1}^{n}[(X_i - m_X - b \times (z_i - m_z)]^2}{n-2}}$$

and $t^{95\%}_{n-2}$ is Student's t-value for (n – 2) degrees of freedom at the 95% confidence level. The worked examples in Section 5.7 illustrate the use of these equations in practice.

5.5.4   Dealing with small data sets

The techniques discussed above may be used to determine the characteristic value of a geotechnical parameter, provided there is sufficient data to justify the assumptions made. Others[28] have argued that statistical methods could be used when the number of data points n is greater than 13 (but we rarely have this much data) and, when the number is less than 13, a pure statistical approach is too pessimistic to be of practical value.

A simple way to estimate the characteristic value from limited knowledge of the ground properties is to assume:[29]

$$X_k = m_X \mp \frac{s_X}{2} \approx \left(\frac{X_{min} + 4X_{mode} + X_{max}}{6}\right) \mp \frac{1}{2}(X_{max} - X_{min})$$

where m_X and s_X are the mean and standard deviation of X; X_{min} and X_{max} are the estimated minimum and maximum values of X; and X_{mode} is the most likely value of X. The term for s_X assumes that X_{max} and X_{min} are three standard deviations above and below the mean value m_X, and hence are extreme values not normally measured in field or laboratory tests.

This formula is valid when assessing characteristic values of independent soil samples, for which their autocorrelation[†] may be neglected. In practice, samples may be considered independent when their locations differ by more than the so-called 'auto-correlation distance', which is typically 0.2–2m vertically (depending on depositional history) and 20–100m horizontally.[30]

An alternative to the above approximation, assuming X_{max} and X_{min} are two standard deviations above and below m_X, is:

$$X_k = m_X \mp \frac{s_X}{2} \approx \left(\frac{X_{min} + 4X_{mode} + X_{max}}{6} \right) \mp \frac{3}{4}\left(X_{max} - X_{min} \right)$$

Example 5.3 illustrates the use of these equations in practice.

One way of improving the outcome of statistical analysis of small data sets is to employ Bayesian updating, which combines prior knowledge about a parameter with the measured values of that parameter via the following equations:[31]

$$m'_X = \frac{m_X + \dfrac{1}{n}\left(\dfrac{s_X}{\sigma_X}\right)^2 \mu_X}{1 + \dfrac{1}{n}\left(\dfrac{s_X}{\sigma_X}\right)^2} \quad \text{and} \quad s'_X = \sqrt{\frac{\dfrac{1}{n}\left(s_X\right)^2}{1 + \dfrac{1}{n}\left(\dfrac{s_X}{\sigma_X}\right)^2}}$$

where m_X and s_X are the *measured* mean and standard deviation of X; μ_X and σ_X are the mean and standard deviation of X *expected* from prior knoweldge; and m'_X and s'_X are the *updated* values of m_X and s_X. The characteristic value of X is then given by:

$$X_k = m'_X \mp \frac{s'_X}{2}$$

Suitable values of μ_X may be obtained, for example, from the literature or from well-established experience (see Section 5.3.6). Values of σ_X are more difficult to obtain, but studies have suggested suitable values for the coefficient of variation σ_X/μ_X (see Section 5.3.1) that can be used in lieu of better data.

[†]'autocorrelation' is the correlation between the elements of a series and others from the same series separated from them by a given interval.

5.5.5   Use or abuse of statistics?

There is something fascinating about science...

> 'In the space of [176] years the Lower Mississippi has shortened itself by
> [242] miles. That is ... a trifle over one mile and a third per year. Therefore,
> any calm person ... can see that in the Old Oolitic Silurian Period, just a
> million years ago next November, the Lower Mississippi was upward of
> [1,300,000] miles long, and stuck out over the Gulf of Mexico like a fishing
> rod ... There is something fascinating about science. One gets such wholesale
> returns of conjecture out of such a trifling investment of fact'
>
> Mark Twain, *Life on the Mississippi* (1874)

We conclude this sub-section with a warning about the naive application of
statistics to geotechnical data:

> 'Attempts by statisticians to tackle geotechnical design have often ended in
> ridicule, and it is very difficult for one person to have sufficient grasp of both
> disciplines that he can use them sensibly.'[32]

Statistical analysis of geotechnical data is a useful adjunct to engineering
judgement of the characteristic value. However, for statistical outcomes to
make sense, the following simple rules should be observed:

- Only use statistics if you have sufficient data
- Use the simplest form of statistics that captures the data trend
- Exclude rogue data points
- Apply simple physical principles to choose which data to exclude
- Analyse data sets independently of each other as well as together
- Be careful to choose appropriate correlations when combining data
- Choose the outcome on the basis of judgement, not the statistics alone

And remember what Benjamin Disraeli once said:

> 'There are lies, damned lies, and statistics.'

5.6   Summary of key points

The selection of characteristic values of relevant ground parameters is
probably the single most important task that a geotechnical engineer
undertakes in design. Although partial factors provide a degree of reliability
(for example, by reducing soil strengths from their expected values), they
cannot compensate for gross errors of judgement in interpretation of the
operational conditions in the ground.

In view of the alarming spread of interpretations of characteristic values that we have discovered in our research for this book, it is essential that engineers acquaint themselves with the fundamental assumptions that underpin Eurocode 7, especially with regard to the selection of ground parameters.

5.7   Worked examples

The worked examples in this chapter illustrate the way in which statistics may be used to determine the characteristic values of various geotechnical parameters: the standard penetration blow count in Thames Gravel (Example 5.1); the undrained strength of London Clay (Example 5.2); and the angle of shearing resistance of Leighton Buzzard Sand (Example 5.3).

Specific parts of the calculations are marked ❶, ❷, ❸, etc., where the numbers refer to the notes that accompany each example.

5.7.1   Standard penetration tests in Thames Gravel

Example 5.1 applies statistical analysis to the data presented in **Figure 5.14** for the site near Gravesend, in Kent.

In Investigation A, where there is no obvious trend for blow count to change with depth, the statistics are based on the blow counts alone and not on the penetration test depths.

In Investigation B, where there appears to be a slight increase in blow count with depth, the analysis is multi-variate and hence considerably more complicated.

Figure 5.17 shows the outcome of the analysis on the data from Investigation B.

Notes on Example 5.1

❶ The blow counts quoted here are plotted as the open white symbols on **Figure 5.14**.

❷ The coefficient of variation of the sample ($V_N = 0.067$) is below the recommended threshold value ($V_N = 0.1$) and hence the calculation should proceed with an assumed standard deviation based on the threshold.

❸ The standard deviation is based on the threshold coefficient of variation.

❹ Student's t-value for the 95% confidence limit has been taken from standard tables for a normal probability density function (see Chapter 2).

❺ The definition of the statistical coefficient is appropriate for calculating a 95% confident mean value when the variance is unknown and assumes that there is no variation with depth.

❻ The characteristic value (19.7) is only slightly less than the mean value of N (20.7), reflecting the lack of variability of this data set.

❼ The blow counts quoted here are plotted as the closed black symbols on **Figure 5.14**.

❽ It is assumed that the value of N varies with depth and hence the statistics are more complicated than for Investigation A. Refer to Section 5.5.3 for an explanation of the statistical procedure used here.

❾ The slope of best-fit regression line gives the rate at which N increases with depth (the slope is negative because depth below ground level has been taken as a negative value).

❿ The standard error is a measure of the variability of the data set and can be used to provide a simple estimate of the 95% confident mean, as discussed below.

Figure 5.17 shows the outcome of the statistical analysis on data from Investigation B. The analysis, which assumes that the blow count N varies according to a normal distribution, confirms the general trend for N to increase with depth, which is visually obvious from the scatter plot and which an experienced geotechnical engineer might assume.

The line labelled 'best fit' gives values of blow count N that have a 50% probability of being exceeded by the *local* value of N at the same depth. This line is readily calculated by any spreadsheet program using linear regression.

The curve labelled 'lower characteristic' gives values of N that have a 95% probability of being exceeded by the *average* value of N over the depths considered (i.e. the 'spatial average'). This is a 95% confident value and is the one most often needed in geotechnical designs which involve a large zone of the ground. The curvature of the lower characteristic line is a consequence of the chosen statistical method and signifies greater uncertainty at the ends of the data set.

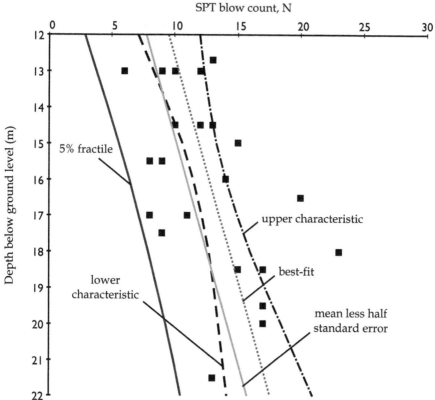

Figure 5.17. Outcome of statistical analysis of standard penetration tests at a site near Gravesend in Kent

The curve labelled 'upper characteristic' gives values of N that have a 95% probability of *not* being exceeded by the average value of N over the depths considered. This is needed when an upper value is critical to the design.

The line labelled '5% fractile' gives values of N that have a 95% probability of being exceeded by the *local* value of N at the same depth. We will call this the 'local' characteristic value. It is needed when the ground is unable to redistribute stresses from highly to lightly loaded zones.

The line labelled 'mean less half standard error' is much easier to calculate than the other lines discussed below, being offset from the best-fit line. It also has the distinct benefit of being linear with depth. Since this line is a reasonable approximation to the lower characteristic line, it can be used as a simple means of obtaining the characteristic line through the data.

Example 5.1
Standard penetration tests in Thames Gravel
Determination of characteristic blow count

<u>Data from investigation A</u>

Consider the results of a series of standard penetration tests (SPTs) in Thames Gravel. The measured blow counts were:

22, 19, 19, 21, 23, 22, 19, 21, 19, 21, 21, 20, 22.

Statistical analysis of data
The number of test results is $n = 13$

The sum and mean are $\sum N = 269$ and $m_N = \dfrac{\sum^N}{n} = 20.7$

The standard deviation is $s_N = \sqrt{\dfrac{\sum (N - m_N)^2}{n - 1}} = 1.38$

and the coefficient of variation $V_N = \dfrac{s_N}{m_N} = 0.067$ ❷

The minimum coefficient of variation is 0.1, hence $V_N = \max\left(V_N, 0.1\right) = 0.1$

and the 'new' $s_N = V_N \times m_N = 2.1$ ❸

With no prior knowledge of blow count (variance unknown)
Student's t-value for 95% confidence limit with (n - 1) degrees of freedom is

$t_{95}(n - 1) = 1.782$ ❹

Hence statistical coefficient is $k_n = t_{95}(n - 1)\sqrt{\dfrac{1}{n}} = 0.494$ ❺

Characteristic blow count is then $N_k = m_N - k_n s_N = 19.7$ ❻

Data from investigation B

Consider the results of the Standard Penetrations Tests from investigation B, given as: (blow count N, depth in m):
(9, -13m), (13, -14.5m), (13, -12.7m), (10, -14.5m), (6, -13m), (12, -14.5m),
(14, -16m), (9, -17.5m), (15, -15m), (20, -16.5m), (23, -18m), (17, -19.5m),
(12, -13m), (12, -14.5m), (8, -15.5m), (11, -17m), (17, -18.5m), (10, -13m),
(12, -14.5m), (9, -15.5m), (8, -17m), (15, -18.5m), (17, -20m), (13, -21.5m) **7**

Statistical analysis of data
The number of tests is $n = 24$

Sum and mean of SPT values are $\sum N = 305$ and $m_N = \dfrac{\sum N}{n} = 12.71$

Sum and mean of depth values $\sum z = -383.2\,m$, $m_z = \dfrac{\sum z}{n} = -15.97\,m$

Sum of squared deviations from mean depth $\sum\limits_{i=1}^{n} \left(z_i - m_z\right)^2 = 139.1\,m^2$

Sum of cross deviations from mean $\sum\limits_{i=1}^{n} \left[\left(N_i - m_N\right)\left(z_i - m_z\right)\right] = -111.8\,m$

Slope of best-fit line b $= \left[\dfrac{\sum\limits_{i=1}^{n} \left[\left(N_i - m_N\right)\left(z_i - m_z\right)\right]}{\sum\limits_{i=1}^{n} \left(z_i - m_z\right)^2} \right] = -0.803\,\dfrac{1}{m}$ **9**

Standard error is $s_e = \sqrt{\dfrac{\sum\limits_{i=1}^{n} \left[N_i - m_N - b\left(z_i - m_z\right)\right]^2}{n-2}} = 3.613$ **10**

5.7.2   Undrained triaxial compression tests on London Clay

Example 5.2 applies statistical analysis to the triaxial test data presented in **Figure 5.11** (see p. 145) for the site at Holborn, London. Only data for London Clay is considered here (i.e. the solid black symbols on **Figure 5.11**). **Figure 5.18** shows the outcome of the analysis.

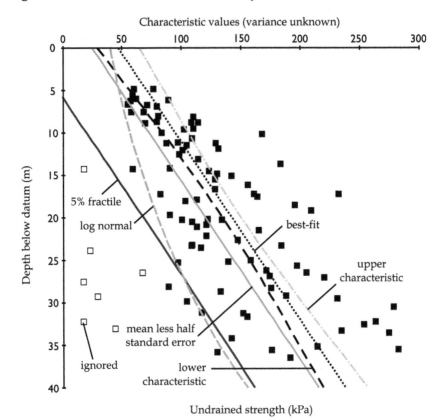

Figure 5.18. Outcome of statistical analysis of undrained triaxial compression tests in London Clay

Notes on Example 5.2

There is much more data shown in **Figure 5.18** than is normally available from typical site investigations in stiff clay. Undrained strengths derived from triaxial compression tests can vary considerably even for samples from the same horizon. Traditional techniques using light cable percussion drilling rigs and U100 tubes do not obtain samples that are truly adequate for strength testing. Since such investigations are common, engineers must be able to judge the soil's characteristic strength from such scattered results.

Example 5.2
Undrained triaxial compression tests on London Clay
Determination of characteristic undrained strength

Assuming normal distribution of strength vs depth

Number of tests is $n = 91$

Sum and mean of strengths $\sum c_u = 12688$ kPa and

$$m_{cu} = \frac{\sum c_u}{n} = 139.43 \text{ kPa}$$

Sum and mean of depths $\sum z = -1763.45$ m and $m_z = \dfrac{\sum z}{n} = -19.38$ m

Sum squared deviations from mean depth $\displaystyle\sum_{i=1}^{n} \left(z_i - m_z\right)^2 = 8000 \text{ m}^2$

Sum cross deviations from mean $\displaystyle\sum_{i=1}^{n} \left[\left(c_{u_i} - m_{cu}\right)\left(z_i - m_z\right)\right] = -38153.7 \dfrac{\text{kN}}{\text{m}}$

Slope of best-fit regression $b = \left[\dfrac{\displaystyle\sum_{i=1}^{n} \left[\left(c_{u_i} - m_{cu}\right)\left(z_i - m_z\right)\right]}{\displaystyle\sum_{i=1}^{n} \left(z_i - m_z\right)^2}\right] = -4.769 \dfrac{\text{kN}}{\text{m}^3}$

Standard error of regression line

$$s_e = \sqrt{\dfrac{\displaystyle\sum_{i=1}^{n} \left[c_{u_i} - m_{cu} - b\left(z_i - m_z\right)\right]^2}{n-2}} = 44.5 \text{ kPa}$$

The line labelled 'best fit' on **Figure 5.18** gives values of c_u which have a 50% probability of being exceeded by the *local* value of c_u at the same depth. This line is readily calculated by spreadsheet programs using linear regression.

The curve labelled 'lower characteristic' gives values of c_u which have a 95% probability of being exceeded by the *average* value of c_u over the depths considered. This is a 95% confident mean value and is the value most often needed in geotechnical designs which involve a large zone of the ground. The curvature of the lower characteristic line is a consequence of the chosen statistical method and signifies greater uncertainty at the ends of the data set.

The curve labelled 'upper characteristic' gives values of c_u which have a 95% probability of *not* being exceeded by the average value of c_u over the depths considered. This is needed when an upper value is critical to the design.

The line labelled '5% fractile' gives values of undrained strength c_u which have a 95% probability of being exceeded by the *local* value of c_u at the same depth. We will call this the 'local' characteristic value. It is needed when the ground is unable to redistribute stresses from highly to lightly loaded zones.

Finally, the line labelled 'mean less half standard error' gives values which are close to the 95% confident mean (lower characteristic) and are much easier to calculate, since the line is offset from the best-fit line and hence has the added benefit of being linear.

5.7.3   Shear box tests on Leighton Buzzard Sand

Example 5.3 considers the determination of the angle of shearing resistance of Leighton Buzzard Sand, based on the results of a small number of laboratory shear box tests. It also illustrates the combination of test results with previous knowledge, using Bayesian updating techniques.

Notes on Example 5.3

❶ The values of shearing resistance were obtained from a series of six laboratory shear box tests.

❷ The 'mean-deviation-from-the-mean' is an alternative measure of variation, which is simpler to calculate than the standard deviation and gives similar values. It can be used to give an 'alternative approximation' of φ_k based on half the mean-deviation-of-the-mean below the mean.

❸ The characteristic value here is the 95% confident mean value with variance unknown.

Example 5.3
Shear box tests on Leighton Buzzard Sand
Determination of characteristic shearing resistance

<u>Statistical analysis of laboratory measurements</u>

Data from laboratory tests

Consider the results of a series of shear box tests on Leighton Buzzard Sand, which measured angles of shearing resistance of: 48.7, 50.8, 50.9, 50.5, 48.3, and 46.3°. **❶**

Statistical analysis

The sum of the results of $n = 6$ tests is $\sum \varphi = 295.5°$

Mean/standard dev.: $m_\varphi = \dfrac{\sum \varphi}{n} = 49.3°$ $\quad s_\varphi = \sqrt{\dfrac{\sum \left(\varphi - m_\varphi\right)^2}{n-1}} = 1.82°$

Coefficient of variation $V_\varphi = \dfrac{s_\varphi}{m_\varphi} = 0.04$

The minimum coefficient of variation is 0.1, hence $V_\varphi = \max\left(V_\varphi, 0.1\right) = 0.1$

and the 'new' $s_\varphi = V_\varphi \times m_\varphi = 4.93°$

Mean deviation from the mean is $MDM_\varphi = \dfrac{\sum \sqrt{\left(\varphi - m_\varphi\right)^2}}{n} = 1.48°$ **❷**

With variance of shearing resistance unknown, Student's t-value for 95% confidence limit with (n - 1) degrees of freedom is $t_{95}(n-1) = 2.015$

Hence statistical coefficient is $k_n = t_{95}(n-1)\sqrt{\dfrac{1}{n}} = 0.823$

Characteristic angle of shearing resistance is $\varphi_k = m_\varphi - k_n s_\varphi = 45.2°$ **❸**

Schneider's approximation $\varphi_k = m_\varphi - \dfrac{s_\varphi}{2} = 46.8°$ **❹**

Alternative approximation $\varphi_k = m_\varphi - \dfrac{MDM_\varphi}{2} = 48.5°$ **❷**

Prior knowledge

Well-established experience

Leigton Buzzard Sand is a dense medium sand with density index $I_D = 80\%$

Critical state angle of shearing resistance is: $\varphi_{crit} = 35°$ **⑤**

At the average expected mean confining pressure $p' = 100kPa$ **⑥**

$$\text{Relative dilatancy index: } I_R = 0.8\left(10 - \ln\left(\frac{p'}{kPa}\right)\right) - 1 = 3.3 \quad ⑤$$

$$\text{Estimated modal angle } \varphi_{mode} = 5° \times I_R + \varphi_{crit} = 51.6° \; ⑤$$

At the lowest expected mean confining pressure $p' = 55kPa$ **⑥**

$$\text{Relative dilatancy index: } I_R = 0.8\left(10 - \ln\left(\frac{p'}{kPa}\right)\right) - 1 = 3.8$$

$$\text{Estimated maximum angle } \varphi_{max} = 5° \times I_R + \varphi_{crit} = 54°$$

At the highest expected mean confining pressure $p' = 220kPa$ **⑥**

$$\text{Relative dilatancy index: } I_R = 0.8\left(10 - \ln\left(\frac{p'}{kPa}\right)\right) - 1 = 2.7$$

$$\text{Estimated minimum angle } \varphi_{min} = 5° \times I_R + \varphi_{crit} = 48.4°$$

Estimates of characteristic value

$$\text{Estimate of mean: } \varphi_{mean} = \left(\frac{\varphi_{min} + 4\varphi_{mode} + \varphi_{max}}{6}\right) = 51.5°$$

$$\text{Estimated characteristic value: } \varphi_k = \varphi_{mean} - \frac{1}{2}\left(\varphi_{max} - \varphi_{min}\right) = 48.7° \; ⑦$$

$$\text{Alternative estimate: } \varphi_k = \varphi_{mean} - \frac{3}{4}\left(\varphi_{max} - \varphi_{min}\right) = 47.3°$$

Bayesian updating

Prior knowledge

Assumed mean value of peak angle of shearing resistance

$\mu_\varphi = \varphi_{mode} = 51.6°$

Assumed coefficient of variation $COV = 0.1$

Hence assumed standard deviation $\sigma_\varphi = \mu_\varphi \times COV = 5.2°$ ⑧

Measured values (from above)

Mean angle of shearing resistance $m_\varphi = 49.3°$

Standard deviation $s_\varphi = 4.9°$

Updated values

Standard deviation $s'_\varphi = \sqrt{\dfrac{\dfrac{1}{n} \times \left(s_\varphi\right)^2}{1 + \dfrac{1}{n} \times \left(\dfrac{s_\varphi}{\sigma_\varphi}\right)^2}} = 1.9°$

Mean value $m'_\varphi = \dfrac{m_\varphi + \dfrac{1}{n} \times \left(\dfrac{s_\varphi}{\sigma_\varphi}\right)^2 \times \mu_\varphi}{1 + \dfrac{1}{n} \times \left(\dfrac{s_\varphi}{\sigma_\varphi}\right)^2} = 49.6°$ ⑨

Schneider's approximation $\varphi_k = m'_\varphi - \dfrac{s'_\varphi}{2} = 48.6°$ ⑩

❹ This characteristic value is half the standard deviation below the mean.

❺ The critical state angle of shearing resistance of Leighton Buzzard Sand, the relative dilatancy index I_R, and the formula for determining φ from I_R are all taken from Bolton's paper [33] on the strength and dilatancy of sands.

❻ The 'modal' value here is calculated using a simple rule-of-thumb that depends on the sand's particle shape, grading, and density. The maximum and minimum values are obtained from the same rule-of-thumb, assuming extreme values for the components A, B, and C.

❼ This characteristic value is a simple weighted average of the minimum, modal, and maximum values.

❽ The coefficient of variation of φ is assumed to be 0.1, based on observed values for a wide range of different soils.

❾ The mean m_φ and standard deviation s_φ from the laboratory measurements are updated using the mean μ_φ and standard deviation σ_φ obtained from previous experience.

❿ This characteristic value is half the updated standard deviation below the updated mean. The full range of values of φ_k obtained in this example vary from 46.8° to 49.6°.

5.8　　Notes and references

1.　Ang, A. H.-S. And Tang, W. H. (2006) *Probability concepts in engineering: emphasis on applications in civil and environmental engineering* (2nd edition), John Wiley and Sons Ltd, 406pp.

2.　Data provided by Stent Foundations (pers. comm., 2007).

3.　BS EN 1997-2: 2006, Eurocode 7 – Geotechnical design, Part 2 – Ground investigation and testing, British Standards Institution, London, 196pp.

4.　Definition from Merriam-Webster's Online Dictionary, www.m-w.com.

5.　Terzaghi, K. and Peck, R. B. (1967) *Soil mechanics in engineering practice* (2nd edition), John Wiley & Sons, Inc, 729pp.

6.　Ang and Tang, ibid.

7. Schneider, H. R. (1997) *Definition and determination of characteristic soil properties*, 12th Int. Conf. Soil Mech. & Fdn Engng, Hamburg: Balkema.

8. See, for example, Phoon, K. K. and Kulhawy, F. H. (1999) 'Characterization of geotechnical variability', *Can. Geotech. J.* 36(4), pp. 612–624.

9. See Schneider, ibid., and Phoon, K. K. and Kulhawy, F. H. (1999) 'Evaluation of geotechnical property variability', *Can. Geotech. J.* 36(4), pp. 625–639.

10. Ground Board of the Institution of Civil Engineers (1991) *Inadequate site investigation*, Thomas Telford.

11. Pearsall, J. (ed.) (1999) *The Concise Oxford Dictionary* (10th edition), Oxford University Press.

12. BS 8002:1994 Code of practice for earth retaining structures, British Standards Institution, London.

13. Padfield, C. J. and Mair, R. J. (1984) *Design of retaining walls embedded in stiff clays*. CIRIA RP104.

14. Pearsall, ibid.

15. Gaba, A. R., Simpson, B., Powrie, W. and Beadman, D. R. (2003) *Embedded retaining walls – guidance for economic design*, London, CIRIA C580.

16. Pearsall, ibid., see definition of 'experience'.

17. See Table 3 of BS 8002, ibid.

18. For example, see Tomlinson, M. J. (2000) *Foundation design and construction*, Prentice Hall; or Bowles, J. E. (1997) *Foundation analysis and design*, McGraw-Hill.

19. See Table 2 of BS 8002, ibid.

20. Data kindly provided by Viv Troughton and Tony Suckling of Stent Foundations (pers. comm., 2007).

21. Data kindly provided by Dr Toh and the Singapore Building and Construction Authority (pers. comm., 2007).

22. Bjerrum , L. (1972) 'Embankments on soft ground', *Proc. Speciality Conf. on Performance of Earth and Earth Supported Structures*, ASCE, 2, pp. 1–54.

23. Suckling T. (2003) 'Driven cast insitu piles, the CPT and the SPT — two case studies', *Ground Engineering*, 36(10), pp. 28–32 (and pers. comm., 2007).

24. Simpson, B. and Driscoll, R. (1998) *Eurocode 7 — a commentary*, BRE.

25. This follows from the central limit theorem — 'one of the most important theorems in probability theory' according to Ang and Tang, ibid., p. 168.

26. Mortensen, J. K., Hansen, G., and Sørensen, B. (1991) 'Correlation of CPT and field vane tests for clay tills', *Danish Geotechnical Society Bulletin*, 7, 62pp.

27. See Ang and Tang, ibid., pp.102–3.

28. Schneider, ibid.

29. Schneider, ibid.

30. Hans Schneider (pers. comm., 2008).

31. Ang and Tang, ibid., pp. 361–2 and Schneider, ibid.

32. Simpson and Driscoll, ibid.

33. Bolton, M. D. (1986) 'The strength and dilatancy of sands', *Géotechnique*, 36(1), pp. 65-78.

Chapter 6

Verification of strength

'It is not possible to fight beyond your strength, even if you strive' – Homer (800–700 BC)[1]

6.1   Basis of design

Verification of strength to Eurocode 7 involves checking that design effects of actions do not exceed their corresponding design resistances.

Verification of strength is expressed in Eurocode 7 by the inequality:

$$E_d \leq R_d \qquad\qquad \text{[EN 1990 exp (6.8)] \& [EN 1997-1 exp (2.5)]}$$

in which E_d = the design effects of actions and R_d = the corresponding design resistance.

This requirement applies to ultimate limit state GEO, defined as:

Failure or excessive deformation of the ground, in which the strength of soil or rock is significant in providing resistance   *[EN 1997-1 §2.4.7.1(1)P]*

and to ultimate limit state STR:

Internal failure or excessive deformation of the structure or structural elements ... in which the strength of structural materials is significant in providing resistance   *[EN 1997-1 §2.4.7.1(1)P]*

Examples of situations where strength is a concern are shown in **Figure 6.1**; from left to right, these include: top, the stem of a cantilever retaining wall must withstand the forces on its back (STR); and a hillside must be strong enough to support its self-weight and other forces acting on it (GEO); middle, the foundation of a footing must be strong enough to support the imposed load on it (GEO); and an embedded retaining wall and its support system must be strong enough to withstand earth pressures over its retained height (STR); and bottom, the ground supporting a pile subject to horizontal loads must be strong enough to prevent excessive horizontal movement (GEO); and, finally, the ground beneath a mass concrete retaining wall must be strong enough to carry the wall's weight and any forces acting on it (GEO).

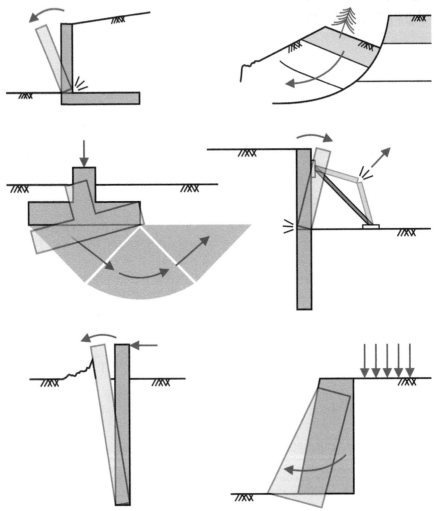

Figure 6.1. *Examples of ultimate limit states of strength*

6.1.1   Effects of actions

'Effects of actions' (or 'action effects') is a general term denoting internal forces, moments, stresses, and strains in structural members — plus the deflection and rotation of the whole structure.   *[EN 1990 §1.5.3.2]*

For most structural designs, verification of limit state STR involves action effects that are independent of the *strength* of the structural materials (see Chapter 2). However, in many geotechnical designs, verification of the STR and GEO limit states involves effects of actions that depend upon the strength of the ground.

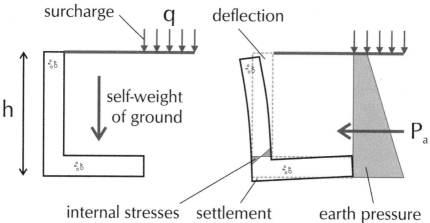

Figure 6.2. *Actions (left) and effects (right) for L-shaped gravity retaining wall*

For example, **Figure 6.2** shows a retaining wall supporting loose soil and an imposed uniform surcharge (q). The earth pressures acting behind the wall produce a horizontal sliding force H_E (an action effect) given by:

$$H_E = P_a = K_a \left(\frac{\gamma h}{2} + q \right) h = \left(\frac{1 - \sin \phi}{1 + \sin \phi} \right) \left(\frac{\gamma h}{2} + q \right) h = f\{h, \gamma, \phi, q\}$$

where h is the wall's height; γ and φ the soil's self-weight and angle of shearing resistance; and K_a is Rankine's active earth pressure coefficient.

This simple example illustrates why the definition of design effects of actions given in the head Eurocode:

$$E_d = E\{F_d; a_d\}$$ *[EN 1990 exp (6.2a, simplified)]*

has to be revised for geotechnical design to:

$$E_d = E\{F_d; X_d; a_d\}$$

where F_d = design actions applied to the structure; X_d = design material properties; and a_d = design dimensions of the structure. (The notation E{...} denotes a function of the enclosed parameters and usually involves multiple parameters of each type listed.)

Put simply, in structural design, effects of actions are generally a function of actions and dimensions only; whereas, in geotechnical design, effects of actions are typically a function of actions, dimensions, *and the strength of the ground*.

The inclusion of X_d in the equation for E_d adds considerable complexity to designs involving geotechnical actions and is one of the reasons for the diversity of design methods used in geotechnical design.

6.1.2   Resistance

'Resistance' is defined as the:

> *capacity of a [member or] component, or cross-section of a [member or] component of a structure, to withstand actions without mechanical failure*
> *[EN 1990 §1.5.2.15] & [EN 1997-1 §1.5.2.7]*

(The words in brackets are omitted in Eurocode 7's definition. The absence of the word 'ground' from either definition appears to be an error – unless we regard the ground as a component of the structure.)

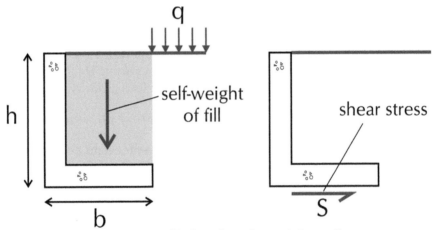

Figure 6.3. Sliding resistance of L-shaped gravity retaining wall

For most structural designs, verification of limit state STR involves resistances that are independent of actions (see Chapter 2). However, in many geotechnical designs, verification of the STR and GEO limit states involves resistances that depend upon *actions*.

For example, **Figure 6.3** illustrates the sliding resistance H_R of the retaining wall shown previously in **Figure 6.2**:

$$H_R = S = \gamma \times h \times b \times \tan \delta = f\{h,b,\gamma,\delta\}$$

where the resistance is a function of the wall's dimensions (h and b), the self-weight of the soil (γ) – *and* the strength of the soil-structure interface (δ, which itself is a function of the soil's drained angle of shearing resistance φ).

Again this example illustrates why the definition of resistance given in the head Eurocode:

$$R_d = \frac{R\{X_d; a_d\}}{\gamma_{Rd}}$$ [EN 1990 exp (6.6, simplified)]

has to be revised for geotechnical design to:

$$R_d = \frac{R\{F_d; X_d; a_d\}}{\gamma_R}$$

where F_d, X_d, and a_d are as defined earlier. Here, R{...} denotes a function of the enclosed parameters and $\gamma_{Rd} = \gamma_R$ = a partial factor on resistance.

In simple terms, in structural design, resistances are generally a function of material strengths and dimensions only; whereas, in geotechnical design, resistances are typically a function of material strengths, dimensions, *and actions, including the self-weight of the ground*.

Once again, the inclusion of F_d in the equation for R_d adds considerable complexity to designs involving geotechnical materials and is another reason for the diversity of design methods used in geotechnical design.

6.2 Introducing reliability into the design

'The word safety is encompassed in the Eurocodes in the word reliability.'[2]

Reliability can be introduced into the design in a number of ways, through the application of suitable partial factors or tolerances, as illustrated in **Figure 6.4**.

In the top half of this diagram, there are three 'channels' that lead into the calculation model: one for actions (left), another for geometrical parameters (centre), and a third for materials properties (right). Certain material properties, such as weight density, have a direct influence on actions, whereas other material properties, such as strength, do not (they do, however, influence the action effects).

Verification occurs in the bottom third of the diagram: the calculation model provides values for effects of actions (left) and resistance (right), which are compared against each other (in the centre).

Partial factors (or tolerances) can be applied to one or more of:
● actions (F) or action effects (E)
● material properties (X) and/or resistances (R)
● geometrical parameters (a)

These factors/tolerances are shown on **Figure 6.4** in the wavy boxes.

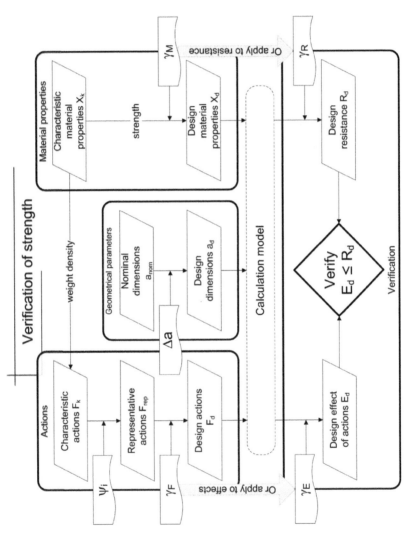

Figure 6.4. Overview of verification of strength

6.2.1 Actions and effects

The calculation of design effects of actions follows the route shown in the left hand channel of **Figure 6.4**:

Characteristic actions → Representative actions → Design actions
→ Design effects of actions

Characteristic actions F_k are calculated according to the rules of Eurocode 1. Characteristic self-weights are calculated as the product of a material's characteristic weight density γ_k and its nominal dimensions a_{nom} (see Chapter 2):

$$F_k = \gamma_k \times a_{nom,1} \times a_{nom,2} \times a_{nom,3}$$

Representative actions F_{rep} are obtained from characteristic actions by multiplying by combination factors $\psi \leq 1.0$ (where $\psi = 1.0$ for permanent actions, see Chapter 2):

$$F_{rep} = \psi F_k$$

The total design action F_d is then obtained as the sum of all the representative actions multiplied by their corresponding partial factors $\gamma_F \geq 1.0$:

$$F_d = \sum_i \gamma_{F,i} \psi_i F_{k,i}$$

The design effects of actions are then obtained from:

$$E_d = E\{F_d; X_d; a_d\} = E\{\gamma_F \psi F_k; X_d; a_d\}$$

Figure 6.5 shows the relative magnitude of actions as appropriate combination factors (1.0 or ψ) and partial factors (γ_G and γ_Q) are applied to them. The diagram assumes arbitrary values for the permanent, leading variable, and accompanying variable actions (G, Q_1, and Q_i respectively). The arrow denotes where design actions enter the calculation model.

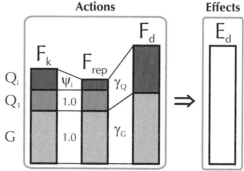

Figure 6.5. Hierarchy of actions and effects when partial factors are applied to actions

Eurocode 7 allows partial factors γ_F to be applied to actions or to their effects, *but typically not to both*. Thus an alternative to the above equation is:

$$E_d = \gamma_E E\left\{\psi F_k; X_d; a_d\right\}$$

where the partial factors γ_E are numerically identical to γ_F.

Figure 6.6 shows the relative magnitude of the actions and effects, as the combination factors are applied to characteristic actions and partial factors to the *effects* of actions. If the calculation model is linear, then the resultant design effects will be identical to those shown in **Figure 6.5**; if the model is non-linear (which is invariably the case in geotechnical engineering), then the resultant design effects will

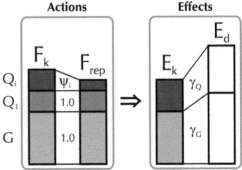

Figure 6.6. Hierarchy of actions and effects when partial factors are applied to action effects

differ. A further complication with this formulation is that permanent and variable effects of actions must be calculated separately to allow different partial factors to be applied to them. Existing computer software is unlikely to have been programmed to do this and hence will need amending to accommodate Eurocode 7.

6.2.2   Material strength and resistance

The calculation of design resistance follows the route shown in the right hand channel of **Figure 6.4**:

> Characteristic material strengths → Design strengths
> → Design resistance

Design material properties X_d are obtained from characteristic material properties X_k by dividing by partial factors $\gamma_M \geq 1.0$:

$$X_d = \frac{X_k}{\gamma_M}$$

The design resistance is then obtained from:

$$R_d = \frac{R\{F_d; X_d; a_d\}}{\gamma_R} = \frac{R\left\{F_d; \dfrac{X_k}{\gamma_M}; a_d\right\}}{\gamma_R}$$

where the partial factor $\gamma_R \geq 1.0$.

It is usual for one of the partial factors γ_M or γ_R to be equal to 1.0 and so the equation above typically reduces to one of two formats, either:

$$R_d = R\left\{F_d; \frac{X_k}{\gamma_M}; a_d\right\} \quad \text{or} \quad R_d = \frac{R\{F_d; X_k; a_d\}}{\gamma_R}$$

Figure 6.7 shows the relative magnitude of material strengths, assuming the first format, as the appropriate partial factors (γ_φ and γ_{cu}) are applied to them. The diagram assumes arbitrary contributions to resistance from a coarse soil with characteristic angle of shearing resistance φ_k and from a fine soil with characteristic undrained shear strength c_{uk}. The arrow denotes the insertion of design material strengths into the calculation model.

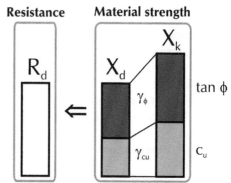

Figure 6.7. Hierarchy of material strengths and resistance when factors are applied to material properties only

Figure 6.8 does likewise for the second format, applying resistance factors (γ_R) instead of material factors. The arrow denotes the insertion of design material strengths into the calculation model. The resultant design resistance R_d will invariably differ from that shown in **Figure 6.7**.

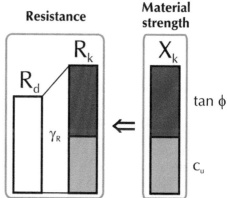

Figure 6.8. Hierarchy of material strengths and resistance when factors are applied to resistance only

6.2.3   Geometry

The calculation of geometrical parameters follows the route shown in the central channel of **Figure 6.4**:

Nominal dimensions → Design dimensions

Design geometrical parameters a_d are obtained from nominal geometrical parameters a_{nom} by adding or subtracting a tolerance Δa:

$$a_d = a_{nom} \pm \Delta a$$

Because it would be impractical to apply tolerances to all the dimensions that affect a design, EN 1990 allows nominal dimensions to be used as design dimensions, i.e. adopting $\Delta a = 0$. The uncertainty in the dimensions is catered for by the values of the partial factors on actions (γ_F) and on material properties (γ_M). The use of $\Delta a > 0$ is reserved for design situations where it is known that small deviations in geometrical parameters have a large impact on the resulting action effects or resistance. See Chapters 11 and 12 for examples where it is normal to assume $\Delta a > 0$.

6.2.4   Verification

Substituting the expressions for effects of actions (Section 6.1.1) and resistance (Section 6.1.2) into that for verification of strength (Section 6.1) gives:

$$E_d = E\{F_d; X_d; a_d\} \le \frac{R\{F_d; X_d; a_d\}}{\gamma_R} = R_d$$

Expanding the terms F_d, X_d, and a_d results in:

$$E\left\{\gamma_F \psi F_k; \frac{X_k}{\gamma_M}; a_{nom} \pm \Delta a\right\} \le \frac{R\left\{\gamma_F \psi F_k; \frac{X_k}{\gamma_M}; a_{nom} \pm \Delta a\right\}}{\gamma_R}$$

which is mathematically equivalent to the process illustrated in **Figure 6.4**.

6.2.5   Partial factors

Partial factors for verification of strength in persistent and transient design situations are specified in EN 1997-1 in Tables A.3 (for actions and effects), A.4 (for soil parameters), and A.5–8 and A.12–14 (for resistance). Some values are amended in the UK National Annexes to ENs 1990 and 1997-1.

The following table lists the partial factors given in EN 1997-1 for general foundations (e.g. slopes, footings, and walls); factors for pile foundations and anchorages are given in Chapters 13 and 14, respectively. The code presents the factors in 'Sets': A1 and A2 apply to actions; M1 and M2 to material properties; and R1 to R3 to resistance. (Values for set R4 are presented in Chapter 13.)

Parameter			Actions or effects		Material properties		Resistance		
			A1	A2	M1	M2	R1	R2	R3
Permanent actions (G)	Unfav'ble	γ_G	1.35	1.0					
	Favourable	$\gamma_{G,fav}$	1.0	1.0					
Variable actions (Q)	Unfav'ble	γ_Q	1.5	1.3					
	Favourable	$\gamma_{Q,fav}$	0	0					
Coefficient of shearing resistance ($\tan \varphi$)		γ_φ			1.0	1.25			
Effective cohesion (c')		$\gamma_{c'}$			1.0	1.25			
Undrained strength (c_u)		γ_{cu}			1.0	1.4			
Unconfined compressive strength (q_u)		γ_{qu}			1.0	1.4			
Weight density (γ)		γ_γ			1.0	1.0			
Bearing resistance (R_v)		γ_{Rv}					1.0	1.4	1.0
Sliding resistance (R_h)		γ_{Rh}					1.0	1.1	1.0
Earth resistance ... retaining structures ... slopes		γ_{Re}					1.0	1.4 1.1	1.0
Prestressed anchorages		γ_a					1.1	1.1	1.0

Partial factors for actions (and effects) in accidental design situations should normally be taken as 1.0 and partial resistance factors (and presumably partial material factors) should be selected according to the particular circumstances of the accidental situation. *[EN 1997-1 §2.4.7.1(3)]*

6.3   Design approaches

During the development of Eurocode 7, it became clear that some countries wanted to adopt a load and material factor approach to the verification of strength (see Section 6.4.2), while others preferred the load and resistance factor approach (see Section 6.4.3).

To accommodate these differing wishes, a compromise was reached whereby each country could choose – through its National Annex – one (or more) of three *design approaches* that should be used within its jurisdiction. The design approaches defined by Eurocode 7 are listed below and explained in more detail in Sections 6.3.1 to 6.3.3. The choices made by each country are discussed in Section 6.3.4.   *[EN 1997-1 §2.4.7.3.4.1(1)P NOTE 1]*

The table below shows which *sets* of partial factor are used in each design approach, depending on the type of structure being designed. In Design Approach 1, *both* Combinations must be checked.

Structure	Partial factor sets used in Design Approach ...			
	1		2	3
	Combination 1	Combination 2		
General	<u>A1</u> & M1 & R1	<u>M2</u> & <u>A2</u> & R1	A1 & R2 & M1	<u>A1</u>* & <u>M2</u> & <u>A2</u>† & R3
Slopes			E1 & R2 & M1	<u>M2</u> & E2 & R3
Piles and anchor- ages	<u>A1</u> & <u>R1</u> & M1	<u>R4</u> & <u>A2</u> & M1	<u>A1</u> & R2 & M1	<u>A1</u>* & <u>M2</u> & <u>A2</u>† & <u>R3</u>

<u><u>doubly underlined</u></u> sets = major partial factors
<u>underlined</u> sets = minor partial factors
*on structural actions
†on geotechnical actions
Sets A1–2 are applied to actions; M1–2 to material properties; R1–4 to resistance; E1–2 to action effects (using values from Sets A1–2).

Of the sixty-three partial factors specified in Annex A of EN 1997-1 for the GEO and STR limit states, just over half are numerically greater than one (the remainder are either unity or zero). In the table above, those sets in which the partial factors are significantly greater than one – and hence introduce reliability into the calculation – are <u>doubly underlined</u>; sets in which only one factor is > 1.0 or the factors are relatively small are <u>singly underlined</u>; and sets in which the factors are all unity are not underlined.[†]

In essence, Design Approach 1 provides reliability by applying different partial factors to two variables in two separate calculations ('Combinations' 1 and 2); whereas Design Approaches 2 and 3 apply factors to two variables simultaneously, in a single calculation – as summarized below.

Structure	Main variable that gets factored in Design Approach ...				
	1		2	3
	Combination 1	Combination 2		
General	Actions	Material properties	Actions (or effects) and resistance	Structural actions (or effects) and material properties
Slopes			Effects of actions and resistance	Structural effects of actions and material properties
Piles and anchorages		Resistance	Actions (or effects) and resistance	Structural actions (or effects) and material properties

[†]This analysis is strictly only valid for the partial factors given in EN 1997-1. Individual countries may change these factors in their National Annexes.

6.3.1   Design Approach 1

The philosophy of Eurocode 7's Design Approach 1 is to check the foundation's reliability in two stages.

First, partial factors are applied to actions (alone), while ground strengths and resistances are left unfactored. This is achieved by employing partial factors from Sets A1, M1, and R1 in what is termed 'Combination 1', as illustrated in **Figure 6.9**. The crosses on the diagram indicate that partial factors in Sets M1 and R1 are all 1.0 and hence ground strengths and

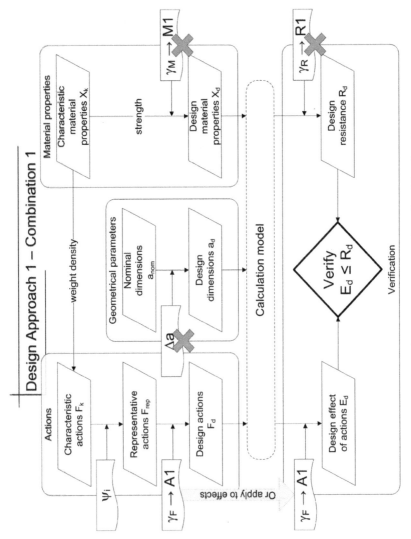

Figure 6.9. Verification of strength to Design Approach 1, Combination 1

resistances are unfactored. The cross that appears against Δa indicates that tolerances are not normally applied to nominal dimensions.

Second, partial factors are applied to ground strengths and variable actions, while non-variable actions and resistances are left unfactored. This is achieved by employing partial factors from Sets A2, M2, and R1 in what is termed 'Combination 2', as illustrated in **Figure 6.10**. Once again, the crosses on the diagram indicate that the partial factors in Sets A2 and R1 are all 1.0 (except those applied to variable actions) and hence non-variable actions and resistance are unfactored.

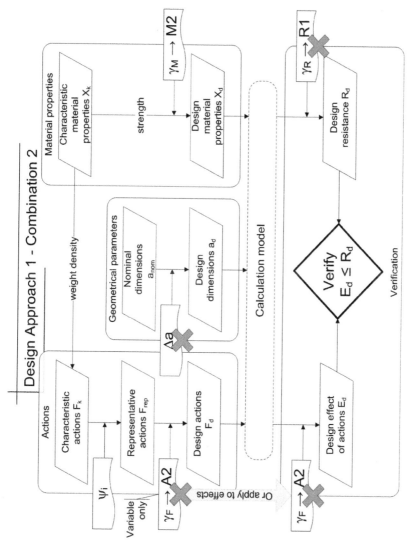

Figure 6.10. Verification of strength to Design Approach 1, Combination 2

(If Design Approach 1 is used to design pile foundations or anchorages, then factors must be applied to resistances instead of material properties. Factors from Set M1 are applied to material properties in both Combinations and factors from Sets R1 and R4 applied to resistances in Combinations 1 and 2 respectively. See Chapters 13 and 14 for a full discussion of this point.)

Numerical values of the partial factors needed in Design Approach 1, for persistent and transient design situations, are:

Design Approach 1			Combination 1			Combination 2		
			A1	M1	R1	A2	M2	R1
Permanent actions (G)	Unfavourable	γ_G	1.35			1.0		
	Favourable	$\gamma_{G,fav}$	1.0			1.0		
Variable actions (Q)	Unfavourable	γ_Q	1.5			1.3		
	Favourable	$\gamma_{Q,fav}$	0			0		
Coeff. of shearing resistance (tan φ)		γ_φ		1.0			1.25	
Effective cohesion (c′)		$\gamma_{c'}$		1.0			1.25	
Undrained strength (c_u)		γ_{cu}		1.0			1.4	
Unconfined compressive strength (q_u)		γ_{qu}		1.0			1.4	
Weight density (γ)		γ_γ		1.0			1.0	
Resistance (R)		γ_R			1.0			1.0

Figure 6.11 shows the relative magnitude of the key parameters when using Combination 1 and **Figure 6.12** using Combination 2.

In Design Approach 1, partial factors are applied early in the calculation process (to actions and material properties), close to the source of uncertainty. This approach is a particular form of material factor design discussed in Section 6.4.2.

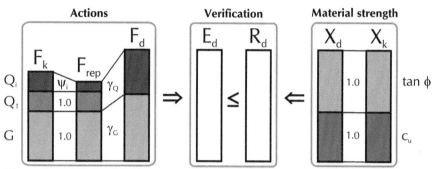

Figure 6.11. *Hierarchy of parameters for Design Approach 1 Combination 1*

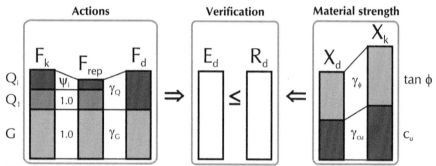

Figure 6.12. *Hierarchy of parameters for Design Approach 1 Combination 2*

6.3.2   Design Approach 2

The philosophy of Eurocode 7's Design Approach 2 is to check the foundation's reliability by applying partial factors to actions or action effects and to resistance simultaneously, while ground strengths are left unfactored.

This is achieved by employing partial factors from Sets A1, M1, and R2, as illustrated in **Figure 6.13**. The crosses indicate that partial factors in Set M1 are all 1.0 and hence ground strengths are unfactored. The cross against Δa indicates that tolerances are not normally applied to nominal dimensions.

(If Design Approach 2 is used in slope stability analyses, then the factors from Set A1 must be applied to action effects and *not* to actions – see Chapter 10 for a full discussion of this point.)

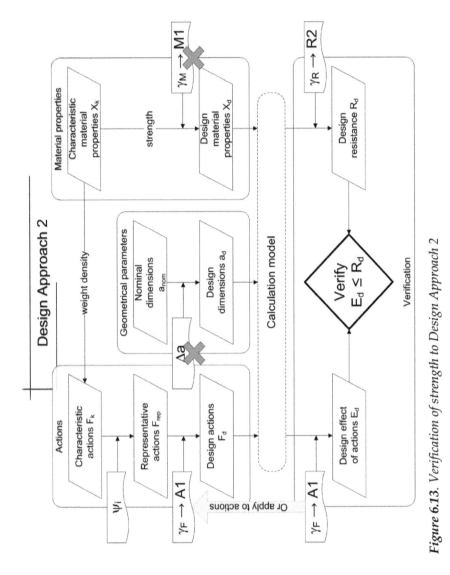

Figure 6.13. Verification of strength to Design Approach 2

The numerical values of the partial factors needed in Design Approach 2, for persistent and transient design situations, are listed in the table below. **Figure 6.14** shows the relative magnitude of the key parameters when using Design Approach 2.

Design Approach 2			A1	M1	R2
Permanent actions (G)	Unfavourable	γ_G	1.35		
	Favourable	$\gamma_{G,fav}$	1.0		
Variable actions (Q)	Unfavourable	γ_Q	1.5		
	Favourable	$\gamma_{Q,fav}$	0		
Material properties (X)		γ_M		1.0	
Bearing resistance (R$_v$)		γ_{Rv}			1.4
Sliding resistance (R$_h$)		γ_{Rh}			1.1
Earth resistance against retaining structures		γ_{Re}			1.4
... in slopes					1.1

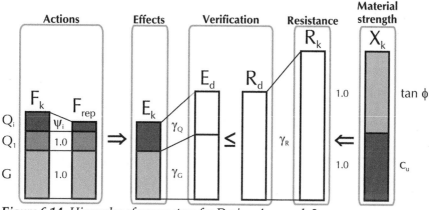

Figure 6.14. Hierarchy of parameters for Design Approach 2

In Design Approach 2, partial factors are applied as late as possible in the calculation process, to action effects and resistances. This is a particular form of load and resistance factor design discussed in Section 6.4.3.

6.3.3 Design Approach 3

The philosophy of Eurocode 7's Design Approach 3 is to check the foundation's reliability by applying partial factors to structural actions and to material properties simultaneously, while geotechnical actions and resistance are left mainly unfactored. (See below for discussion of the difference between structural and geotechnical actions.)

This is achieved by using partial factors from Sets A1 or A2 (on structural and geotechnical actions respectively), M2, and R3, as illustrated in **Figure 6.15**. The crosses on the diagram indicate that the partial factors in Set R3 are all 1.0 (except for those applied to tensile pile resistance) and hence most resistances are unfactored. The cross against Δa indicates that tolerances are not normally applied to nominal dimensions.

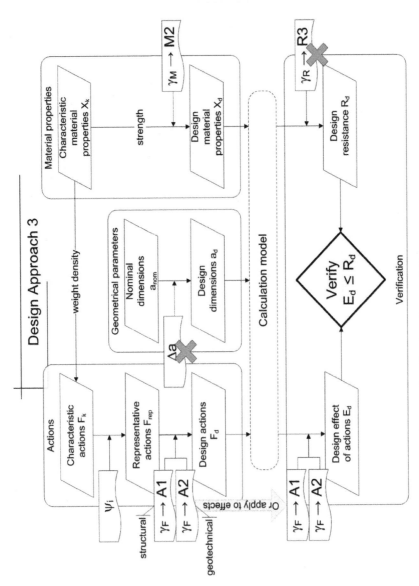

Figure 6.15. Verification of strength to Design Approach 3

If Design Approach 3 is used in slope stability analyses, then factors from Set A2 are applied to *all* actions, not just to geotechnical ones. Numerical values of the partial factors needed in Design Approach 3, for persistent and transient design situations, are:

Design Approach 3			A1	A2	M2	R3
Permanent actions (G)	Unfavourable	γ_G	1.35	1.0		
	Favourable	$\gamma_{G,fav}$	1.0	1.0		
Variable actions (Q)	Unfavourable	γ_Q	1.5	1.3		
	Favourable	$\gamma_{Q,fav}$	0	0		
Coefficient of shearing resistance (tan φ)		γ_φ			1.25	
Effective cohesion (c′)		$\gamma_{c'}$			1.25	
Undrained strength (c_u)		γ_{cu}			1.4	
Unconfined compressive strength (q_u)		γ_{qu}			1.4	
Weight density (γ)		γ_γ			1.0	
Resistance (R) (except for pile shaft in tension)		γ_R				1.0
Pile shaft resistance in tension		$\gamma_{R,st}$				1.1

Figure 6.16 shows the relative magnitude of the key parameters when using Design Approach 3. In this approach, partial factors are applied early in the calculation process (i.e. to actions and material properties), but – unlike Design Approach 1 – in a single stage. Design Approach 3 is a form of material factor design discussed in Section 6.4.2.

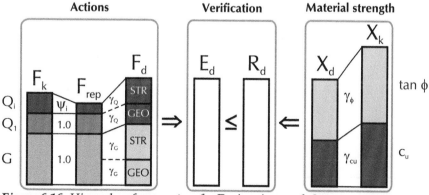

Figure 6.16. Hierarchy of parameters for Design Approach 3

An important feature of Design Approach 3 is the distinction between structural and geotechnical actions — larger factors are applied to the former than to the latter, suggesting greater uncertainty in their values. A geotechnical action is defined as an:

> *action transmitted to the structure by the ground, fill[, standing water†] or*
> *groundwater.*   *[EN 1990 §1.5.3.7] and [EN 1997-1 §1.5.2.1]*

Eurocode 7 does not explicitly define what a structural action is, but by implication it is an action that is not geotechnical.

Figure 6.17 shows some design situations where the distinction between structural and geotechnical actions is not straightforward. Strict interpretation of the definition above results in traffic action applied to the gravity wall being classified as a geotechnical action!

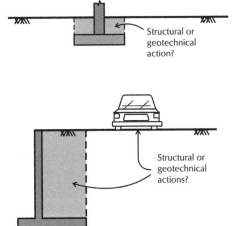

The table below summarizes several alternative definitions for 'geotechnical action' and applies them to the design situations shown in **Figure 6.17**. None is without problems. Fortunately, as Section 6.3.4 below shows, this issue has become largely academic because Design Approach 3 is almost exclusively being used for the design of slopes, where all actions (including traffic) should be treated as geotechnical.

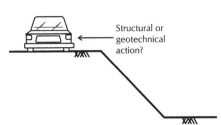

Figure 6.17. Situations where definition of geotechnical action is unclear

---

†Text in brackets omitted in EN 1990.

Definition of geotechnical action	Concrete in wall/ footing	Backfill on ...		Traffic ...	
		Footing	Wall	Behind wall	On slope crest
Transmitted to structure by ground?	No (STR)	Yes? (GEO)	Yes? (GEO)	Yes (GEO)	Yes? (GEO)
Designers' Guide†	STR	STR	GEO	STR	GEO
Soil or rock?	No (STR)	Yes (GEO)	Yes (GEO)	No (STR)	No (STR)
Below ground level	Yes (GEO)	Yes (GEO)	Yes (GEO)	No (STR)	No (STR)
Magnitude uncertain?	No (GEO)	No (GEO)	No (GEO)	Yes (STR)	Yes (STR)

†Choice made in Eurocode 7 Designers' Guide[3]
?It is unclear whether this should be Yes or No

6.3.4   Choice of design approach by different European countries

Eurocode 7 Part 1 allows each country to specify in its National Annex which design approach must be used within its jurisdiction. The choices made by the countries within CEN[4] are summarized in **Figure 6.18** (for slopes) and **Figure 6.19** (for other geotechnical structures).

[EN 1997-1 §2.4.7.3.4.1(1)P NOTE 1]

As **Figure 6.18** shows, the most popular design approach (DA) for slopes is DA3 (adopted by 65% of CEN countries), followed by DA1 (25%). Only one country (Spain) has chosen DA2, while Ireland permits any of the design approaches to be used. As discussed in Chapter 9, DAs 1 and 3 produce almost identical results when applied to slope stability, so almost all of Europe has adopted a common approach for this issue.

Figure 6.19 illustrates the choice of design approach for structures other than slopes. DA2 has been adopted by 55% of CEN countries, followed by DA1 (30%), and then DA3 (10%). Ireland permits any of the design approaches to be used. This division between the various European countries reflects their respective traditions and practice in geotechnical engineering, rather than any judgment regarding the philosophical and engineering merits of each design approach.

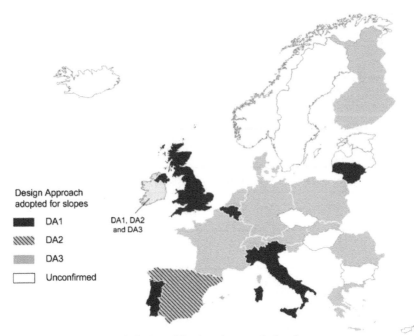

Figure 6.18. *National choice of Design Approach for slopes*

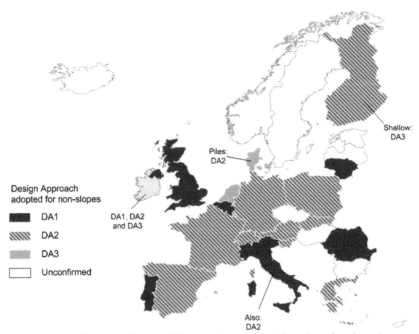

Figure 6.19. *National choice of Design Approach (other than for slopes)*

6.4   Alternative ways of dealing with design uncertainty

The following sub-sections briefly review alternative ways of dealing with uncertainties in geotechnical design.

6.4.1   Allowable or working stress design (ASD or WSD)

Early codes of practice for geotechnical design ensured reliability in calculations by the introduction of a 'factor of safety', applied in some cases to the structure's dimensions (e.g. to the width of a footing or the embedded depth of a retaining wall) and in other cases to the structure's load-bearing capacity (e.g. to ultimate bearing capacity or passive earth pressure).

According to Meyerhof,[5] the concept of a factor of safety was first used in geotechnical design by Bélidor and Coulomb in the 18th Century. During the first half of the 20th Century, it was common practice in Europe and North America to utilize a single, 'global' factor of safety F in geotechnical design, defined as the ratio of the ultimate resistance of the foundation Q_{ult} to the applied load P:

$$F = \frac{Q_{ult}}{P}$$

Re-arranging this equation enables the allowable load P_a to be determined:

$$P_a = \frac{Q_{ult}}{F}$$

Values of F typically vary between 1.3 and 3.0, depending on the structure and the type of failure considered.

6.4.2   Load and strength factor design

In the middle of the 20th Century, first Taylor[6] and then Brinch Hansen[7] introduced the concept of partial factors into geotechnical design, with separate factors being applied to different types of load, shear strength parameters for soils, and on the components of pile capacity. The partial factors recommended by Brinch Hansen, summarized in the table below, are not too dissimilar to the partial factors that have found their way into Eurocode 7 (see Section 6.2.5) and were chosen to give about the same design outcomes as conventional global factors of safety.

The philosophy of material factor design is to apply partial factors as close as possible to the source of uncertainty.

Parameter		Partial factor
Loads	Dead loads, soil weight	1.0
	Live and environmental loads	1.5
	Water pressures, accidental loads	1.0
Shear strength	Friction (tan φ)	1.2
	Cohesion (c'), slopes, earth pressures	1.5
	Cohesion (c'), spread foundations	1.7
	Cohesion (c'), piles	2.0
Ultimate pile capacities	Load tests	1.6
	Dynamic formulas	2.0
Deformations		1.0

This approach has been used in Denmark since the early 1960s and was first codified in the 1965 Danish *Code of practice for foundation engineering*.[8] It has since been widely adopted in European practice, for example in BS 8002.[9]

6.4.3   Load and resistance factor design (LRFD)

In northern America, an alternative approach to geotechnical design based on limit state principles has become popular, recently being incorporated into AASHTO's bridge design specifications,[10] the American Petroleum Institute's recommended practice for offshore structures,[11] and the Canadian Foundation Engineering Manual.[12]

The philosophy of the 'load and resistance factor design' (LRFD) method is to apply partial factors to the outcomes of the design calculations, i.e. to action effects and resistance.

The fundamental equation that must be satisfied in the LRFD method (using AASHTO's formulation) is:

$$\sum_i \eta_i \gamma_i Q_i \le \varphi R_n = R_r$$

where η_i = a load modifier (which is numerically between 0.95 and 1.0) accounting for ductility, redundancy, and operational importance; γ_i = a load factor, typically ≥ 1.0; Q_i = a force effect; φ = a resistance factor ≤ 1.0; R_n = the nominal resistance; and R_r = the factored resistance.

Since the LRFD resistance factor φ is typically less than one, its reciprocal is directly analogous to the Structural Eurocode's resistance factor γ_R. Indeed, the equation above may be re-written with Eurocode-like symbols as:

$$E_d = \sum_i \gamma_{F,i} E_i \leq \frac{R_k}{\gamma_R} = R_d$$

Although the LRFD method has many features in common with Eurocode 7's Design Approach 2, the resistance factors it employs are different, as can be seen by comparing the table below with the values given in Section 6.3.2.

Structure	Partial factor on...	Value	Reciprocal
Overall stability	Earth slope resistance (φ)	0.65–0.75	1.33–1.54
Spread foundation	Bearing resistance (φ_b)	0.45–0.55	1.82–2.22
	Sliding resistance (φ_r)	0.80–0.90	1.11–1.25
	Passive earth pressure (φ_{ep})	0.50	2.0
Drilled shaft	Side resistance (φ_{stat})	0.45–0.60	1.67–2.22
	Tip resistance (φ_{stat})	0.40–0.50	2.0–2.5
	Uplift resistance (φ_{up})	0.35–0.45	2.22–2.86
Driven pile	Skin friction/end bearing (φ_{stat})	0.25–0.45	2.22–4.0
	Uplift resistance (φ_{up})	0.20–0.40	2.5–5.0

6.5   Summary of key points

Limit states STR and GEO involve failure or excessive deformation of the structure or ground, in which the strength of the structure or ground is significant in providing resistance.

Verification of these limit states is demonstrated by satisfying the inequality:
$$E_d \leq R_d$$
in which E_d = design effects of actions and R_d = design resistance. This equation is applied using one of three Design Approaches, chosen in the National Annex.

6.6   Notes and references

1. Homer (800–700 BC), *The Iliad*.

2. European Commission (2003) *Guidance paper L: Application and use of Eurocodes*, CEN.

3. Frank, R., Bauduin, C., Kavvadas, M., Krebs Ovesen, N., Orr, T., and Schuppener, B. (2004) *Designers' guide to EN 1997-1: Eurocode 7: Geotechnical design – General rules*, London: Thomas Telford.

4. Schuppener, B. (2007) *Eurocode 7: Geotechnical design - Part 1: General rules – its implementation in the European Member states*, Proc. 14th European Conf. on Soil Mechanics and Geotechnical Engineering, Madrid.

5. Meyerhof, G. G. (1994) *Evolution of safety factors and geotechnical limit state design* (2nd Spencer J. Buchanan Lecture), Texas A&M University.

6. Taylor, D. W. (1948) *Fundamentals of soil mechanics*, New York: J. Wiley.

7. Brinch Hansen, J. (1956) *Limit design and safety factors in soil mechanics*, Danish Geotechnical Institute, Bulletin No. 1.

8. DS 415: 1965, Code of practice for foundation engineering, Dansk Ingeniørforening.

9. BS 8002: 1994 Code of practice for earth retaining structures, British Standards Institution, London.

10. AASHTO LRFD Bridge Design Specifications (4th Edition, 2007), American Association of State Highway and Transportation Official.

11. American Petroleum Institute (2003) *Recommended practice for planning, designing and constructing fixed offshore platforms – Load and Resistance Factor Design*, American Petroleum Institute.

12. Canadian Geotechnical Society (2006) *Foundation Engineering Manual*, Canadian Geotechnical Society.

Chapter 7

Verification of stability

'Give me a lever long enough and a fulcrum on which to place it, and I shall move the world' — Archimedes (c. 287-212 BC)

7.1   Basis of design

Verification of stability involves checking that destabilizing effects of actions do not exceed the corresponding stabilizing effects, plus any resistance that enhances those stabilizing effects.

Verification of stability is expressed in Eurocode 7 by the inequality:

$$E_{d,dst} \leq E_{d,stb} + R_d$$

in which $E_{d,dst}$ = the design effect of destabilizing actions, $E_{d,stb}$ = the design effect of stabilizing actions, and R_d = any design resistance that helps to stabilize the structure.

This requirement applies to ultimate limit state EQU, defined as:

[loss] of equilibrium of the structure or the ground considered as a rigid body, [where ... minor variations ... in actions ... are significant, and†] in which the strengths of structural materials and the ground are insignificant in providing resistance   *[EN 1990 §6.4.1(1)P] and [EN 1997-1 §2.4.7.1(1)P]*

to ultimate limit state UPL, defined as:

loss of equilibrium of the structure or the ground due to uplift by water pressure (buoyancy) or other vertical actions   *[EN 1997-1 §2.4.7.1(1)P]*

and to ultimate limit state HYD, defined as:

hydraulic heave, internal erosion and piping in the ground caused by hydraulic gradients   *[EN 1997-1 §2.4.7.1(1)P]*

---

†the words in brackets are omitted from Eurocode 7's definition.

Examples of situations where stability is a concern are shown in **Figure 7.1**. From left to right, these include, top: overturning moments on a wind turbine must not exceed restoring moments from the foundation (EQU); uplift forces on a basement due to effective stress rebound must not exceed the weight of the structure and any shear along its walls (UPL); and heave caused by inflow of water due to differential water levels around an embedded wall must be acceptable (HYD). And, bottom: the overturning moment on a water-retaining cofferdam must not exceed the restoring moment due to the cofferdam's weight (EQU); uplift caused by pore pressures beneath the base of a cutting must not exceed the weight of ground in that cutting (UPL); and erosion of soil caused by large hydraulic gradients must be avoided (HYD).

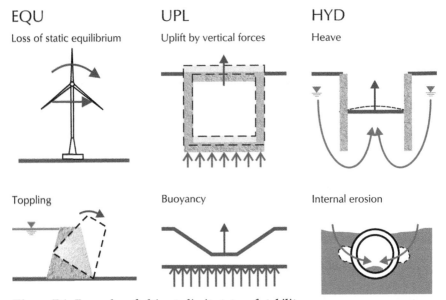

Figure 7.1. Examples of ultimate limit states of stability

7.2   Introducing reliability into the design

Reliability is introduced into design against loss of stability by applying partial factors (or tolerances) to:
- destabilizing actions (F_{dst})
- stabilizing actions (F_{stb})
- material properties (X) or resistances (R)
- geometrical parameters (a)

These factors/tolerances are shown on **Figure 7.2** in the wavy boxes.

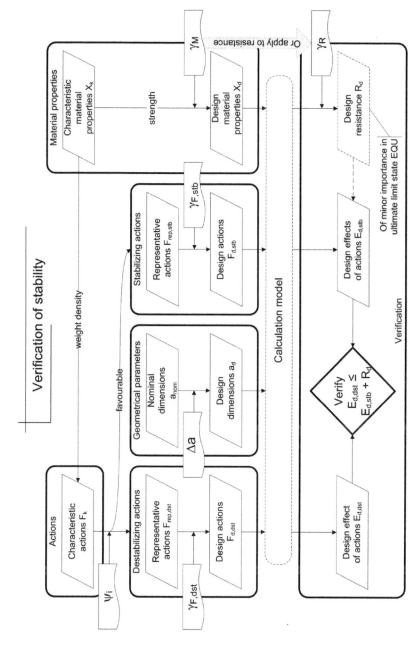

Figure 7.2. Overview of verification of stability

Destabilizing design actions $F_{d,dst}$ are obtained from destabilizing characteristic actions ($F_{k,dst}$) by first multiplying by combination factors ψ (where appropriate, see Chapter 2) and then multiplying by partial factors $\gamma_{F,dst}$ = 1.1–1.5 (as shown in the far left hand channel of **Figure 7.2**):

$$F_{d,dst} = \sum_i \gamma_{F,dst,i} \psi_i F_{k,dst,i}$$

Stabilizing design actions $F_{d,stb}$ are obtained from stabilizing characteristic actions ($F_{k,stb}$) by multiplying by combination factors ψ (where appropriate, see Chapter 2) and then multiplying by $\gamma_{F,stb}$ = 0–0.9 (middle right channel):

$$F_{d,stb} = \sum_i \gamma_{F,stb,i} \psi_i F_{k,stb,i}$$

Design material properties X_d are obtained from characteristic material properties (X_k) by dividing by partial factors γ_M = 1.0–1.4 or, alternatively, design resistances R_d are obtained from characteristic resistances (R_k) by dividing by γ_R = 1.0-1.4 (far right channel):

$$X_d = \frac{X_k}{\gamma_M} \text{ or } R_d = \frac{R_k}{\gamma_R}$$

Finally, design dimensions a_d are obtained from nominal dimensions a_{nom} by adding or subtracting a tolerance Δa (middle left hand channel):

$$a_d = a_{nom} \pm \Delta a$$

7.2.1   Partial factors

Partial factors for verification of stability are specified in EN 1997-1 in Tables A.1 and A.2 for EQU, A.15 and A.16 for UPL, and A.17 for HYD. Some

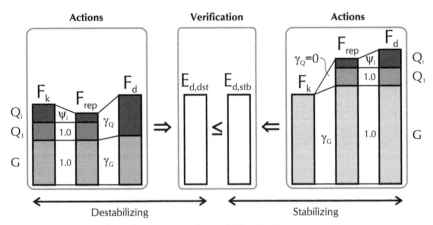

Figure 7.3. *Hierarchy of parameters in stability limit states*

values are amended in the UK National Annexes to BS ENs 1990 and 1997-1. The following table summarizes the factors that must be used, for persistent and transient design situations, and **Figure 7.3** shows the relative magnitude of the key parameters when checking the EQU, UPL, and HYD limit states.

Parameter			Ultimate limit state		
			EQU	UPL	HYD
Permanent actions (G)	Destabilizing	$\gamma_{G,dst}$	1.1 Bridges 1.05	1.0 [1.1]	1.35
	Stabilizing	$\gamma_{G,stb}$	0.9 Bridges 0.95	0.9	0.9
Variable actions (Q)	Destabilizing	$\gamma_{Q,dst}$	1.5 Road 1.35 Pedest. 1.35 Rail 1.4-1.7 Wind 1.7 Thermal 1.55	1.5	1.5
	Stabilizing	$\gamma_{Q,stb}$	0	0†	0
Coefficient of shearing resistance (tan φ)		γ_φ	1.25 [1.1]	1.25	-†
Effective cohesion (c′)		$\gamma_{c'}$	1.25 [1.1]	1.25	
Undrained strength (c_u)		γ_{cu}	1.4 [1.2]	1.4	
Unconfined compressive strength (q_u)		γ_{qu}	1.4 [1.2]	1.4†	
Weight density (γ)		γ_γ	1	1.0†	
Tensile pile resistance (R_{st})		γ_{st}		1.4 [*]	-†
Anchorage resistance (R_a)		γ_a		1.4 [*]	

†Inferred value (not given explicitly in EN 1997-1)
Values in [brackets] from UK National Annex to BS EN 1997-1
Underlined values from UK National Annex to BS EN 1990 (A1)
*Values to be determined according to ultimate limits states STR/GEO
-Not applicable

7.3   Loss of static equilibrium

Ultimate limit state EQU is defined as:

> *[loss] of equilibrium of the structure or the ground considered as a rigid body,*
> *[where ... minor variations ... in actions ... are significant, and†] in which the*
> *strengths of structural materials and the ground are insignificant in*
> *providing resistance.*   *[EN 1990 §6.4.1(1)P] and [EN 1997-1 §2.4.7.1(1)P]*

Verification of static equilibrium is expressed in the Eurocodes by the inequality:

$$E_{d,dst} \leq E_{d,stb}(+R_d)$$ *[EN 1990 exp (6.7)] and [EN 1997-1 exp (2.4,modified)]*

in which $E_{d,dst}$ = the design effect of destabilizing actions, $E_{d,stb}$ = the design effect of stabilizing actions, and R_d = any design resistance that helps to stabilize the structure. The term in brackets (i.e. R_d) is absent in EN 1990.‡

EN 1997-1 notes that 'if any shearing resistance ... is included [in EQU], it should be of minor importance'. The inclusion of R_d in the inequality above appears to contradict this note and the latter part of the definition of EQU ('strengths ... are insignificant in providing resistance'). If resistance *is* insignificant, why is it included in the inequality? If resistance is *significant*, then limit states STR and GEO (see Chapter 6) ought to control the design.
                                                                 [EN 1997-1 §2.4.7.2(2)P NOTE 1]

EN 1997-1 also notes that EQU:

> *... is mainly relevant in structural design. In geotechnical design, EQU*
> *verification will be limited to rare cases, such as a rigid foundation bearing*
> *on rock.*   *[EN 1997-1 §2.4.7.2(2)P NOTE 1]*

Consider, therefore, the mass concrete dam bearing on rock that is shown in **Figure 7.4**, which would qualify as a rare case where EQU is relevant in geotechnical design. For simplicity, we will treat the dam as a rectangular block of height H and breadth B, whereas in reality its shape would be more complicated.

---

†the words in brackets are omitted in Eurocode 7's definition.

‡The symbol T_d is used in Eurocode 7, but this is likely to be changed to R_d in a future corrigendum.

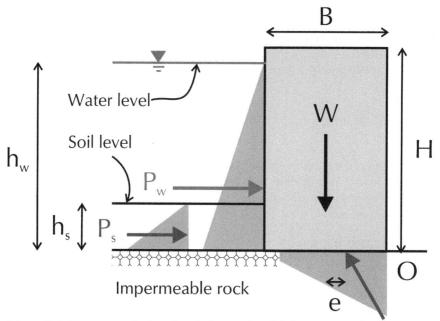

Figure 7.4. *Mass concrete dam founded on rock, retaining water and soil*

The dam retains a mixture of water (to a height h_w) and soil (to height h_s). The soil is considerably weaker than the underlying rock.

The total horizontal thrust P on the back of the dam is given by:

$$P = P_w + P_s = \frac{1}{2}\gamma_w h_w^2 + \frac{1}{2}K_a\left(\gamma_s - \gamma_w\right)h_s^2$$

where γ_w and γ_s are the weight densities of water and the soil and K_a is the soil's active earth pressure coefficient, which is assumed here (for simplicity) to be given by:

$$K_a = \frac{1 - \sin\varphi}{1 + \sin\varphi}$$

with φ the soil's angle of shearing resistance.

The destabilizing (i.e. overturning) moment that the earth and water pressures produce about the dam's toe O is:

$$M = \frac{1}{6}\gamma_w h_w^3 + \frac{1}{6}K_a\left(\gamma_s - \gamma_w\right)h_s^3$$

Counteracting these actions is the dam's weight W, which provides a maximum (i.e. ultimate) sliding resistance P_{ult}:

$$P_{ult} = \gamma_c BH \times \tan \delta$$

where γ_c is the concrete's weight density and δ the angle of interface friction between the dam and the underlying rock. The stabilizing (i.e. restoring) moment of W about the dam toe O defines the limiting (i.e. ultimate) value of M:

$$M_{ult} = \gamma_c BH \times \frac{B}{2}$$

Traditionally, the stability of the dam against rigid-body motion would be checked using appropriate factors of safety F_s for sliding and F_o for overturning. The allowable thrust P_a and allowable overturning moment M_a are then given by:

$$P_a \leq \frac{P_{ult}}{F_s} \text{ and } M_a \leq \frac{M_{ult}}{F_o}$$

If we substitute the equations for P, P_{ult}, M, and M_{ult} into these inequalities and assume the water level rises to the top of the dam (i.e. $h_w = H$), we obtain, after re-arrangement:

$$\frac{B}{H} \geq \left(\frac{F_s}{2}\right)\left(\frac{\gamma_w}{\gamma_c}\right)\left(1 + K_a\left(\frac{\gamma_s - \gamma_w}{\gamma_c}\right)\left(\frac{h_s}{H}\right)^2\right) \times \left(\frac{1}{\tan \delta}\right)$$

and

$$\frac{B}{H} \geq \sqrt{\left(\frac{F_o}{3}\right)\left(\frac{\gamma_w}{\gamma_c}\right)\left(1 + K_a\left(\frac{\gamma_s - \gamma_w}{\gamma_c}\right)\left(\frac{h_s}{H}\right)^3\right)}$$

Figure 7.5 illustrates the consequences of the above for a dam retaining various depths of soil, assuming $\gamma_s = 20\text{kN/m}^3$, $\gamma_c = 24\text{kN/m}^3$, $\gamma_w = 10\text{kN/m}^3$, $\varphi = 25°$ (hence $K_a = 0.406$), and $\delta = 35°$.

The lines labelled 'Traditional' are based on $F_s = 1.5$ for sliding and $F_o = 2.0$ for overturning. As **Figure 7.5** shows, overturning governs the minimum width of the dam for all heights of retained soil. (Of course, had we assumed a smaller value of interface friction δ then sliding might have governed the dam's width.)

An additional line for $F_o = 3.0$, which ensures the resultant force lies within the middle third of the foundation (i.e. e < B/6), is also shown on **Figure 7.5**. This middle-third rule, which prevents tension occurring between the dam and the founding rock, is more critical than the requirement for $F_o = 2$.

Eurocode 7 requires us to verify the stability of rigid bodies according to limit state EQU. We can do this by replacing F_s and F_o in the inequalities above by the ratio $\gamma_{G,dst}/\gamma_{G,stb}$ (= 1.1/0.9 = 1.21) and by applying a material factor γ_φ (= 1.25) to tan φ and tan δ, reducing φ to 20.5° (and consequently *increasing* K_a to 0.482) and δ to 29.2°. As **Figure 7.5** shows, sliding governs the minimum width of the dam for all values of h_s/H – a result that is somewhat counter-intuitive, since EQU deals with static equilibrium!

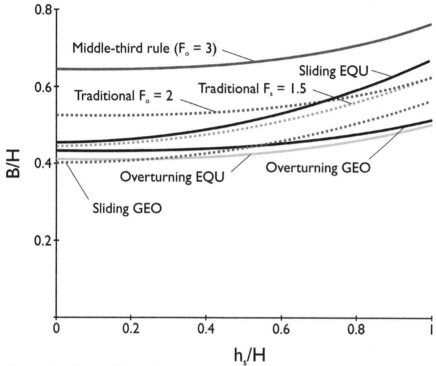

Figure 7.5. *Sizing of dam of **Figure 7.4** to avoid sliding and overturning*

A fundamental principle of limit state design is that all limit states are of equal importance. Consequently, we should check all design situations for all possible limit states. We can check the situation shown in **Figure 7.5** for limit states GEO and STR (see Chapter 6) by replacing F_s and F_o in the inequalities above by the ratio $\gamma_G/\gamma_{G,fav}$ (= 1.35 for Design Approach 1, Combination 1, and 1.0 for Combination 2) and by applying a material factor γ_φ (= 1.0 or 1.25) to tan φ and tan δ. As **Figure 7.5** shows, overturning governs the minimum width of the dam for values of $h_s/H < 0.5$ and sliding governs for $h_s/H > 0.5$. (This change-over depends on the value of interface friction δ assumed.) Once again, this result is counter-intuitive, since GEO and STR deal with the strength of materials!

Many engineers would instinctively consider overturning to be controlled by an EQU limit state and sliding by GEO. The simple calculation above demonstrates that Eurocode 7's partial factors lead to the opposite conclusion in some situations. We believe that sliding should be governed by limit state GEO and EQU should only be used to guard against overturning when any actions from the ground are minor.

It is generally unwise to allow tension to occur either within the structure or between its base and the ground. To prevent this from happening, the resultant force must remain within the middle third of the foundation base. Satisfying this condition – which is not required by Eurocode 7 – is likely to govern the design.

7.4   Uplift

Ultimate limit state UPL is defined as:

> *loss of equilibrium of the structure or the ground due to uplift by water*
> *pressure (buoyancy) or other vertical actions*   *[EN 1997-1 §2.4.7.1(1)P]*

Because uplift involves predominantly vertical actions, verification of stability against uplift is expressed in Eurocode 7 by the inequality:

$$V_{d,dst} = G_{d,dst} + Q_{d,dst} \leq G_{d,stb} + R_d \qquad \text{[EN 1997-1 exp (2.8)]}$$

in which V_d = the design vertical action, G_d = design permanent actions, Q_d = design variable actions, and R_d = any design resistance that helps to stabilize the structure. The subscripts 'dst' and 'stb' denote destabilizing and stabilizing components, respectively. This inequality is merely a more specific version of the general equation given in Section 7.1:

$$E_{d,dst} \leq E_{d,stb} + R_d$$

Eurocode 7 Part 1 allows resistance to uplift to be treated as a stabilizing permanent vertical action, thereby simplifying the expression above to:

$$G_{d,dst} + Q_{d,dst} \leq G_{d,stb}$$

However, doing so will lead to a different outcome to that obtained with the previous equation, since in the first case the resistance is obtained by dividing material strengths by their appropriate partial factors (e.g. $\gamma_\varphi = 1.25$ or $\gamma_{cu} = 1.4$), whereas in the second case the resistance is multiplied by the partial factor on stabilizing permanent actions ($\gamma_{G,stb} = 0.9$). Since it gives a more conservative result, we believe it may be better to treat resistance explicitly as resistance and not as a favourable action.

Destabilizing design vertical actions $V_{d,dst}$ are obtained from destabilizing characteristic permanent ($G_{k,dst}$) and variable ($Q_{k,dst}$) actions by first multiplying by combination factors ψ where appropriate (see Chapter 2) and then multiplying by partial factors γ_G and γ_Q greater than or equal to 1.0:

$$V_{d,dst} = \sum_j \gamma_{G,dst,j} G_{k,dst,j} + \sum_i \gamma_{Q,dst,i} \psi_i Q_{k,dst,i}$$

Stabilizing design vertical actions $V_{d,stb}$ are obtained from stabilizing characteristic permanent actions ($G_{k,stb}$) by multiplying by a partial factor less than or equal to 1.0:

$$V_{d,stb} = \sum_j \gamma_{G,stb,j} G_{k,stb,j}$$

There is no term for variable actions in this expression, since it would be unsafe to include it (mathematically, $\gamma_{Q,stb} = 0$).

The design resistance R_d, if any, that helps to stabilize the structure can be obtained in one of two ways: either directly from design material properties X_d with a resistance factor $\gamma_R = 1$ or from characteristic material properties with $\gamma_R > 1$. See Chapter 6 for further discussion of the way design resistance may be calculated.

Finally, design dimensions a_d are are obtained from nominal dimensions a_{nom} by adding or subtracting a tolerance Δa. Eurocode 7 does not give any specific recommendations for the value of Δa to use in uplift verifications.

Consider the depressed highway shown in **Figure 7.6**, which is subject to uplift owing to a naturally high water table outside the constructed section.

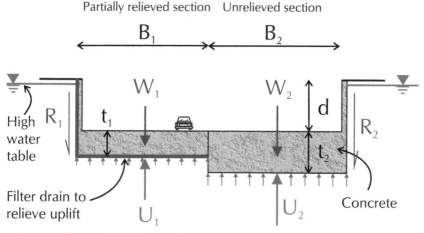

Figure 7.6. Depressed highway subject to uplift

Two options are shown in **Figure 7.6** for dealing with the uplift pressures beneath the highway: in the 'partially relieved' section, a filter drain is installed to relieve water pressures and allow the thickness t_1 of the road base to be minimized; in the 'unrelieved' section, no drain is installed and the base thickness t_2 must be selected to balance the uplift force. The partially relieved section will require ongoing maintenance to ensure the drains continue to function correctly in the future.

For the situations shown in **Figure 7.6**, the destabilizing design vertical action is given by:

$$V_{d,dst} = \gamma_{G,dst} U_k = \gamma_{G,dst} \times \gamma_w \times (d + t_i) \times B_i$$

and the stabilizing design vertical action by:

$$V_{d,stb} = \gamma_{G,stb} W_k = \gamma_{G,stb} \times \gamma_{ck} \times t_i \times B_i$$

where d, t, and B are as defined on **Figure 7.6**; γ_w and γ_{ck} = weight densities of water and concrete, respectively; and we have assumed that the water table has risen to the ground surface and ignored the weight of the side walls.

In addition, the resistance R caused by effective earth pressures acting on the side wall of the section helps to stabilize the structure:

$$R = \frac{1}{2} K_a \times (\gamma_k - \gamma_w) \times (d + t_i)^2 \times \tan(\delta)$$

$$= \frac{1}{2} \beta \times (\gamma_k - \gamma_w) \times (d + t_i)^2$$

where γ_k = the weight density of the soil adjacent to the highway; δ = the angle of interface friction between the wall and the ground; and K_a is the soil's active earth pressure coefficient, given by:

$$K_a = \frac{1 - \sin\varphi}{1 + \sin\varphi}$$

where φ is the soil's angle of shearing resistance. By assuming active earth pressure conditions, we avoid over-estimating the favourable effect of R. If we also assume that the angle of interface friction is given by $\delta = (2/3)\varphi$, then the factor β becomes:

$$\beta = \left(\frac{1 - \sin\varphi}{1 + \sin\varphi} \right) \times \tan\left(\frac{2}{3}\varphi \right)$$

Figure 7.7 shows the value of β based on various assumptions about the soil's angle of shearing resistance φ. The curve labelled 'characteristic' assumes $\varphi = \varphi_{k,inf}$, the soil's 'inferior' characteristic angle of shearing resistance. The maximum value of β in this case occurs when $\varphi_{k,inf} = 27.3°$.

The curve labelled 'inferior design' assumes that $\varphi = \varphi_{d,inf}$, the soil's inferior design angle of shearing resistance, given by:

$$\varphi_{d,inf} = \tan^{-1}\left(\frac{\tan\varphi_{k,inf}}{\gamma_\varphi}\right)$$

where the partial factor $\gamma_\varphi = 1.25$. The maximum value of β in this case occurs when $\varphi_{k,inf} = 32.8°$. As **Figure 7.7** shows, the value of β is smaller for the characteristic curve when $\varphi_{k,inf} > 29.9°$. In other words, applying a partial factor to the soil's characteristic angle of shearing resistance *increases* the resistance available along the wall when $\varphi_{k,inf}$ is greater than about 30°![1]

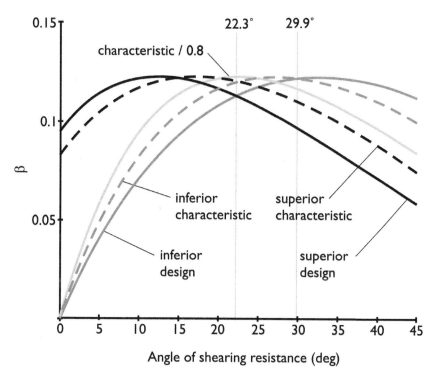

Figure 7.7. *Variation of favourable resistance with angle of shearing resistance*

The UK National Annex to BS EN 1997-1[2] attempts to deal with this anomaly by noting:

The value of the partial factor for soil parameters should be taken as the reciprocal of the specified value if such a reciprocal value produces a more onerous effect …

The curve labelled 'characteristic/0.8' on **Figure 7.7** shows the effect of entering $\gamma_\phi = 0.8 \, (= 1/1.25)$ in the previous equation for $\phi_{d,inf}$. However, there is a serious problem with this approach, as illustrated in **Figure 7.8** and explained below.

Figure 7.8 shows a hypothetical normal (aka Gaussian) probability density function for the soil's angle of shearing resistance ϕ, assuming a mean value $\mu_\phi = 30°$ and standard deviation $\sigma_\phi = 3°$ (giving a coefficient of variation $\delta_\phi = \sigma_\phi/\mu_\phi = 0.1$).

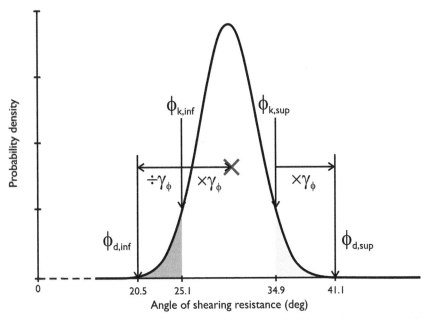

Figure 7.8. Difference between 'inferior' and 'superior' design values of strength

The inferior characteristic angle of shearing resistance $\phi_{k,inf}$ may be estimated from:

$$\phi_{k,inf} = \mu_\phi - \kappa\sigma_\phi = 30° - 1.645 \times 3° = 25.1°$$

where κ is a statistical coefficient assumed here to be 1.645. (See Chapter 5 for a full discussion of this equation and the choice of a suitable value for κ.)

The inferior design angle of shearing resistance $\phi_{d,inf}$ is then given by:

$$\phi_{d,inf} = \tan^{-1}\left(\frac{\tan\phi_{k,inf}}{\gamma_\phi}\right) = \tan^{-1}\left(\frac{\tan 25.1°}{1.25}\right) = 20.5°$$

and is shown on **Figure 7.8**.

A design angle of shearing resistance φ_d obtained from:

$$\varphi_d = \tan^{-1}\left(\frac{\tan\varphi_{k,inf}}{\gamma_{\varphi,sup}}\right) = \tan^{-1}\left(\frac{\tan 25.1°}{0.8}\right) = 30.3°$$

is an entirely meaningless value, falling at an arbitrary position (shown by the cross) within the probability density function for φ. Instead, the calculation of the superior design angle of shearing resistance $\varphi_{k,sup}$ should be based on:

$$\varphi_{k,sup} = \mu_\varphi + \kappa\sigma_\varphi = 30° + 1.645 \times 3° = 34.9°$$

which then gives:

$$\varphi_{d,sup} = \tan^{-1}\left(\frac{\tan\varphi_{k,sup}}{\gamma_{\varphi,sup}}\right) = \tan^{-1}\left(\frac{\tan 34.9°}{0.8}\right) = 41.1°$$

Returning now to **Figure 7.7**, the curve labelled 'superior design' assumes that $\varphi = \varphi_{d,sup}$, with $\varphi_{k,sup} = \varphi_{k,inf} + 10°$. (The difference of 10° selected here for illustration only.) This curve gives the most pessimistic values of β when $\varphi_{k,inf}$ > 22.3°.

When establishing the resistance provided by retaining walls against uplift, it is not obvious whether a high or a low characteristic angle of shearing resistance will give the most conservative result. Instinctively, we might expect that a weaker soil would give lower resistance but, because of the interaction between strength and earth pressures, this is not always true.

7.5 Hydraulic failure

Ultimate limit state HYD is defined in Eurocode 7 as:

hydraulic heave, internal erosion and piping in the ground caused by hydraulic gradients   *[EN 1997-1 §2.4.7.1(1)P]*

The following sub-sections consider each of these phenomena in turn.

7.5.1 Hydraulic heave

Verification of stability against hydraulic heave is expressed in Eurocode 7 by two different (but supposedly equivalent) inequalities. One is given in terms of forces and weights:

$$S_{d,dst} \leq G'_{d,stb}$$ *[EN 1997-1 exp (2.9b, modified)]*

in which $S_{d,dst}$ = the design seepage force destabilizing a column of soil and $G'_{d,stb}$ = the design submerged weight of that soil column.

The other is expressed in terms of stresses and pressures:

$$u_{d,dst} \leq \sigma_{d,stb}$$ [EN 1997-1 exp (2.9a, modified)]

in which $u_{d,dst}$ = the design total pore water pressure that is destabilizing the soil column and $\sigma_{d,stb}$ = the (stabilizing) design total stress that resists that pore pressure. Regrettably, Eurocode 7 does not specify how partial factors should be applied in the verification of HYD, which can lead to an apparent disparity between these two equations.

Applying Terzaghi's principle of effective stress[3], we can rearrange the latter equation as follows:

$$\sigma_{d,stb} - u_{d,dst} = \sigma_d' \geq 0$$

which merely states that the design effective stress at the base of the soil column must not become negative.

Consider the embedded retaining wall shown in **Figure 7.9**, under which water flows owing to a difference in water level across the wall. Experiments have shown[4] that a block of soil (shaded) of width d/2 is susceptible to piping failure if the hydraulic gradient over the depth of embedment d exceeds a critical value i_{crit}.

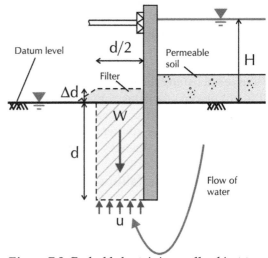

Figure 7.9. *Embedded retaining wall subject to piping due to heave*

For the purposes of this example, the datum has been taken at formation level on the left hand side of the wall. With this assumption, the total head h is given by Bernouilli's equation (total head = elevation + pressure + kinetic heads):

$$h = z + \frac{u}{\gamma_w} + \frac{v^2}{2g}$$

where z is the elevation above the datum level; u the pore water pressure; v the water velocity; and g the acceleration due to gravity. In situations involving groundwater, the kinetic head is usually negligible in comparison with the other heads and can be ignored. Hence:

$$h \approx z + \frac{u}{\gamma_w}$$

The total head acting over the base of the shaded soil column shown in **Figure 7.9** may be approximated by:

$$h \approx -d + \frac{(H+d)+d}{2} = \frac{H}{2}$$

where H is the height of the retained water above formation level. This assumes that the head loss caused by seepage into the excavation is equal on both sides of the wall.

With the above assumptions, the characteristic hydraulic gradient i_k through the shaded region is:

$$i_k = h/d = H/2d$$

The characteristic seepage force destabilizing the soil column (per unit run of wall) is given by:

$$S_k = \gamma_w \times i_k \times volume = \gamma_w i_k \left(\frac{d^2}{2}\right)$$

where γ_w is the weight density of water. Since this is a permanent destabilizing action, its design value is:

$$S_{d,dst} = \gamma_{G,dst} \gamma_w i_k \left(\frac{d^2}{2}\right)$$

where $\gamma_{G,dst}$ is the partial factor (= 1.35) on destabilizing permanent actions.

The characteristic submerged weight of the soil column (per unit run of wall) is given by:

$$G'_k = \gamma'_k \times volume = (\gamma_k - \gamma_w)\left(\frac{d^2}{2}\right)$$

where γ_k' is the soil's characteristic submerged weight density and γ_k its characteristic total weight density. Since this is a permanent stabilizing action, its design value is:

$$G'_{d,stb} = \gamma_{G,stb}(\gamma_k - \gamma_w)\left(\frac{d^2}{2}\right)$$

where $\gamma_{G,stb}$ is the partial factor (= 0.9) on stabilizing permanent actions.

Substituting these expressions into $S_{d,dst} \leq G'_{d,stb}$ and simplifying produces:

$$i_k \leq \left(\frac{\gamma_{G,stb}}{\gamma_{G,dst}}\right)\left(\frac{\gamma_k - \gamma_w}{\gamma_w}\right) = \left(\frac{0.9}{1.35}\right) i_{crit} = \frac{i_{crit}}{1.5} \approx 0.67$$

where i_{crit} is the critical hydraulic gradient (numerically ≈ 1). For this situation, we conclude that the partial factors specified for limit state HYD are equivalent to a global factor of 1.5 on the critical hydraulic gradient i_{crit}. Traditional advice suggests a 'safety factor of 1.5 to 2 is desired against piping or heaving' for this situation[5] and (more onerously) 'a factor of safety ... of 4 to 5 is considered adequate'[6].

The characteristic pore water pressure acting on the underside of the shaded block is:

$$u_{k,dst} = \gamma_w \left(h + d \right) = \gamma_w \left(\frac{H}{2} + d \right)$$

and, since this is a permanent destabilizing action, its design value is:

$$u_{d,dst} = \gamma_{G,dst} \gamma_w \left(\frac{H}{2} + d \right) = \gamma_{G,dst} \gamma_w \left(i_k + 1 \right) d$$

The characteristic vertical total stress acting on the same plane is:

$$\sigma_{k,stb} = \gamma_k d$$

and, since this is a permanent stabilizing action, its design value is:

$$\sigma_{d,stb} = \gamma_{G,stb} \gamma_k d$$

Substituting these expressions into $u_{d,dst} \leq \sigma_{d,stb}$ and simplifying produces:

$$i_k \leq \left(\frac{\gamma_{G,stb}}{\gamma_{G,dst}} \right) \left(\frac{\gamma_k - \gamma_w}{\gamma_w} \right) + \left(\frac{\gamma_{G,stb}}{\gamma_{G,dst}} \right) - 1 = \frac{i_{crit}}{1.5} - \left(\frac{1}{3} \right) \approx 0.33 \approx \frac{i_{crit}}{3.0}$$

since $i_{crit} \approx 1$. For this situation, we conclude that the partial factors specified for limit state HYD are equivalent to a global factor of 3.0 on the critical hydraulic gradient i_{crit}. Thus for the same situation, the verification based on pore pressure and total stress is more onerous than the one based on seepage force and submerged weight.[7]

The total stress approach gives a traditional factor of safety of 3, which is within the range of values (1.5–4.0) suggested in the literature, whereas the seepage force approach gives a safety factor of 1.5, which is at the lower limit of the traditionally recommended values. (Note that, to avoid piping, the characteristic hydraulic gradient should be even smaller than calculated here — see Section 7.5.3.) This demonstrates the problem that, when net forces or pressures are used (as in the seepage force approach), an unintended reduction in reliability can result. We recommend using gross forces and pressures (as in the total stress approach) wherever possible.

7.5.2   Internal erosion

Internal erosion — within a soil stratum, at the interface between strata, or at the soil-structure interface — occurs when large hydraulic gradients carry particles away. If the erosion continues, collapse of the structure follows.

Filter protection at the ground's free surface and suitable filter criteria must be used to prevent internal erosion and to minimize material transport. Non-cohesive soils that satisfy the filter criteria should be used and, in some cases, in multiple layers that ensure a stepwise change in particle size distribution. EN 1997-1 does not provide detailed rules for filter design, for which the reader should refer to any well-established text on the subject.[8]

If the filter criteria are not satisfied, Eurocode 7 requires verification that the design value of the hydraulic gradient i_d is 'well below' the critical hydraulic gradient, taking account of the direction of flow, the grain size distribution and shape of grains, and stratification of the soil.† *[EN 1997-1 §10.4(5)P and (6)P]*

In other words:

$$i_d \ll i_{crit} = \frac{\gamma - \gamma_w}{\gamma_w}$$

Regrettably, Eurocode 7 does not give a partial factor for determining i_d, but based on previous experience this should be at least equal to 4.0.

In order to reduce the potential for particles to be moved through filter materials, much lower design hydraulic gradients are required to resist internal erosion than to resist heave.

7.5.3   Piping

Piping is a form of internal erosion that begins, for example, at the surface of a reservoir and then regresses until a pipe-shaped discharge tunnel is formed in the soil, between the soil and the foundation, or at the interface between cohesive and non-cohesive strata. Failure occurs when the upstream end of the eroded tunnel reaches the bottom of the reservoir.

Areas susceptible to piping must be inspected regularly during periods of extremely unfavourable hydraulic conditions, such as floods, so that

---

† §10.4(5)P wrongly says 'it shall be verified that the critical hydraulic gradient is well below the design value of the gradient at which soil particles begin to move' — this is a known error that will be corrected in a future corrigendum.

measures can be taken immediately to mitigate the conditions (using materials stored in the vicinity).

Prevention of piping must be achieved by providing sufficient resistance against internal soil erosion in the areas where water outflow may occur. Joints or interfaces between the structure and the ground, which may provide preferential seepage paths, must be considered when determining hydraulic conditions in those outflow areas.

Piping is a special case of internal erosion and is of particular significance in the design of embankment dams and/or dams founded on granular fine soils. It is extremely important that hydraulic gradients and the potential for seepage to concentrate in particular areas of the structure is avoided. To mitigate against such problems, careful control of the earthworks is required and seepage paths should be controlled using suitably permeable materials and carefully designed and constructed filter drainage systems.

7.6   Summary of key points

Limit states EQU, UPL, and HYD involve failure of the ground due to an imbalance of forces when the resistance of the ground does not govern.

Verification of these limit states is demonstrated by ensuring:

$$E_{d,dst} \leq E_{d,stb} + R_d$$

where $E_{d,dst}$ = destabilizing design effects of actions, $E_{d,stb}$ = stabilizing design effects, and R_d = design resistance.

The differences between ultimate limit states EQU, UPL, and HYD are determined by which terms dominate the equation above. In EQU, the resistance is of minor importance; in UPL, only vertical actions are considered; and HYD focuses on microscopic rather than macroscopic stability.

7.7   Worked examples

The worked examples in this chapter look at the static equilibrium of a wind turbine subject to wind loads (Example 7.1); equilibrium of a double-walled cofferdam subject to water loads (Example 7.2); uplift of a box caisson (Example 7.3); uplift of the basement of a buried structure (Example 7.4); hydraulic stability of a weir (Example 7.5); and piping due to heave of an embedded retaining wall (Example 7.6).

Specific parts of the calculations are marked ❶, ❷, ❸, etc., where the numbers refer to the notes that accompany each example.

7.7.1 Wind turbine

Example 7.1 considers the foundation design for the wind turbine[9] shown in **Figure 7.10**, which is subject to a permanent vertical force V_{Gk} and imposed variable horizontal force H_{Qk} and moment M_{Qk}. The base, which is square on plan, is set at a depth D below ground surface.

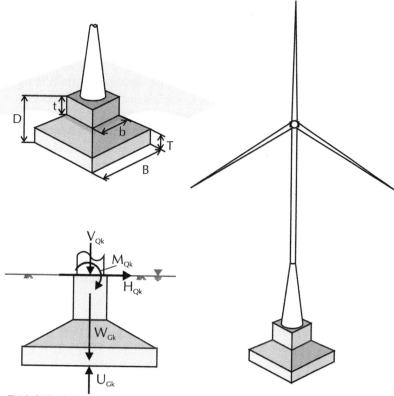

Figure 7.10. Wind turbine subject to overturning moments

The vertical loads are relatively low, resulting in small bearing pressures. Because the turbine is sitting on rock, the bearing resistance of the ground will be high. This example is therefore likely to be governed by the EQU limit state, since the properties of the ground are insignificant to the stability of the wind turbine.

Notes on Example 7.1

❶ Strength and stiffness properties are not relevant in limit state EQU in this example.

❷ The self-weight of the foundation (concrete and backfill sitting on it) is a stabilizing permanent action. The other stabilizing action is the self-weight of the turbine, V_{Gk}.

❸ Pore water pressures acting on the underside of the foundation are destabilizing. Here we take a conservative estimate of ground water level (at ground level) and treat the uplift force as a permanent action. Alternatively, a lower water table could be assumed and some component of the uplift force considered variable.

❹ Partial factors > 1.0 are applied to destabilizing actions (thereby increasing their design values) and factors < 1.0 are applied to stabilizing actions (decreasing their values) . Stabilizing variable actions are ignored.

❺ The design destabilizing moment comprises: 1) the moment caused by the variable horizontal action H_{Qk} multiplied by the distance D of its point of action from the foundation base; 2) the moment applied at the top of the foundation; and 3) the moment owing to uplift from ground water.

❻ The stabilizing moment is a function of the foundation weight W_{Gk} and the imposed load from the turbine V_{Gk}, multiplied by their lever arm B/2.

❼ The design destabilizing moment is more than the design stabilizing moment (degree of utilization > 100%), which means that this design does not meet the requirements of EN 1997-1.

❽ IEC 61400 is a commonly used design standard for wind turbines. A key difference from Eurocode 7 is that it applies a single partial factor to permanent and variable destabilizing loads.

❾ The original calculation on which this Example 7.1 is based treats the uplift from groundwater as a negative stabilizing force, rather than as a positive destabilizing force. Combined with lower partial factors, IEC 61400 implies the design is satisfactory and hence some power engineers may regard the design to EN 1997-1 as too conservative.

❿ The traditional lumped factor of safety for this foundation is 1.17 if the water thrust is treated as a destabilizing force, and 1.32 if is treated as a negative stabilizing force. The perceived safety of the structure clearly depends on the assumption made about the effect of water pressures.

Example 7.1
Wind turbine
Verification of stability against overturning (EQU)

Design situation

Consider the foundation of a wind turbine, which is required to carry imposed forces in two directions, vertical V_{Gk} = 2000kN (permanent) and horizontal H_{Qk} = 1500kN (variable), and an imposed moment M_{Qk} = 50000kNm (variable), all of which act at the top of the foundation.

The base is square with width B = 15m at its bottom and b = 5.5m at its top. The narrowest part of the base has thickness t = 1500mm and the widest part T = 600mm. The underside of the foundation is at depth D = 3.0m. The characteristic weight density of reinforced concrete is γ_{ck} = 25$\dfrac{kN}{m^3}$ (as per EN 1991-1-1 Table A.1). The backfill that sits on top of the base has characteristic weight density $\gamma_{k,f}$ = 18$\dfrac{kN}{m^3}$. Groundwater is at ground level with characteristic weight density γ_w ≡ 9.81$\dfrac{kN}{m^3}$. ❶

Actions

The area of the base at its top is A_t = b × b = 30.3 m^2 and at its bottom A_b = B × B = 225 m^2

The volume of concrete in the foundation is then:

$$V_c = \left(A_b \times T\right) + \left[\frac{A_b + A_t}{2} \times (D - t - T)\right] + \left(A_t \times t\right) = 295.2\, m^3$$

The volume of backfill on top of the foundation is

$$V_f = \left(A_b \times D\right) - V_c = 379.8\, m^3$$

Thus, the characteristic self-weight of the foundation (concrete plus backfill) is then: ❷

  concrete $W_{Gk,c}$ = $\gamma_{ck} \times V_c$ = 7381 kN

  backfill $W_{Gk,f}$ = $\gamma_{k,f} \times V_f$ = 6836 kN

  total W_{Gk} = $W_{Gk,c} + W_{Gk,f}$ = 14217 kN

The characteristic uplift force from groundwater pressure acting on the underside of the base is:

$$U_{Gk} = \gamma_w \times D \times B \times B = 6622\,kN \; ❸$$

Effects of actions
Partial factors on destabilizing permanent and variable actions are $\gamma_{G,dst} = 1.1$ and $\gamma_{Q,dst} = 1.5$ and on stabilizing permanent actions

$$\gamma_{G,stb} = 0.9 \quad ❹$$

The design destabilizing moment about the toe is: ❺

$$M_{Ed,dst} = \gamma_{Q,dst} \times \left(H_{Qk} \times D + M_{Qk} \right) + \gamma_{G,dst} \times U_{Gk} \times \frac{B}{2} = 136\,MNm$$

The design stabilizing moment about the toe is: ❻

$$M_{Ed,stb} = \gamma_{G,stb} \times \left(W_{Gk} + V_{Gk} \right) \times \frac{B}{2} = 109\,MNm$$

Verification of stability against overturning

The degree of utilization is $\Lambda_{EQU} = \dfrac{M_{Ed,dst}}{M_{Ed,stb}} = 125\,\% \; ❼$

The design is acceptable if Λ_{EQU} is ≤ 100%

Partial factors on unfavourable and favourable actions for abnormal loads are $\gamma_f = 1.1$ and $\gamma_{f.fav} = 0.9$ respectively. **⑧**

The design destabilizing moment about the toe is:

$$M_{Ed,dst} = \gamma_f \times \left(H_{Qk}D + M_{Qk}\right) = 60 \text{ MNm}$$

The design stabilizing moment about the toe is: **⑨**

$$M_{Ed,stb} = \gamma_{f,fav} \times \left(W_{Gk} + V_{Gk} - U_{Gk}\right) \times \frac{B}{2} = 65 \text{ MNm}$$

The degree of utilization is $\Lambda_{EQU} = \dfrac{M_{Ed,dst}}{M_{Ed,stb}} = 93\%$

The design is acceptable if Λ_{EQU} is ≤ 100%

Traditional lumped factor of safety

If water thrust is regarded as a destabilizing force, then

$$F = \frac{\left(W_{Gk} + V_{Gk}\right) \times \dfrac{B}{2}}{\left(H_{Qk} \times D\right) + M_{Qk} + \left(U_{Gk} \times \dfrac{B}{2}\right)} = 1.17$$

If water thrust is regarded as a (negative) stabilizing force, then

$$F = \frac{\left(W_{Gk} + V_{Gk} - U_{Gk}\right) \times \dfrac{B}{2}}{H_{Qk} \times D + M_{Qk}} = 1.32 \quad \textbf{⑩}$$

7.7.2   Concrete dam

Example 7.2 considers the design of the mass concrete dam shown in **Figure 7.11**, which is sitting on a permeable rock and retaining a height h of water.

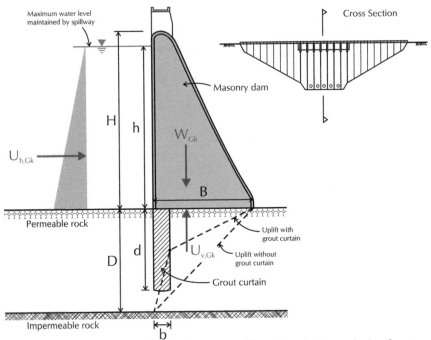

Figure 7.11. Concrete dam founded on permeable rock, retaining a body of water

The pore pressure beneath the dam can be obtained from an analysis[10] of the flow beneath a structure with a single sheet pile which is resting on the surface of an infinite depth of porous material.

Figure 7.12 gives the solution for the dam of **Figure 7.11** (with dimensions B = 2.6m, H = 5m, h = 4.5m, d = 2m, and D = ∞) for three particular cases: with no cutoff; with sheet piles at the dam's heel; and with sheet piles at the dam's toe.

The pore pressure u shown in **Figure 7.12** is based on the following equation:[11]

$$\frac{u}{\gamma_w h} = \frac{1}{\pi} \cos^{-1}\left(\frac{\lambda_1 d \pm \sqrt{d^2 + x^2}}{\lambda_2 d} \right)$$

where γ_w is the weight density of water; h is the height of retained water; d is the depth of the sheet pile; x is the horizontal distance from the dam's heel; and the intermediate variables λ_1 and λ_2 are given by:

$$\left.\begin{array}{c}\lambda_1\\\lambda_2\end{array}\right\} = \frac{1}{2d}\left(\sqrt{d^2 + a^2} \mp \sqrt{d^2 + (B-a)^2}\right)$$

where 'a' is the distance of the sheet pile from the heel; and B is the width of the dam.

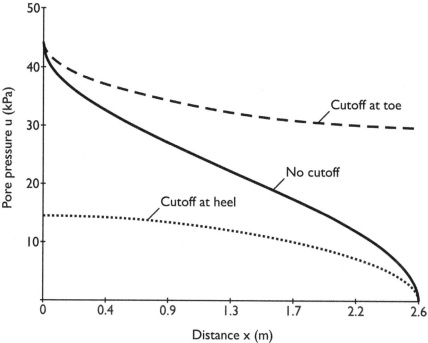

Figure 7.12. Pore water pressure beneath the dam of Figure 7.11

Notes on Example 7.2

❶ The weight density of concrete here is 24 kN/m³ because we assume it is unreinforced (for nominally reinforced concrete, γ_{ck} = 24 kN/m³). This example considers three different scenarios: with no cutoff, with sheet piles at the heel of the dam, and with sheet piles at its toe.

❷ Assuming the dam is broadly triangular in shape and of uniform weight density.

Example 7.2
Mass concrete dam
Verification of static equilibrium (EQU)

Design situation

Consider a mass concrete dam of width $B = 2.6$m and height $H = 5$m which is founded on permeable rock. The height of water retained by the dam is maintained by a spillway at $h = 4.5$m above the foundation. A grout curtain or sheet pile cut-off wall of depth $d = 2$m is used to reduce uplift pressures on the base of the dam. The characteristic weight density of mass concrete is

$$\gamma_k = 24\frac{kN}{m^3} \text{ (as per EN 1991-1-1) and of water } \gamma_w \equiv 9.81\frac{kN}{m^3}. \ ❶$$

Dam with no cutoff

Actions

The characteristic self-weight of dam is given approximately by:

$$W_{Gk} = \gamma_k \times \left(\frac{H \times B}{2}\right) = 156\frac{kN}{m} \ ❷$$

and its moment about the dam toe is: $M_{W,Gk} = W_{Gk} \times \left(\frac{2 \times B}{3}\right) = 270.4\frac{kNm}{m}$

Characteristic thrust from water behind dam: $P_{w,Gk} = \dfrac{\gamma_w \times h^2}{2} = 99.3\frac{kN}{m}$

and its moment about the dam toe is: $M_{Pw,Gk} = P_{w,Gk} \times \left(\dfrac{h}{3}\right) = 149\frac{kNm}{m}$

Integrating the pressure diagram above to obtain the characteristic uplift on the dam, we obtain (for no cutoff): $U_{Gk} = 57.4\frac{kN}{m} \ ❸$

This value obtained by assuming a linear fall in pressure across the dam is:

$$U_{Gk} = \frac{\gamma_w \times h \times B}{2} = 57.4\frac{kN}{m}$$

Taking moments of the pressure diagram about the toe, we get the characteristic overturning moment due to uplift: $M_{U,Gk} = 93.3\frac{kNm}{m}$

Effects of actions

According to EN 1997-1, partial factors on destabilizing and stabilizing permanent actions are $\gamma_{G,dst} = 1.1$ and $\gamma_{G,stb} = 0.9$ respectively.

The design destabilizing moment about the toe is: ④

$$M_{Ed,dst} = \gamma_{G,dst} \times \left(M_{Pw,Gk} + M_{U,Gk} \right) = 266.5 \frac{kNm}{m}$$

The design stabilizing moment about the toe is: ⑤

$$M_{Ed,stb} = \gamma_{G,stb} \times M_{W,Gk} = 243.4 \frac{kNm}{m}$$

Verification of stability against overturning

The degree of utilization is $\boxed{\Lambda_{EQU} = \dfrac{M_{Ed,dst}}{M_{Ed,stb}} = 109\%}$ ⑥

The design is unacceptable if the degree of utilization is > 100%
The traditional lumped factor of safety for this design based on gross

moments is: $F = \dfrac{M_{W,Gk}}{M_{Pw,Gk} + M_{U,Gk}} = 1.12$

but based on net moments (i.e. deducting the uplift from the self-weight of

the dam) is: $F = \dfrac{M_{W,Gk} - M_{U,Gk}}{M_{Pw,Gk}} = 1.19$

Dam with cutoff at heel

Actions

The self-weight of the dam and thrust from water behind dam are unchanged from above. Integrating the pressure diagram to obtain the characteristic uplift on the dam with a cutoff at its heel, we get:

$$U_{Gk} = 28.3 \frac{kN}{m} ⑦$$

Taking moments about the toe, the characteristic overturning moment due to

uplift is then: $M_{U,Gk} = 43.1 \frac{kNm}{m}$

Effects of actions

The design destabilizing moment about the toe is:

$$M_{Ed,dst} = \gamma_{G,dst} \times \left(M_{Pw,Gk} + M_{U,Gk} \right) = 211.3 \frac{kNm}{m}$$

The design stabilizing moment about the toe is:

$$M_{Ed,stb} = \gamma_{G,stb} \times M_{W,Gk} = 243.4 \frac{kNm}{m}$$

The degree of utilization is $\boxed{\Lambda_{EQU} = \dfrac{M_{Ed,dst}}{M_{Ed,stb}} = 87\%}$ ⑧

The design is unacceptable if the degree of utilization is > 100%
The traditional lumped factor of safety based on gross moments is:

$$F = \frac{M_{W,Gk}}{M_{Pw,Gk} + M_{U,Gk}} = 1.41 \quad \text{but on net } F = \frac{M_{W,Gk} - M_{U,Gk}}{M_{Pw,Gk}} = 1.53$$

Dam with cutoff at toe

Actions

The self-weight of the dam and thrust from water behind dam are unchanged from above. Integrating the pressure diagram to obtain the characteristic uplift on the dam with a cutoff at its heel, we get:

$$U_{Gk} = 86.5\frac{kN}{m}$$ ⑨

Taking moments about the toe, the characteristic overturning moment due to uplift is then: $M_{U,Gk} = 118.8\dfrac{kNm}{m}$

Effects of actions

The design destabilizing moment about the toe is:

$$M_{Ed,dst} = \gamma_{G,dst} \times \left(M_{Pw,Gk} + M_{U,Gk}\right) = 294.6\frac{kNm}{m}$$

The design stabilizing moment about the toe is:

$$M_{Ed,stb} = \gamma_{G,stb} \times M_{W,Gk} = 243.4\frac{kNm}{m}$$

Verification of stability against overturning

The degree of utilization is $\boxed{\Lambda_{EQU} = \dfrac{M_{Ed,dst}}{M_{Ed,stb}} = 121\%}$ ⑩

The design is unacceptable if the degree of utilization is > 100%
The traditional lumped factor of safety based on gross moments is:

$$F = \frac{M_{W,Gk}}{M_{Pw,Gk} + M_{U,Gk}} = 1.01 \quad \text{but on net: } F = \frac{M_{W,Gk} - M_{U,Gk}}{M_{Pw,Gk}} = 1.02$$

❸ This value is obtained by integrating the equation for pore pressure given above, with a = 0m and d = 0.0001m ≈ 0m. The subsequent value for the moment is obtained by integrating the same equation, having first multiplied by the distance from the toe.

❹ The destabilizing moment is the sum of the moments from the horizontal water thrust and the uplift beneath the dam.

❺ The stabilizing moment comes from the self-weight of the dam.

❻ The degree of utilization exceeds 100% and hence limit state EQU is *not* verified. The traditional lumped factor of safety against overturning in this case is between 1.12 and 1.19, depending on how that safety factor is calculated.

❼ This value is obtained by integrating the pore pressure equation, with a = 0m (sheet pile at heel) and d = 2m.

❽ The degree of utilization drops to 87% with the cutoff at the heel and limit state EQU is verified. The traditional lumped factor of safety is between 1.41 and 1.53.

❾ This value is obtained by integrating the pore pressure equation, with a = B (sheet pile at toe) and d = 2m.

❿ The degree of utilization rises to 121% with the cutoff at the toe and limit state EQU is once again *not* verified. The traditional lumped factor of safety is approximately 1.0.

7.7.3 Box caisson

Example 7.3 considers the design of the box caisson shown in **Figure 7.13**, which is sitting on a river bed and whose stability against uplift is purely a function of the caisson's weight and any uplift pressures from the water.

The purpose of this example is to consider how the UPL limit state is applied. In reality, the caisson would not sit directly on the river bed but be sunk into it. For simplicity in this example, the caisson is not buried.

(It could be argued that this worked example is not a geotechnical design situation, since it does not involve geotechnical parameters. However, box caissons usually appear in books on geotechnical engineering design, so we have included it for completeness.)

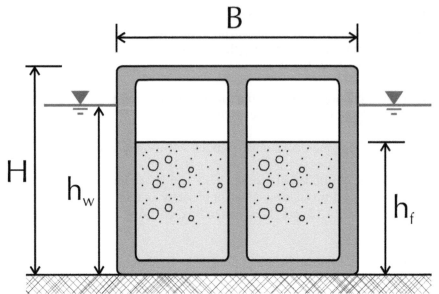

Figure 7.13. *Box caisson subject to uplift forces*

Notes on Example 7.3

❶ The partial factors given here are taken from EN 1997-1 Annex A4 and provide no margin on destabilizing permanent actions. (Different values are specified in the UK National Annex, as discussed under ❸ below.)

❷ The design is (just) verified against uplift according to EN 1997-1. The equivalent lumped factor of safety of 1.13 would probably be regarded by most engineers as too low.

❸ The UK National Annex to BS EN 1997-1 increases the partial factor on destabilizing permanent actions to 1.1 and UPL is no longer avoided. The depth of ballast must be increased to meet the requirements of the UK National Annex.

❹ An acceptable design is achieved by increasing the depth of ballast to 3.9m. The traditional factor of safety is then 1.26, which is a more acceptable level of reliability.

❺ The Eurocodes allow floods to be treated as accidental design conditions, for which the partial factors reduce to 1.0. So, although the destabilizing water uplift is increased by the flood, a depth of ballast of 3.9m still produces a satisfactory design against uplift.

Example 7.3
Box caisson
Verification of stability against uplift (UPL)

Design situation
Consider a box caisson of width $B = 15\,m$ and height $H = 7\,m$ which is to form a crossing over a river. The depth of water in the river is $h_w = 5.7\,m$. The caisson is to be filled with ballast, with characteristic weight density $\gamma_k = 18\,\dfrac{kN}{m^3}$, up to a height $h_f = 2.2\,m$. The characteristic weight density of water is $\gamma_w \equiv 9.81\,\dfrac{kN}{m^3}$ and of reinforced concrete $\gamma_{ck} = 25\,\dfrac{kN}{m^3}$ (as per EN 1991-1-1). The thickness of the caisson's walls, roof, and floor are assumed to be $t = 0.4\,m$.

Actions
The characteristic destabilizing water uplift beneath the caisson is:
$$U_{Gk} = \gamma_w \times B \times h_w = 838.8\,\frac{kN}{m}$$

The characteristic self-weight of the structure is approximately:
$$W_{Gk_1} = \gamma_{ck} \times [3 \times t \times (H - 2t) + 2 \times t \times B] = 486\,\frac{kN}{m}$$

The characteristic self-weight of the ballast (ignoring the volume taken up by the caisson's walls) is approximately:
$$W_{Gk_2} = \gamma_k \times (B - 3t) \times \left(h_f - t\right) = 447.1\,\frac{kN}{m}$$

Hence $W_{Gk} = \displaystyle\sum W_{Gk} = 933.1\,\dfrac{kN}{m}$

Effects of actions - persistent and transient design situations
According to EN 1997-1, partial factors on destabilizing and stabilizing permanent actions are $\gamma_{G,dst} = 1$ and $\gamma_{G,stb} = 0.9$ respectively. ❶

The design destabilizing vertical action is $V_{d,dst} = \gamma_{G,dst} \times U_{Gk} = 838.8\,\dfrac{kN}{m}$

The design stabilizing vertical action is $V_{d,stb} = \gamma_{G,stb} \times W_{Gk} = 839.8\,\dfrac{kN}{m}$

Verification of stability against uplift

The degree of utilization is $\boxed{\Lambda_{UPL} = \dfrac{V_{d,dst}}{V_{d,stb}} = 100\,\%}$ **②**

The design is unacceptable if the degree of utilization is > 100%

The traditional lumped factor of safety for this design is $F = \dfrac{W_{Gk}}{U_{Gk}} = 1.11$

Verification to BS EN 1997-1

The UK National Annex to BS EN 1997-1 increases the value of the partial factor for destabilizing actions to $\gamma_{G,dst} = 1.1$. The design destabilizing

vertical action is then increased to: $V_{d,dst} = \gamma_{G,dst} \times U_{Gk} = 922.6\,\dfrac{kN}{m}$ **③**

The design stabilizing vertical action is $V_{d,stb} = \gamma_{G,stb}\,W_{Gk} = 839.8\,\dfrac{kN}{m}$

The degree of utilization is $\boxed{\Lambda_{UPL} = \dfrac{V_{d,dst}}{V_{d,stb}} = 110\,\%}$

The design is unacceptable if the degree of utilization is > 100%

Re-design to BS EN 1997-1

To obtain a satisfactory design to BS EN 1997-1, we need to increase the depth of ballast to $h_f = 2.6m$. Its self-weight is then given by:

$$W_{Gk_2} = \gamma_k \times (B - 3t) \times \left(h_f - t\right) = 546.5\,\frac{kN}{m}$$

Hence $W_{Gk} = \sum W_{Gk} = 1032.5\,\dfrac{kN}{m}$

The design stabilizing vertical action is $V_{d,stb} = \gamma_{G,stb} \times W_{Gk} = 929.2\,\dfrac{kN}{m}$

The degree of utilization is $\boxed{\Lambda_{UPL} = \dfrac{V_{d,dst}}{V_{d,stb}} = 99\,\%}$

The design is unacceptable if the degree of utilization is > 100%

Traditional lumped factor of safety for this design is $F = \dfrac{W_{Gk}}{U_{Gk}} = 1.23$ **④**

Design for accidental design situation

The extreme water level for this river is estimated to be $h_w = H = 7\,m$

The characteristic uplift force is increased to $U_{Gk} = \gamma_w B h_w = 1030\,\dfrac{kN}{m}$

Partial factors for accidental design situations are reduced to $\gamma_{G,dst} = 1$ and $\gamma_{G,stb} = 1$

The design vertical actions are then:

destabilizing $V_{d,dst} = \gamma_{G,dst} U_{Gk} = 1030\,\dfrac{kN}{m}$

stabilizing $V_{d,stb} = \gamma_{G,stb} W_{Gk} = 1032.5\,\dfrac{kN}{m}$

The degree of utilization is $\boxed{\Lambda_{UPL} = \dfrac{V_{d,dst}}{V_{d,stb}} = 100\,\%}$ ❺

The design is unacceptable if the degree of utilization is > 100%

The traditional lumped factor of safety for this design is $F = \dfrac{W_{Gk}}{U_{Gk}} = 1$

7.7.4   Basement subject to uplift

Example 7.4 considers the design against uplift of the basement shown in **Figure 7.14**, with both one and two levels of basement.

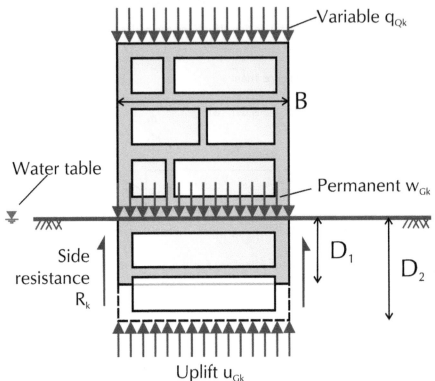

Figure 7.14. Basement subject to uplift forces

In this example, the resistance from friction on the basement walls is significant. Since it is a stabilizing action, this resistance should be an 'inferior' value, which may, counter-intuitively, come from 'superior' values of soil strength. The example also considers the use of tension piles to increase the stabilizing forces. The design of the individual piles should follow the methods described in Chapter 13.

Notes on Example 7.4

❶ Although a variable component of loading is identified, it is a stabilizing force and will not be used when verifying the relevant ultimate limit state. The subscript 'sup' refers to superstructure and the subscript 'sub' refers to substructure.

❷ There are no variable destabilizing actions, so only the 1.1 factor is used on the permanent component. The variable component of the stabilizing force is not included.

❸ The value of β_k (= 0.113) is the product of the characteristic earth pressure coefficient $K_{a,k}$ (= 0.238) and the characteristic coefficient of interface friction $\tan \delta_k$ (with δ_k = 25.3°). Both K_a and δ depend on φ.

❹ When $\tan \varphi_k$ is divided by the partial factor γ_φ, the design angle of shearing resistance becomes φ_d = 32°. This increases the value of K_a to 0.307 but decreases the value of δ_k to 21.3°, resulting in a *larger* value of β (= 0.12). This is a less severe condition than for characteristic conditions (see ❸).

❺ This value of β is calculated from the superior value of $\varphi_{k,sup}$ (= 45°), divided by a superior partial factor $\gamma_{\varphi,sup}$ (= 0.8), producing $\varphi_{d,sup}$ = 51.3°. This decreases the value of K_a to 0.123 but increases the value of δ_k to 34.2°, resulting in a *smaller* value of β (= 0.084). This is more severe than for ❸ and ❹ and hence will be used for design.

❻ The design resistance for the one-storey basement is verified. The traditional global factor of safety is 1.22.

❼ The design resistance for the two-storey basement is *not* verified. The traditional global factor of safety is 0.87.

❽ Additional resistance may be provided by piles. The horizontal effective stress σ'_h acting on the piles is calculated from $\sigma'_h = K_s \sigma'_v$, where σ'_v is the vertical effective stress along the pile. For simplicity, we have assumed K_s is independent of material properties.

❾ The value of design resistance for the UPL limit state is calculated by applying a material factor (γ_φ = 1.25) to the tangent of the characteristic angle of interface friction between the soil and the pile δ_k. In this instance, δ_k should be an inferior value (since this minimizes the pile resistance). It could be argued that the design resistance should be treated as a stabilizing action, in which case the characteristic resistance would be multiplied by $\gamma_{G,stb}$ = 0.9. If this approach was adopted, it would be less conservative than the approach we have adopted in these calculations.

❿ With the addition of piles, the design resistance of the two-storey basement is now verified. The traditional global factor of safety is 1.34.

Example 7.4
Basement subject to uplift
Verification of stability against uplift (UPL)

<u>One-storey basement</u>

<u>*Design situation*</u>
Consider a three-storey building which applies a self-weight loading at
foundation level estimated to be w_{Gk} = 30kPa (permanent) and carries
imposed loads on its floors and roof amounting to q_{Qk} = 15kPa (variable). The
building is to be supported by a one-storey basement of width B = 18m and
depth D = 4.5m. The basement walls are t_w = 400mm thick, its floors
t_f = 250mm thick, and its base slab t_b = 500mm thick. The characteristic

weight density of reinforced concrete is γ_{ck} = 25$\dfrac{kN}{m^3}$, as per EN 1991-1-1.

The ground profile comprises 20m of dense sand and groundwater levels are
close to ground level. The sand's characteristic weight density is γ_k = 19$\dfrac{kN}{m^3}$,

its angle of shearing resistance φ_k = 38°, and its 'superior' angle of shearing
resistance $\varphi_{k,sup}$ = 45°. The weight density of water should be taken as

γ_w = 9.81$\dfrac{kN}{m^3}$.

<u>*Actions*</u>
The characteristic water pressure acting on the underside of the basement is
u_k = $\gamma_w \times$ D = 44.1 kPa, giving a resultant destabilizing action underneath the

basement of U_{Gk} = $u_k \times$ B = 795$\dfrac{kN}{m}$. Characteristic actions from the

super-structure are $W_{Gk,sup}$ = $w_{Gk} \times$ B = 540$\dfrac{kN}{m}$ (permanent) and

$Q_{Qk,sup}$ = $q_{Qk} \times$ B = 270$\dfrac{kN}{m}$ (variable). ❶

Characteristic self-weight of the sub-structure (basement) is:

from the walls $W_{Gk,w} = 2 \times t_w \times D \times \gamma_{ck} = 90\dfrac{kN}{m}$

from the floors $W_{Gk,f} = t_f \times \left(B - 2t_f\right) \times \gamma_{ck} = 109.4\dfrac{kN}{m}$

from the base slab $W_{Gk,b} = t_b \times \left(B - 2t_f\right) \times \gamma_{ck} = 218.8\dfrac{kN}{m}$

total weight $W_{Gk,sub} = W_{Gk,w} + W_{Gk,f} + W_{Gk,b} = 418\dfrac{kN}{m}$ ❶

Total self-weight of the building is $W_{Gk} = W_{Gk,sup} + W_{Gk,sub} = 958\dfrac{kN}{m}$.

Effects of actions
Partial factors on destabilizing permanent and variable actions are
$\gamma_{G,dst} = 1.1$ and $\gamma_{Q,dst} = 1.5$ and on stabilizing permanent actions
$\gamma_{G,stb} = 0.9$. Thus the destabilizing vertical action is

$V_{d,dst} = \gamma_{G,dst} \times U_{Gk} = 874.1\dfrac{kN}{m}$ and the stabilizing vertical action

$V_{d,stb} = \gamma_{G,stb} \times W_{Gk} = 862.3\dfrac{kN}{m}$. ❷

Material properties
The sand's characteristic angle of shearing resistance is $\varphi_k = 38°$, giving an

active earth pressure coefficient $K_{a,k} = \dfrac{1 - \sin\left(\varphi_k\right)}{1 + \sin\left(\varphi_k\right)} = 0.238$ and angle of

wall friction $\delta_k = \dfrac{2}{3}\varphi_k = 25.3°$. Thus $\beta_k = K_{a,k}\tan\left(\delta_k\right) = 0.113$ ❸ .

The partial factor on the coefficient of shearing resistance $\gamma_\varphi = 1.25$ gives

a design angle of shearing resistance $\varphi_d = \tan^{-1}\left(\dfrac{\tan\left(\varphi_k\right)}{\gamma_\varphi}\right) = 32°$. Thus the

active earth pressure coefficient increases to $K_{a,d} = \dfrac{1 - \sin\left(\varphi_d\right)}{1 + \sin\left(\varphi_d\right)} = 0.307$

while the angle of wall friction reduces to $\delta_d = \dfrac{2}{3}\varphi_d = 21.3°$. Thus

$$\beta_{d,inf} = K_{a,d} \tan(\delta_d) = 0.12 \quad \text{④}$$

We need to check that a lower β is not obtained with the superior angle of shearing resistance $\varphi_{k,sup} = 45°$ and partial factor $\gamma_{\varphi,sup} = \dfrac{1}{\gamma_\varphi} = 0.8$.

The superior design angle of shearing resistance is then

$$\varphi_{d,sup} = \tan^{-1}\left(\frac{\tan(\varphi_{k,sup})}{\gamma_{\varphi,sup}}\right) = 51.3°, \text{ giving}$$

$$K_{a,d,sup} = \frac{1 - \sin(\varphi_{d,sup})}{1 + \sin(\varphi_{d,sup})} = 0.123 \quad \text{and} \quad \delta_{d,sup} = \frac{2}{3}\varphi_{d,sup} = 34.2°.$$

Hence $\beta_{d,sup} = K_{a,d,sup} \times \left(\tan(\delta_{d,sup})\right) = 0.084 \quad \text{⑤}$
Thus $\beta_d = \min(\beta_{d,inf}, \beta_{d,sup}) = 0.084$

Resistance
The average vertical effective stress down the basement wall is:

$$\sigma'_v = \frac{(\gamma_k - \gamma_w) \times D}{2} = 20.7 \text{ kPa}$$

The characteristic resistance along the basement walls is given by:

$$R_k = \beta_k \times \frac{(\gamma_k - \gamma_w) \times D^2}{2} = 10.5 \frac{kN}{m}$$

The design resistance along the basement walls is given by:

$$R_d = \beta_d \times \frac{(\gamma_k - \gamma_w) \times D^2}{2} = 7.8 \frac{kN}{m}$$

Verification of stability against uplift

The degree of utilization is $\boxed{\Lambda_{UPL} = \dfrac{V_{d,dst}}{V_{d,stb} + R_d} = 100\%}$ ⑥

The design is unacceptable if the degree of utilization is > 100%

The traditional lumped factor of safety for this design situation is:

$$F = \frac{W_{Gk,sup} + W_{Gk,sub} + R_k}{U_{Gk}} = 1.22 \quad \text{⑥}$$

Two-storey basement

Design situation

The building considered above is now required to have a two-storey basement with depth $D = 7.5m$.

Actions

The characteristic water pressure acting on the underside of the basement is $u_k = \gamma_w \times D = 73.6\ kPa$, giving a resultant destabilizing action underneath the basement of $U_{Gk} = u_k \times B = 1324\ \dfrac{kN}{m}$. Characteristic actions from the super-structure remain $W_{Gk,sup} = 540\ \dfrac{kN}{m}$ (permanent) and $Q_{Qk,sup} = 270\ \dfrac{kN}{m}$ (variable). The characteristic self-weight of the two floors is now $W_{Gk,f} = 2t_f \times \left(B - 2t_f\right) \times \gamma_{ck} = 218.8\ \dfrac{kN}{m}$ and of the walls $W_{Gk,w} = 2 \times t_w \times D \times \gamma_{ck} = 150\ \dfrac{kN}{m}$, resulting in a total weight of sub-structure $W_{Gk,sub} = W_{Gk,w} + W_{Gk,f} + W_{Gk,b} = 588\ \dfrac{kN}{m}$. Hence the total self-weight of the building is now $W_{Gk} = W_{Gk,sup} + W_{Gk,sub} = 1128\ \dfrac{kN}{m}$.

Effects of actions

The destabilizing vertical action is $V_{d,dst} = \gamma_{G,dst} \times U_{Gk} = 1456.8\ \dfrac{kN}{m}$ and the stabilizing vertical action $V_{d,stb} = \gamma_{G,stb} \times W_{Gk} = 1014.8\ \dfrac{kN}{m}$.

Material properties

Are unchanged.

Resistance

The average vertical effective stress down the basement wall is:

$$\sigma'_v = \frac{\left(\gamma_k - \gamma_w\right) \times D}{2} = 34.5\ kPa$$

The characteristic resistance along the basement walls is given by:

$$R_k = \beta_k \times \frac{\left(\gamma_k - \gamma_w\right) \times D^2}{2} = 29.1\,\frac{kN}{m}$$

The design resistance along the basement walls is given by:

$$R_d = \beta_d \times \frac{\left(\gamma_k - \gamma_w\right) \times D^2}{2} = 21.6\,\frac{kN}{m}$$

Verification of stability against uplift

The degree of utilization is $\boxed{\Lambda_{UPL} = \dfrac{V_{d,dst}}{V_{d,stb} + R_d} = 141\%}$ ❼

The design is unacceptable if the degree of utilization is > 100%
The traditional lumped factor of safety for this design situation is:

$$F = \frac{W_{Gk,sup} + W_{Gk,sub} + R_k}{U_{Gk}} = 0.87 \;\; ❼$$

Additional resistance from tension piles

To overcome the shortfall in stabilizing actions and resistance, the basement will be held down by n = 4 rows of tension piles, each pile L = 10m long and d = 450mm in diameter. The piles will be installed using a contiguous flight auger. Pile rows will be spaced at s = 5m spacing along the building (i.e. into the plane of the drawing). An earth pressure coefficient K_s = 1 is assumed in order to determine skin friction along the pile shaft. ❽

The average vertical effective stress along the pile shafts is:

$$\sigma'_{v,pile} = \left(\gamma_k - \gamma_w\right)\left(D + \frac{L}{2}\right) = 114.9\ kPa$$

The characteristic resistance of each pile is given by:

$$R_{k,pile} = \pi \times d \times L \times \sigma'_{v,pile} \times \tan\left(\delta_k\right) \times K_s = 768.8\ kN$$

and the design resistance by:

$$R_{d,pile} = \pi \times d \times L \times \sigma'_{v,pile} \times \frac{\tan\left(\delta_k\right)}{\gamma_\varphi} \times K_s = 615.1\ kN$$

Hence the total characteristic resistance is now:

$$R_k = R_k + \left(\frac{n}{s}\right) \times R_{k,pile} = 644.2\,\frac{kN}{m}$$

and the total design resistance is:

$$R_d = R_d + \left(\frac{n}{s}\right) \times R_{d,pile} = 513.7 \frac{kN}{m} \quad ⑨$$

Verification of stability against uplift (with tension piles)

The degree of utilization is $\Lambda_{UPL} = \dfrac{V_{d,dst}}{V_{d,stb} + R_d} = 95\%$ ⑩

The design is unacceptable if the degree of utilization is > 100%

The traditional lumped factor of safety for this design situation is:

$$F = \frac{W_{Gk,sup} + W_{Gk,sub} + R_k}{U_{Gk}} = 1.34 \quad ⑩$$

7.7.5   Stability of weir against hydraulic failure

Example 7.5 considers the design of the weir shown in **Figure 7.15** against hydraulic failure at its toe when water emerges with too large an exit gradient.

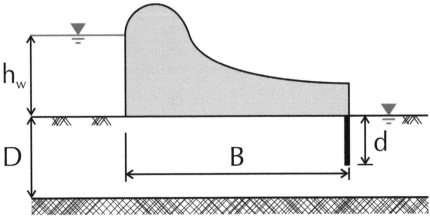

Figure 7.15. *Weir subject to large hydraulic gradients at downstream toe*

In order to assess the exit hydraulic gradient at the toe of the weir, we have chosen to use an analytical solution, rather than carry out a flow net or numerical solution. This has the advantage of enabling the solution of a range of problems without drawing a flow net for each change in geometry.

Notes on Example 7.5

❶ The exit gradient can be determined in a number of ways (for example, by drawing a flow net) – this equation has the advantage of providing a convenient analytical solution.[12]

❷ The exit gradient is calculated by substituting the values of B, h_w, and d into Khosla's equation ❶.

❸ For limit state HYD, partial factors are given in EN 1997-1 Annex A5.

❹❺ Eurocode 7 provides two expressions, 2.9(a) and 2.9(b), for verifying HYD. The former compares the design destabilizing pore water pressure to the design stabilizing vertical stress; the latter compares the design destabilizing seepage force to the design stabilizing submerged weight. Both equations are applied at the base of the column of soil under consideration.

Example 7.5
Stability of weir against hydraulic failure
Verification of stability against piping (HYD)

Design situation
Consider a weir of width B = 15m which is retaining free water of height
h_w = 5m A cut-off wall of depth d = 3.2m helps to reduce the hydraulic

gradient at the downstream end of the weir (the 'exit' gradient). The weir is
founded on a permeable stratum of characteristic weight density

γ_k = 19.5 $\dfrac{kN}{m^3}$. An impermeable stratum is located at depth D = ∞. The

characteristic weight density of water is γ_w ≡ 9.81 $\dfrac{kN}{m^3}$.

Calculation model
For the situation where D = ∞, the exit gradient can be calculated from

Khosla's equation: ❶

$$i_E\left(B, h_w, d\right) = \frac{h_w}{\pi \times d \times \sqrt{\dfrac{1 + \sqrt{1 + \left(\dfrac{B}{d}\right)^2}}{2}}}$$

where B is the width of the weir, h_w the height of retained water, and d the
depth of cut-off.

Actions
The characteristic exit gradient for our design situation is:

$$i_k = i_E\left(B, h_w, d\right) = 0.29 \quad ❷$$

At depth d on the downstream side of the cutoff, we have the characteristic
destabilizing pore water pressure: $u_k = \gamma_w \times \left(1 + i_k\right) \times d$ = 40.6 kPa and the

characteristic stabilizing vertical stress: $\sigma_k = \gamma_k \times d$ = 62.4 kPa

The characteristic seepage force on the downstream side of the cutoff is
(assuming a plan area A = 1m^2) $S_k = \gamma_w \times i_k \times d \times A$ = 9.2 kN
The characteristic submerged weight of the downstream side of the cutoff
is: $G'_k = \left(\gamma_k - \gamma_w\right) \times d \times A$ = 31 kN

Effects of actions

Partial factors on destabilizing and stabilizing permanent actions are

$\gamma_{G,dst} = 1.35$ and $\gamma_{G,stb} = 0.9$ respectively. ❸

Using EN 1997-1 exp. 2.9(a): ❹

The design destabilizing pore water pressure is:

$$u_{d,dst} = \gamma_{G,dst} \times u_k = 54.8 \text{ kPa}$$

The design stabilizing vertical total stress is:

$$\sigma_{d,stb} = \gamma_{G,stb} \times \sigma_k = 56.2 \text{ kPa}$$

Using EN 1997-1 exp. 2.9(b): ❺

The design seepage force is $S_{d,dst} = \gamma_{G,dst} \times S_k = 12.4 \text{ kN}$

The design submerged weight is $G'_{d,stb} = \gamma_{G,stb} \times G'_k = 27.9 \text{ kN}$

Verification of stability against hydraulic heave using EN 1997-1 exp. 2.9(a)

The degree of utilization is $\boxed{\Lambda_{HYD} = \dfrac{u_{d,dst}}{\sigma_{d,stb}} = 98 \%}$ ❻

The design is unacceptable if the degree of utilization is > 100%

Verification of stability against hydraulic heave using EN 1997-1 exp. 2.9(b)

The degree of utilization is $\boxed{\Lambda_{HYD} = \dfrac{S_{d,dst}}{G'_{d,stb}} = 44 \%}$ ❼

The design is unacceptable if the degree of utilization is > 100%

Traditional factor of safety against piping

The soil's critical hydraulic gradient is $i_{crit} = \dfrac{\gamma_k - \gamma_w}{\gamma_w} = 1$

Factor of safety on hydraulic gradient is $F = \dfrac{i_{crit}}{i_k} = 3.38$ ❽

❻❼ The degree of utilization using expression 2.9(a) is close to 100%, whereas using 2.9(b) it is less than 50%. Eurocode 7 does not explicitly state where the partial factors should be applied, which leads to the discrepancy between these expressions, which was not anticipated by the authors of the standard.

❽ The traditional global factor of safety for this situation is 3.38. Recommended values for the global factor are between 1.5 and over 4.0. Generally, where the consequences of piping failure may have serious effects, then higher global factors are adopted. It is appears that when Eq. 2.9(a) is used, equivalent traditional factors of safety are between 3.0 and 4.5 but, when Eq. 2.9(b) is used, it is closer to 1.5. We conclude that Eq. 2.9(b) does not provide a sufficient level of reliability.

7.7.6   Piping due to heave (HYD)

Example 7.6 considers the design of an embedded retaining wall against piping due to heave, as shown in **Figure 7.16**.

The wall forms part of a temporary cofferdam to allow the construction of a bridge pier in the dry. The river water level remains reasonably constant at H = 1.9m above bed level and the water level inside the cofferdam is being kept at or just below bed level by pumping. In order to reduce seepage into the cofferdam and to overcome

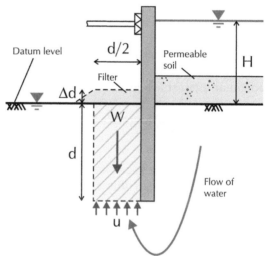

Figure 7.16. Embedded retaining wall subject to piping due to heave

the potential for piping as water seeps upwards into the excavation, it is proposed to embed the sheet pile wall into the ground by d = 6m. The example investigates the application of the two procedures recommended in EN 1997-1 for hydraulic failure to assess the suitability of this depth of penetration.

Notes on Example 7.6

❶ Text books suggest it is acceptable to consider the upward seepage forces over a width equivalent to half the embedment of the wall.

❷ To assess the excess pressure head at the toe of the wall, an acceptable approximation is to assume that half the total head loss is achieved on each side of the wall. This is only an approximation. In reality, it will vary across the base of the soil column being considered and a detailed flow analysis would be required if the result was critical to the design.

❸ Partial factors for limit state HYD given in Annex A to EN 1997-1.

❹ Stabilizing and destabilizing forces on the soil column are calculated here using Expression 2.9(b) of EN 1997-1, with appropriate partial factors.

❺ Expression 2.9(b) suggests the design has ample reliability against heave.

❻ Stabilizing and destabilizing pressures on the soil column are calculated here using Expression 2.9(a). This expression suggests the design is only just adequate against heave.

❼ Expression 2.9(a) suggests the design just has enough reliability to avoid heave.

❽ Eurocode 7 is not specific about where the partial factors $\gamma_{G,dst}$ and $\gamma_{G,stb}$ should be applied. In this alternative calculation, the *excess* pore water pressure is multiplied by $\gamma_{G,dst}$ and the *effective* stress at the base of the soil column multiplied by $\gamma_{G,stb}$.

❾ The alternative calculation gives an identical result to Expression 2.9(b), see ❺.

❿ This calculation of heave results in a traditional global factor of safety of 4.63, which is more than adequate for this problem according to some authors (who suggest F should be 1.5–2.0) but only just adequate according to others (who suggest F ≥ 4). See the discussion in Section 7.5.1 for details of these recommended values for F.

Example 7.6
Piping due to heave
Verification of stability against hydraulic failure (HYD)

Design situation

Consider a sheet pile wall cofferdam that retains water at a height $H = 1.9m$ above its formation level. The walls of the cofferdam are embedded $d = 6m$ below formation. The characteristic weight density of the foundation soil is $\gamma_k = 17\dfrac{kN}{m^3}$ and of water is $\gamma_w \equiv 9.81\dfrac{kN}{m^3}$. The stability of the zone of soil of width $\dfrac{d}{2} = 3m$ next to the embedded part of the wall must be checked for hydraulic failure by piping. ❶

Actions

Average pressure head at base of soil column: $h_a = \dfrac{H}{2} = 0.95\,m$ ❷

Average hydraulic gradient through soil column: $i_k = \dfrac{h_a}{d} = 0.158$

Static pore pressure at wall toe: $u_0 = \gamma_w \times d = 58.9\,kPa$

Excess pore pressure at wall toe: $\Delta u = \gamma_w \times h_a = 9.3\,kPa$

Total pore pressure at wall toe: $u = u_0 + \Delta u = 68.2\,kPa$

Total vertical stress at wall toe: $\sigma_v = \gamma_k \times d = 102\,kPa$

Effective vertical stress at wall toe: $\sigma'_v = \sigma_v - u_0 = 43.1\,kPa$

Self-weight of soil column is: $W_k = \gamma_k \times d \times \dfrac{d}{2} = 306\dfrac{kN}{m}$

Submerged weight of soil column: $W'_k = \left(\gamma_k - \gamma_w\right) \times d \times \dfrac{d}{2} = 129.4\dfrac{kN}{m}$

Characteristic seepage force: $S_k = \gamma_w \times i_k \times d \times \dfrac{d}{2} = 28\dfrac{kN}{m}$

Effects of actions

Partial factors on destabilizing permanent and variable actions are $\gamma_{G,dst} = 1.35$ and $\gamma_{Q,dst} = 1.5$ and on stabilizing permanent actions $\gamma_{G,stb} = 0.9$. ❸

Verification of stability against piping using seepage force and submerged weight

Design destabilizing seepage force: $E_{d,dst} = \gamma_{G,dst} \times S_k = 37.7\,\frac{kN}{m}$ **④**

Design stabilizing weight: $E_{d,stb} = \gamma_{G,stb} \times W'_k = 116.5\,\frac{kN}{m}$ **④**

The degree of utilization is: $\Lambda_{HYD} = \dfrac{E_{d,dst}}{E_{d,stb}} = 32\,\%$ **⑤**

The design is unacceptable if the degree of utilization is > 100%

Verification of stability against piping using pore pressure and total stress

Design destabilizing pore pressure is: $E_{d,dst} = \gamma_{G,dst} \times u = 92\ kPa$ **⑥**

Design stabilizing total stress is: $E_{d,stb} = \gamma_{G,stb} \times \sigma_v = 92\ kPa$ **⑥**

The degree of utilization is: $\Lambda_{HYD} = \dfrac{E_{d,dst}}{E_{d,stb}} = 100\,\%$ **⑦**

The design is unacceptable if the degree of utilization is > 100%

Alternatively...

Design destabilizing pore pressure is: $E_{d,dst} = \gamma_{G,dst} \times \Delta u = 12.6\ kPa$ **⑧**

Design stabilizing total stress is: $E_{d,stb} = \gamma_{G,stb} \times \sigma'_v = 39\ kPa$ **⑧**

The degree of utilization is: $\Lambda_{HYD} = \dfrac{E_{d,dst}}{E_{d,stb}} = 32\,\%$ **⑨**

The design is unacceptable if the degree of utilization is > 100%

Traditional lumped factor of safety

Terzaghi & Peck defined the factor of safety against piping as:

$$F = \frac{W'_k}{S_k} = 4.63$$ **⑩**

The design is unacceptable if F is < 1.5-2.0

7.8   Notes and references

1. See Orr, T. L. L. (2005) 'Evaluation of uplift and heave designs to Eurocode 7', *Proc. Int. Workshop on the Evaluation of Eurocode 7*, Trinity College, Dublin, pp. 147–158.

2. NA to BS EN 1997-1: 2004, UK National Annex to Eurocode 7: Geotechnical design – Part 1: General rules, British Standards Institution.

3. Terzaghi, K. (1936) 'The shearing resistance of saturated soils', *1st Int. Conf. on Soil Mechanics*, 1, pp. 54–56.

4. Terzaghi, K., and Peck, R. B. (1967) *Soil mechanics in engineering practice* (2nd edition), John Wiley & Sons, Inc.

5. See p. 178 of Reddi, L. N. (2003), *Seepage in soils*, John Wiley & Sons.

6. Harr, M. E. (1962) *Groundwater and seepage*, McGraw-Hill.

7. Orr, ibid.

8. See, for example, Fell, R., MacGregor, P., Stapledon, D., and Bell, G. (2005) *Geotechnical engineering of dams*, Leiden, Netherlands, A. A. Balkema Publishers.

9. Wind turbine design data kindly provided by Donald Cook and Chris Hoy of Donaldson Associates, Glasgow (pers. comm., 2007).

10. See Harr, ibid., who gives details of the work of Khosla et al.

11. See Harr, ibid., p. 108.

12. See Harr, ibid., p. 111 or Reddi, ibid., p. 150, who both refer to earlier work by Khosla, Bose, and Taylor in 1954.

Chapter 8

Verification of serviceability

'The ability to quote is a serviceable substitute for wit' — W. Somerset Maugham (1874-1965)

8.1 Basis of design

Serviceability limit states are defined as:

> *States that correspond to conditions beyond which specified service requirements for a structure or structural member are no longer met*
> [EN 1990 §1.5.2.14]

Verification of serviceability involves checking that design effects of actions (e.g. settlements) do not exceed their corresponding design limiting values (i.e. limiting settlements).

Verification of serviceability is expressed in Eurocode 7 by the inequality:

$$E_d \leq C_d \qquad \text{[EN 1990 exp (6.13)] \& [EN 1997-1 exp (2.10)]}$$

in which E_d = the design effects of actions and C_d = the limiting design value of the relevant serviceability criterion.

Examples of situations where serviceability is a concern are shown in **Figure 8.1** overleaf. From left to right, these include: (top) the settlement, s, of a pad footing due to its self-weight and any imposed loads must not exceed the project-specific limiting settlement and the differential settlement, Δs, of foundations for a framed structure must be within specified limits; (middle) the horizontal deflection, δ, of a retaining wall due to unbalanced earth pressures must be within specified limits and vibration due to machinery must not cause discomfort to neighbours; and (bottom) the rate at which water is pumped from a basement must be sufficient to prevent heave underneath the basement and the capacity of the pump, on the downstream side of a dam, must be sufficient to remove water flowing underneath the dam.

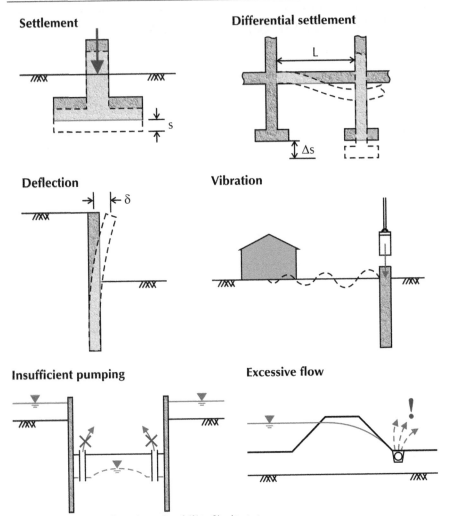

Figure 8.1. Examples of serviceability limit states

8.1.1   Effects of actions

'Effects of actions' (or 'action effects') is a general term denoting internal forces, moments, stresses, and strains in structural members – plus the deflection and rotation of the whole structure.   *[EN 1990 §1.5.3.2]*

For serviceability limit states, the effects of actions are the various forms of foundation movement shown in **Figure 8.2**: settlement (s), rotation (θ), angular strain (α), and tilt (ω) – as well as differential settlement (δs), relative deflection (Δ), and relative rotation or angular distortion (β). The ratio Δ/L

is known as the deflection ratio. Most of these terms are already in widespread use in geotechnical practice.[1]

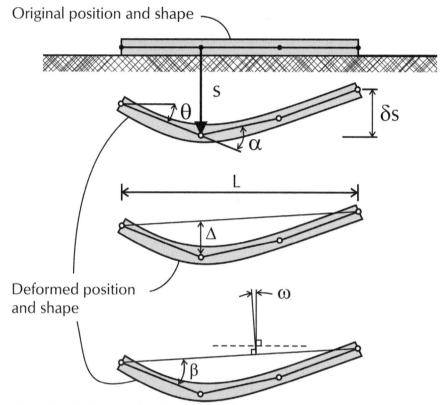

Original position and shape

Deformed position and shape

Figure 8.2. Definition of various forms of foundation movement

8.1.2   Limiting serviceability criteria

Annex H of EN 1997-1 also provides guidance on allowable deformations in open-framed structures, infilled frames, and load bearing and continuous brick walls (see table below). These guidelines apply to normal structures, not to extraordinary structures or to structures subject to markedly non-uniform loading.

Movement			Maximum movement to avoid limit state	
			Serviceability	Ultimate
Settlement		s	50 mm*	-
Relative rotation	sagging	β	1/2000–1/300†	1/150
	hogging		1/4000–1/600‡	1/300

*Larger values may be acceptable if relative rotations and tilt are tolerable
†1/500 is acceptable for many structures
‡1/1000 is acceptable for many structures

8.2   Introducing reliability into the design

Reliability is introduced into design against loss of serviceability by selecting suitable limiting values of displacement, as illustrated in **Figure 8.3**.

Partial factors for serviceability limit states are normally taken as 1.0.

[EN 1997-1 §2.4.8(2)]

Hence the equation for verification of serviceability (see Section 8.1) becomes:

$$E_d = E\left\{F_{rep}; X_k; a_{nom}\right\} \leq C_d$$

and no partial factors are introduced in the flow diagram of **Figure 8.3**.

The combination factors ψ shown on **Figure 8.3** are applied to accompanying variable actions only and are those specified for the characteristic, frequent, or quasi-permanent combinations (see Chapter 2), i.e. ψ = ψ_2.

In ultimate limit state (ULS) verifications, combinations of actions for permanent and transient design situations employ combination factors ψ = ψ_0. Since ψ_0 is numerically greater than ψ_2 for most actions, representative actions are usually larger for ultimate than for serviceability limit states.

It is important to recognize that actions and material properties may vary during the structure's design life and hence serviceability limit states may need to be checked at various times. Of critical importance in verifying serviceability is appropriate selection of the limiting effects of actions. These must represent a realistic assessment of what is necessary for the long term performance of the structure, rather than overly conservative limits which simplify structural analysis.

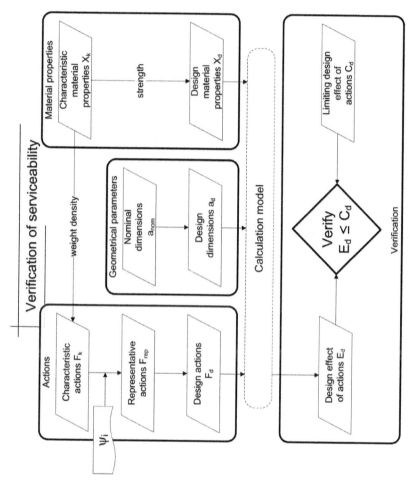

Figure 8.3. Overview of verification of serviceability

8.3   Simplified verification of serviceability

In traditional geotechnical practice, serviceability limit states have been avoided by a variety of means, such as: for foundations, limiting the bearing pressures underneath the foundation to 'allowable' (conservative) values; for piles, by applying large safety factors to base and shaft capacities; and for embedded retaining walls, using 'mobilization factors' to reduce the passive earth pressure assumed to achieve moment equilibrium.

All these methods are fundamentally the same. They attempt to reduce foundation movement by ensuring failure has a sufficiently remote possibility of occurring. Eurocode 7 acknowledges that deformations can be kept within required serviceability limits provided 'a sufficiently low fraction of the ground strength is mobilised'; a value of the deformation is not needed; and comparable experience exists with similar ground, structures, and method of application. Unfortunately, the standard does not specify (other than for spread footings, discussed below) how low this fraction should be.   *[EN 1997-1 §2.4.8(4)]*

Eurocode 7 also gives emphasis to problems associated with heave and vibration, highlighting particular features that must be considered when assessing these issues.   *[EN 1997-1 §6.6.4 and 6.6.4]*

For compression piles in medium to dense soils and for tension piles, serviceability limit states will normally be prevented by verifying that an ultimate limit state will not occur.   *[EN 1997-1 §7.6.4.1 NOTE]*

For conventional structures founded on clays, Eurocode 7 requires settlements to be calculated explicitly whenever the ratio of the characteristic bearing resistance R_k to the applied serviceability loads E_k is less than three. In addition, if this ratio is less than two, those calculations should take account of the ground's non-linear stiffness. By implication, if the ratio R_k/E_k is greater than or equal to three, then the serviceability limit state may be deemed to have been verified by this ultimate limit state calculation.
[EN 1997-1 §6.6.2(16)]

Hence verification of serviceability may also be demonstrated by satisfying the inequality:

$$E_k \leq \frac{R_k}{\gamma_{R,SLS}}$$

in which E_k = the characteristic effects of actions, R_k = the characteristic resistance to those actions, and $\gamma_{R,SLS}$ = a partial resistance factor ≥ 3.

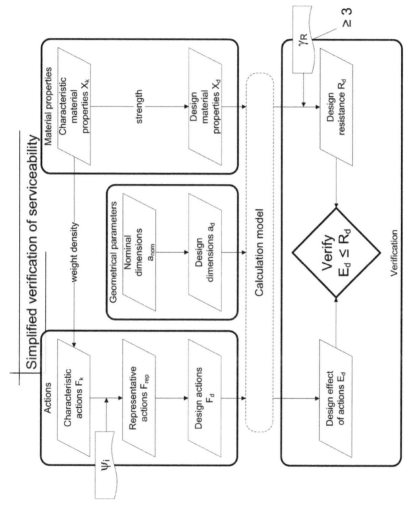

Figure 8.4. Alternative verification of serviceability

Figure 8.4 shows how the flow chart of Chapter 6 (dealing with verification of strength) can be modified to verify serviceability instead. Partial factors on actions, materials properties, effects of actions, and resistance are replaced by a single partial factor on resistance equal to 3.0. The use of this alternative verification obviates the need to establish limiting values of movement C_d.

8.4   Methods to determine settlement

Annex F of EN 1997-1 presents two methods to evaluate the settlement of spread foundations, which are discussed in Chapter 10.   *[EN 1997-1 §F.1(1)]*

Eurocode 7 Part 2 provides an eclectic selection of example methods to determine foundation settlement from common field tests:

Field test	Annex	Example method (and country* of origin)	
Cone and piezocone tests (CPT and CPTU)	D.1	Correlation between drained Young's modulus and q_c	SWE
	D.3	Semi-empirical method for calculating settlement of spread foundations in coarse soil	NLD
	D.4	Correlation between oedometer modulus and q_c	FRA
	D.5	Establishing the stress-dependent oedometer modulus from CPT results	DEU
Pressuremeter test (PMT)	E.2	Semi-empirical method for calculating settlement of spread foundations from Menard pressuremeter	FRA
Standard penetration test (SPT)	F.3	Empirical direct method for calculation of settlement of spread foundations in granular soil	GBR
Dynamic probing test (DP)	G.3	Establishing the stress-dependent oedometer modulus from DP results	DEU
Weight sounding test (WST)	H.1	Correlation with effective angle of shearing resistance and drained Young's modulus	SWE
Flat dilatometer test (DMT)	J	Correlations to determine one-dimensional tangent modulus	ITA
Plate loading test (PLT)	K.2	Plate settlement modulus	GBR
	K.3	Coefficient of sub-grade reaction	SWE
	K.4	Settlement of spread foundations in sand	SWE

*DEU = Germany, FRA = France, GBR = United Kingdom, ITA = Italy, NLD = Netherlands, SWE = Sweden

8.5   Summary of key points

Serviceability limit states involve conditions beyond which specified service requirements are no longer met. Verification of these limit states may be demonstrated by satisfying the inequality:

$$E_d \leq C_d$$

or, for conventional structures founded on clay:

$$E_k \leq \frac{R_k}{\gamma_{R,SLS}}$$

where the symbols are as defined earlier in this chapter.

In essence, the introduction of Eurocode 7 makes little or no change to the way that geotechnical structures are checked for serviceability conditions.

8.6   Worked examples

A worked example that illustrates the verification of serviceability is included in Chapter 10, *Design of footings*.

8.7   Notes and references

1. Burland, J. B., and Wroth, C. P. (1975), 'Settlement of buildings and associated damage', *Proc. Conf. on Settlement of Structures*, Cambridge, pp. 611–654.

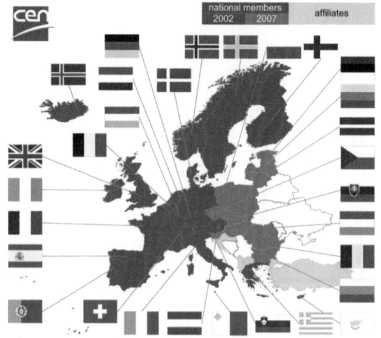

Plate 1. *Countries whose national standards bodies are members of the European Committee for Standardization*

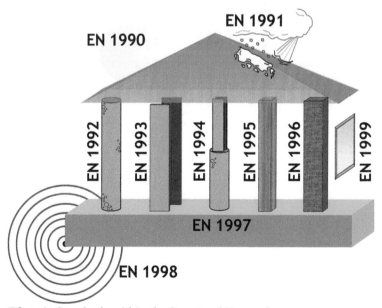

Plate 2. *Standards within the Structural Eurocodes programme*

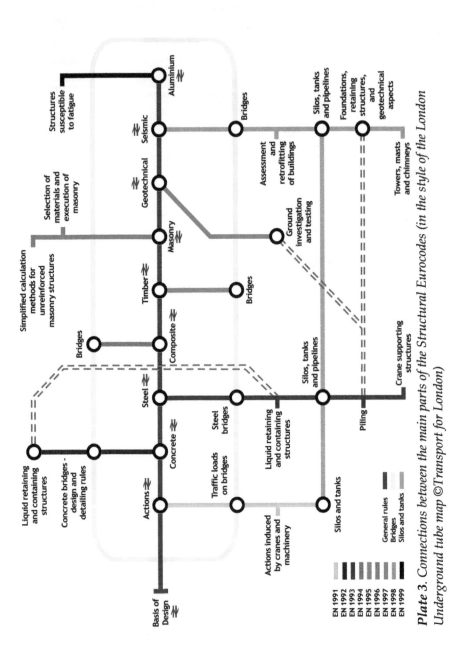

Plate 3. Connections between the main parts of the Structural Eurocodes (in the style of the London Underground tube map ©Transport for London)

Plate 4. Connections between Eurocode 7 and associated European and International standards (based on the National Rail schematic map, ©Association of Train Operating Companies)

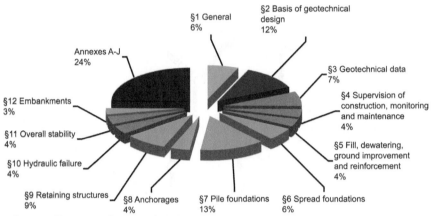

Plate 5. Contents of Eurocode 7 Part 1

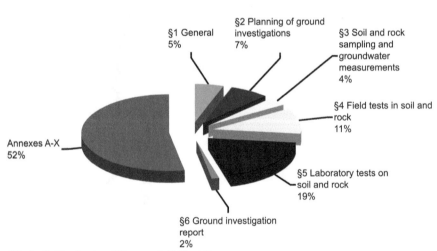

Plate 6. Contents of Eurocode 7 Part 2

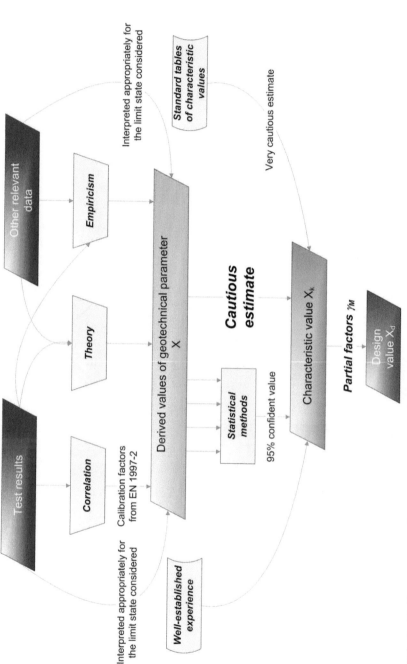

Plate 7. Summary of ground characterization

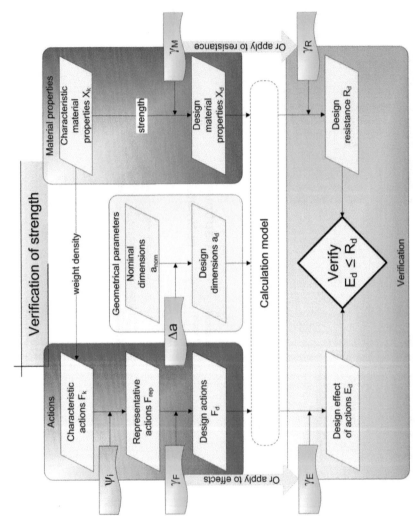

Plate 8. Verification of strength

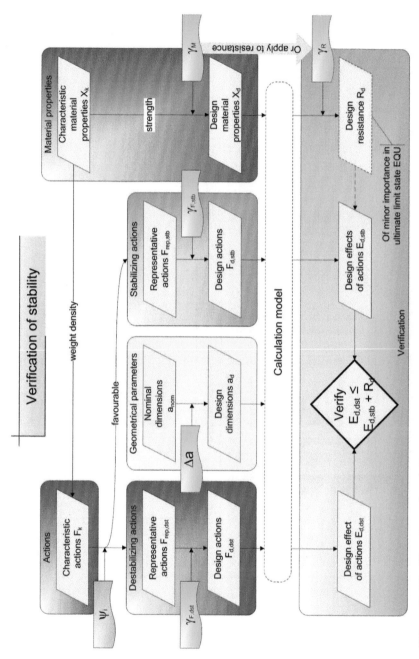

Plate 9. Verification of stability

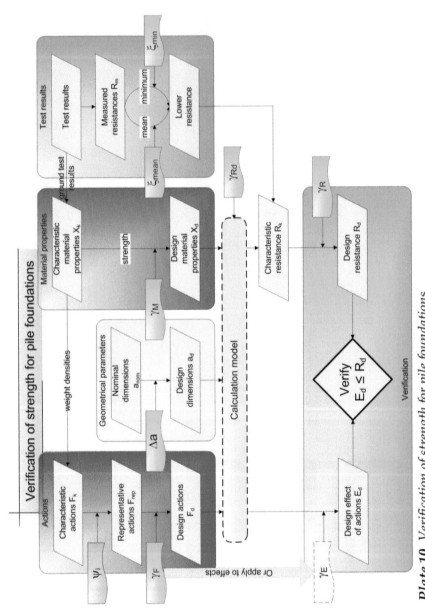

Plate 10. Verification of strength for pile foundations

Chapter 9

Design of slopes and embankments

The design of slopes and embankments is covered by Sections 11 ('Overall stability') and 12 ('Embankments') of Eurocode 7 Part 1, whose contents are as follows:

§x.1 General (2 paragraphs in §11.1 and 2 paragraphs in §12.1)

§x.2 Limit states (2 and 2)

§x.3 Actions and design situations (6 and 8)

§x.4 Design and construction considerations (11 and 13)

§x.5 Ultimate limit state design (26 and 7)

§x.6 Serviceability limit state design (3 and 4)

§11.7 Monitoring (2) and §12.7 Supervision and monitoring (5)

where 'x' is 11 or 12 as appropriate.

Section 11 of EN 1997-1 covers the overall stability of and movements in the ground in relation to foundations, retaining structures, natural slopes, embankments, and excavations. Section 12 covers the design of embankments for small dams and infrastructure.

9.1 Ground investigation for slopes and embankments

Annex B.3 of Eurocode 7 Part 2 provides outline guidance on the depth of investigation points for embankments and cuttings, as illustrated in **Figure 9.1**. (See Chapter 4 for guidance on the spacing of investigation points.)

The recommended minimum depth of investigation, z_a, for embankments and dams (left-hand side of diagram) is the greater of:

$$0.8h \leq z_a \leq 1.2h \text{ and } z_a \geq 6m$$

and for cuttings (right-hand side) is the greater of:

$$z_a \geq 0.4d \text{ and } z_a \geq 2m$$

The depth z_a may be reduced to 2m if the embankment or cutting is built on competent strata[†] with 'distinct' (i.e. known) geology. With 'indistinct'

[†]i.e. weaker strata are unlikely to occur at depth, structural weaknesses such as faults are absent, and solution features and other voids are not expected.

geology, at least one borehole should go to at least 5m. If bedrock is encountered, it becomes the reference level for z_a.   *[EN 1997-2 §B.3(4)]*

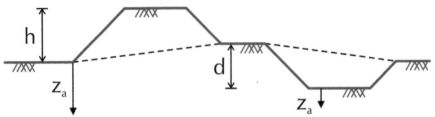

Figure 9.1. *Recommended depth of investigation for embankments and cuttings*

Greater depths of investigation may be needed for very large or highly complex projects or where unfavourable geological conditions are encountered.   *[EN 1997-2 §B.3(2)NOTE & B.3(3)]*

For many sites where slope stability is an issue, the natural ground is not level and deeper investigations may be required. The simple guidelines above should be read with that in mind and the required depth of investigation determined from the expected conditions.

9.2   Design situations and limit states

Limit states for slopes and embankments typically involve loss of overall stability of the ground and associated structures, excessive movement, loss of serviceability, or disruption of drains in an embankment dam.

Section 12 includes an extensive list of additional limit states for embankments, including: internal erosion; surface erosion or scour; deformations leading to loss of serviceability; damage to adjacent structures; problems with transition zones; effects of freezing and thawing; degradation of base course materials; deformations due to hydraulic actions; and changes in environmental conditions.

Examples of situations where slope stability is a concern are shown in **Figure 9.2.**[1] From left to right, these include: (top) translational slab sheet slide and block slide on a weak stratum; (middle) circular and non-circular slides; and (bottom) a large slide encompassing an otherwise adequate structure.

Each design situation must consider the construction process; previous or continuing ground movements; the effects of the slope or embankment on existing structures or slopes; vibrations; climatic variations; vegetation and its removal; human or animal activity; variations in water content or pore-water pressure; and wave action.   *[EN 1997-1 §11.2(2)P]*

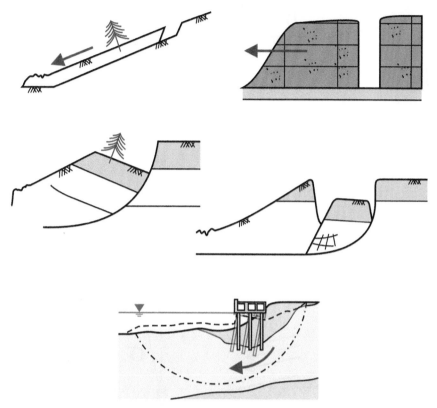

Figure 9.2. Examples of ultimate limit states for slopes

9.3   Basis of design

This book does not attempt to provide complete guidance on the design of slopes and embankments, for which the reader should refer to any well-established text on the subject.[2]

9.4   Stability of an infinitely long slope

Consider the classical problem of determining the stability of an infinitely long slope, as illustrated in **Figure 9.3**. If the slope is underlain by a permeable stratum, as shown in the top half of **Figure 9.3**, any water in the slope flows vertically downwards into the underlying stratum and, as a result, pore water pressures do not build up in the soil. The stability of this slope is identical to that of a dry slope with the same gradient β.

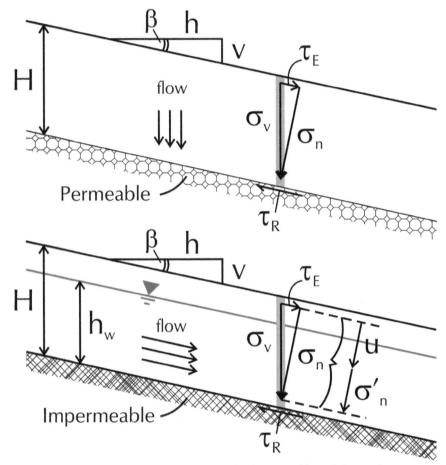

Figure 9.3. *Confined soil slopes overlying (top) permeable and (bottom) impermeable strata*

Instead, if the slope is underlain by an impermeable stratum, as shown in the bottom half of **Figure 9.3**, then any water in the slope flows parallel to the underlying stratum, resulting in a hydraulic gradient parallel to the soil/rock interface.

9.4.1   Closed-form solution using global factor of safety

The global factor of safety against sliding F for both these situations is given by the 'infinite slope expression':[3]

$$F = \frac{c' + (1 - r_u)\gamma H \cos^2 \beta \tan \varphi}{\gamma H \sin \beta \cos \beta} \quad \text{with } r_u = \frac{\gamma_w h_w}{\gamma H}$$

where c′ and φ are the soil's effective cohesion and angle of shearing resistance; γ and $γ_w$ are the weight densities of the soil and water; r_u is a pore pressure parameter; β is the slope angle; and the dimensions H and h_w are defined in **Figure 9.3**. For a dry slope or a slope on a permeable stratum, r_u = 0.

According to BS 6031,[4] first-time slides with a good standard of investigation should be designed with a safety factor between 1.3 and 1.4; slides involving entirely pre-existing slip surfaces should be designed with F ≈ 1.2.

The infinite slope expression can be re-arranged to give the stability number N necessary to provide a given factor of safety:

$$N = \frac{c'}{γH} = F\sin β \cos β - (1 - r_u)\cos^2 β \tan φ$$

Figure 9.4 shows the values of N required to achieve various global factors of safety F, in the range 1.0 to 1.5, for a 1:3 slope (β = 18.4°) with r_u = 0.5. Similar charts can be developed for other values of r_u and β.

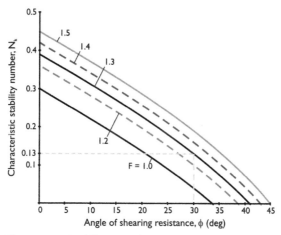

Figure 9.4. Stability number required to achieve various global factors of safety (F) against sliding for a 1:3 infinite slope with r_u = 0.5

An example serves to illustrates how the chart is used. Imagine a 2.5m thick slope comprising soil with a weight density γ = 20 kN/m³ and angle of shearing resistance φ = 30°. From **Figure 9.4**, the stability number needed to achieve a factor of safety F = 1.3 is approximately N = 0.13. Hence the minimum effective cohesion c′ that the soil must possess is:

$$c' = N\gamma H = 0.13 \times 20kN / m^3 \times 2.5m = 6.5kPa$$

9.4.2   Closed-form solution using partial factors from Eurocode 7

As discussed in Chapter 6, verification of strength according to Eurocode 7 involves demonstrating that design effects of actions E_d do not exceed the corresponding design resistance to those actions R_d:

$$E_d \leq R_d$$

where the design effects are given by:

$$E_d = \sigma_{vd} \sin\beta\cos\beta = \gamma_G\gamma_k H \sin\beta\cos\beta$$

and the design resistance by:

$$R_d = \frac{c'_d + (1 - r_u)\sigma_{vd}\cos^2\beta\tan\varphi_d}{\gamma_{Re}}$$

$$= \frac{\left(\dfrac{c'_k}{\gamma_c}\right) + (1 - r_u)(\gamma_G\gamma_k H)\cos^2\beta\left(\dfrac{\tan\varphi_k}{\gamma_\varphi}\right)}{\gamma_{Re}}$$

The subscripts d and k denote design and characteristic values, respectively.

Combining these expressions and re-arranging them produces the 'characteristic stability number' N_k:

$$N_k = \frac{c'_k}{\gamma_k H} = (\gamma_G\gamma_c\gamma_{Re})\sin\beta\cos\beta - \left(\frac{\gamma_G\gamma_c}{\gamma_\varphi}\right)(1 - r_u)\cos^2\beta\tan\varphi_k$$

In deriving this expression, the partial factor γ_G was applied to the characteristic vertical action ($\sigma_{vk} = \gamma_k H$), which appears in the equations for both E_d and R_d. Implicitly, we have assumed that the vertical action is unfavourable wherever it appears in this formulation. Alternatively, we could apply the partial factor to the characteristic action effects, taking account of whether those effects are favourable (using $\gamma_{G,fav}$) or unfavourable (using γ_G). This leads to a different expression for N_k:

$$N_k = \frac{c'_k}{\gamma_k H} = (\gamma_G\gamma_c\gamma_{Re})\sin\beta\cos\beta - \left(\frac{\gamma_{G,fav}\gamma_c}{\gamma_\varphi}\right)(1 - r_u)\cos^2\beta\tan\varphi_k$$

The groupings of factors that appear in these equations are summarized below with their values for each of Eurocode 7's Design Approaches.

Figure 9.5 compares the stability numbers required to achieve the reliability demanded by each Design Approach. The diagram is drawn for a 1:3 slope with $r_u = 0.5$ and is directly analogous to **Figure 9.4**.

The curve for Design Approach 1 (DA1) has two distinct segments that join at $\varphi_k = 11°$. For $\varphi_k \geq 11°$, the value of N_k is controlled by Combination 2 and the slope has an equivalent global factor of safety $F = 1.25$ ($= \gamma_\varphi$). For $\varphi_k < 11°$, the value of N_k is controlled by Combination 1 and the global safety factor increases from $F = 1.25$ at $\varphi_k = 11°$ to $F = 1.35$ as $\varphi_k \to 0°$. The safety factor jumps in value to $F = 1.4$ when $\varphi_k = 0°$ since γ_c is replaced by γ_{cu} in the expression for N_k.

Individual partial factor or partial factor 'grouping'	Design Approach			
	1		2	3
	Combination 1	Combination 2		
γ_G	1.35	1.0	1.35	1.0*
$\gamma_{G,fav}$	1.0	1.0	1.0	1.0
γ_Q	1.5	1.3	1.5	1.3*
$\gamma_\varphi = \gamma_c$	1.0	1.25	1.0	1.25
γ_{cu}	1.0	1.4	1.0	1.4
γ_{Re}	1.0	1.0	1.1	1.0
$\gamma_G \times \gamma_c \times \gamma_{Re}$	1.35	1.25	1.485	1.25
$\gamma_G \times \gamma_{cu} \times \gamma_{Re}$	1.35	1.4	1.485	1.4
$\gamma_G \times \gamma_c / \gamma_\varphi$	1.35	1.0	1.35	1.0
$\gamma_{G,fav} \times \gamma_c / \gamma_\varphi$	1.0	1.0	1.0	1.0
γ_Q / γ_G	1.11	1.3	1.11	1.3

*Factor from Set A2 on geotechnical actions

A second curve for Design Approach 1 is shown on **Figure 9.5** with the label 'DA1 (using $\gamma_{G,fav}$)'. This represents the alternative assumption which applies $\gamma_{G,fav} = 1.0$ to favourable action effects. This curve lies above that for DA1 throughout and its equivalent global factor of safety is constant at $F = 1.35$.

The curve for Design Approach 2 (DA2) lies above that for DA1 when $\varphi_k < 24°$, but below it for larger values of φ_k. Along this curve, the equivalent global factor of safety changes continuously from F = 1.485 at φ_k = 0° to F = 1.0 at φ_k = 43° (not shown on **Figure 9.5**).

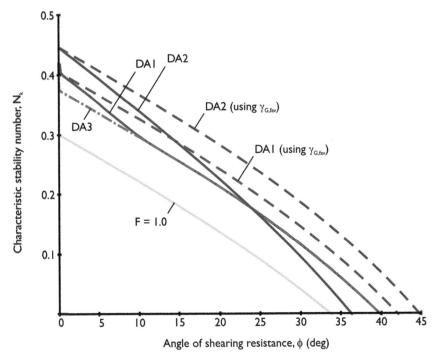

Figure 9.5. *Stability number required to verify limit state GEO for a 1:3 infinite slope with r_u = 0.5 according to Eurocode 7's Design Approaches*

A second curve for Design Approach 2 is shown on **Figure 9.5** with the label 'DA2 (using $\gamma_{G,fav}$)', which applies $\gamma_{G,fav}$ = 1.0 to favourable action effects. This curve lies above that for DA2 throughout and its equivalent global factor of safety is constant at F = 1.485.

The curve for Design Approach 3 (DA3) lies below that for DA1 when φ_k < 11°, but is co-incident with it for larger values of φ_k. Along this curve, the equivalent global factor of safety is constant at F = 1.25.

Hence the relative conservatism of the three Design Approaches is DA2 > DA1 ≥ DA3 for φ_k < 24° and DA1 ≥ DA3 > DA2 for φ_k > 24°.

Figure 9.5 shows quite clearly that Design Approach 2 does not provide a consistent level of reliability as the soil's angle of shearing resistance changes. Furthermore, DA2 produces a level of reliability that is much greater than that traditionally required of infinite slopes and therefore may be considered uneconomical.

Design Approaches 1 and 3 give identical results over the practical range of shearing resistance ($\varphi > 18°$).

9.4.3   Design charts for infinitely long slopes

Based on the preceding analysis, we have developed a series of design charts that allow infinitely long slopes to be designed according to Eurocode 7. The charts cover the verification of limit state GEO using Design Approach 1 and are presented in full in Appendix A. An example of one of these charts in shown in **Figure 9.6**.

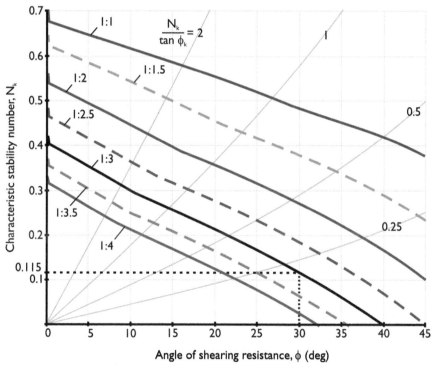

Figure 9.6. *Design chart for verifying limit state GEO using Design Approach 1, for an infinitely long slope with $r_u = 0.5$*

An example illustrates how to use this chart. Imagine a 2.5m thick slope which is required to stand at a gradient of 1:3. The slope comprises soil with characteristic weight density $\gamma_k = 20$ kN/m^3 and characteristic angle of shearing resistance $\varphi_k = 30°$. From **Figure 9.6**, the characteristic stability number needed to verify a 1:3 slope is $N_k = 0.115$. Hence the characteristic effective cohesion c'_k that the soil must possess is:

$$c'_k = N_k \gamma_k H \geq 0.115 \times 20 kN \, / \, m^3 \times 2.5m = 5.75 kPa$$

This is slightly less than was obtained in Section 9.4.1, because the equivalent global factor of safety for Design Approach 1 is 1.25 for $\varphi_k = 30°$ (and was 1.3 in the previous example).

9.5   Stability of a finite slope (based on method of slices)

Bromhead[5] defines the factor of safety used in limit equilibrium methods as the ratio of the mobilized shear strength to the actual shear strength available. This is akin to applying a partial factor to material strength and hence those Design Approaches that apply partial factors to material properties are highly suited to the solution of slope stability problems.

Methods for the limit equilibrium analysis of slopes range from simple translational sliding along a flat plane (as discussed in Section 9.4) to movement along complex shaped sliding surfaces involving curved and planar surfaces. The simplest forms of curved surface involve circular arcs and the discussion below concentrates on such surfaces, which typically compare the overturning moment M_O about a point of rotation to the restoring moment M_R:

$$F = \frac{M_R}{M_O}$$

In order to take into account varying conditions along the slip surface and within the sliding mass, it is usual to split the sliding mass into slices and consider the stability of each slice in turn (see **Figure 9.7**). The results for each slice are summed to give both the overall values for the slope and the factor by which the soil strength needs to be reduced in order to induce failure.

In drained (effective stress) analyses, the available shear strength along the failure plane is a function of the effective normal stress and any effective cohesion acting on that plane. In undrained (total stress) analyses, the available shear strength is a function of the undrained strength alone, which simplifies the analysis considerably.

Figure 9.7 shows a typical slope with the key features of a circular slip analysis indicated. The overturning moment M_O is defined as:

$$M_O = \sum_i \left\{ \left(W_i + Q_i \right) \times x_i \right\}$$

where W_i is the self-weight of slice 'i'; Q_i is any applied surcharge acting on that slice; and x_i is the slice's lever arm about the point of rotation, O. In traditional calculations, the applied surcharge Q_i is included in the weight term W_i. However, because Eurocode 7 applies different partial factors to permanent and variable actions (i.e. to self-weight and applied surcharge, respectively, in this instance), we have separated the terms in this book.

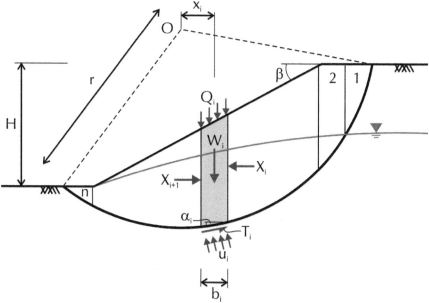

Figure 9.7. Key features of circular slip analysis

It should be noted that x can take either positive or negative values and so M_O has components which are either unfavourable, i.e. increase the overturning moment (positive x); or favourable, i.e. decrease the overturning moment (negative x).

9.5.1   Undrained analysis based on total stresses

In total stress analyses, the restoring moment M_R is defined as:

$$M_R = r \times \sum_i \left\{ c_{u,i} \times l_i \right\}$$

where r is the radius of the slip circle; l_i is the length of the slip surface beneath 'i'; and $c_{u,i}$ is the undrained shear strength along the base of that slice.

The factor of safety F is defined as:

$$F = \frac{M_R}{M_O} = \frac{\sum_i \left\{ c_{u,i} \times l_i \right\}}{\sum_i \left\{ \left(W_i + Q_i \right) \sin \alpha_i \right\}}$$

where α_i is the angle between the base of the slice and the horizontal, i.e.:

$$\alpha_i = \sin^{-1} \left(x_i / r \right)$$

and the other terms are as defined above. This is known as the 'Conventional Method' of analysis.

9.5.2 Eurocode 7 implementation of the Conventional Method

As discussed in Chapter 6, verification of strength according to Eurocode 7 involves demonstrating that design effects of actions E_d do not exceed the corresponding design resistance R_d. i.e.:

$$E_d \le R_d$$

In slope stability analysis, the overturning moment M_O is an action effect and the restoring moment M_R is the resistance to that effect. Hence Eurocode 7 requires the design of slopes and embankments to satisfy:

$$\frac{E_d}{R_d} = \frac{M_{Ed}}{M_{Rd}} = \frac{\sum_i \left\{ \left(W_{d,i} + Q_{d,i} \right) \sin \alpha_i \right\}}{\sum_i \left\{ c_{ud,i} \times l_i \right\}} \le 1$$

where $W_{d,i}$ = the design self-weight of slice 'i'; $Q_{d,i}$ = any imposed surcharge acting on that slice; $c_{ud,i}$ = the design undrained shear strength along the base of the slice; and the other terms are as defined above.

This equation can be re-written in terms of characteristic parameters, as follows:

$$\frac{E_d}{R_d} = \left(\gamma_G \gamma_{cu} \gamma_{Re} \right) \frac{\sum_i \left\{ \left(W_{k,i} + \left(\gamma_Q / \gamma_G \right) Q_{k,i} \right) \sin \alpha_i \right\}}{\sum_i \left\{ c_{uk,i} \times l_i \right\}}$$

The values of the partial factor grouping in this equation are given on page 269 for each Design Approach defined in Eurocode 7.

9.5.3 Drained analysis based on effective stresses

Slopes made of coarse soils and long-term slopes made of fine soils must be analysed in terms of effective stresses. The normal stresses acting on the slip plane and on the sides of the slices make the governing equations indeterminate and closed-form solutions impossible to obtain.

Techniques such as Bishop's Routine Method for circular slips (a.k.a. the 'Simplified Method')[6] make simplifying assumptions about interslice forces that allow iteration to produce a global factor of safety F for the slope. An advantage of Bishop's Routine Method is that it fully satisfies moment equilibrium and uses vertical equilibrium to derive the effective normal forces on each slice. Usually, the vertical interslice force for each slice is set to zero – while the horizontal interslice force does not appear in the equations, since only vertical equilibrium is considered. The full form of the method considers the sum of the interslice forces in both vertical and horizontal directions to be zero.[7] Alternative solutions have been derived for non-circular slip surfaces.[8]

The factor of safety given by Bishop's Routine Method is:

$$F = \sum_i \left\{ \frac{\left(c'_i b_i + (W_i + Q_i - u_i b_i)\tan\phi_i\right)\sec\alpha_i}{1 + \tan\alpha_i\left(\tan\phi_i/F\right)} \right\} \Big/ \sum_i \left\{ (W_i + Q_i)\sin\alpha_i \right\}$$

where b_i is the breadth of slice 'i'; u_i the pore water pressure; c'_i the effective cohesion; ϕ_i the angle of shearing resistance along the base of that slice; and the other variables are as defined earlier.

Note that, if ϕ is set to zero and c' to c_u throughout, this equation reduces to that given earlier for the Conventional Method based on total stresses (see Section 9.5.1):

$$F = \sum_i \left\{ c_{u,i} b_i \sec\alpha_i \right\} \Big/ \sum_i \left\{ (W_i + Q_i)\sin\alpha_i \right\}$$

9.5.4   Eurocode 7 implementation of Bishop's Routine Method

As discussed above, verification of strength according to Eurocode 7 involves demonstrating that design effects of actions E_d are not greater than the corresponding design resistance R_d. i.e.:

$$E_d \le R_d \quad \text{or} \quad \frac{E_d}{R_d} = \frac{M_{Ed}}{M_{Rd}} = \Lambda_{GEO} \le 1$$

where the ratio of the moments Λ_{GEO} (called the 'degree of utilization' in Chapter 6) is given by:

$$\Lambda_{GEO} = \frac{\sum_i \left\{ (W_{d,i} + Q_{d,i})\sin\alpha_i \right\}}{\sum_i \left\{ \dfrac{\left(c'_{d,i} b_i + (W_{d,i} + Q_{d,i} - u_{d,i} b_i)\tan\phi_{d,i}\right)\sec\alpha_i}{1 + \tan\alpha_i \tan\phi_{d,i}\left(\Lambda_{GEO}\right)} \right\}}$$

Here the subscripts 'd' denote design values (i.e. after partial factors have been applied). When $M_{Ed} \rightarrow M_{Rd}$, so $\Lambda_{GEO} = M_{Rd}/M_{Ed} \rightarrow 1$ and this term can therefore be eliminated from the bottom of the equation.

This equation can be re-written in terms of characteristic parameters as:

$$\Lambda_{GEO} = \frac{\left(\gamma_G \gamma_c \gamma_{Re}\right) \sum_i \left\{\left(W_{k,i} + \left(\gamma_Q/\gamma_G\right)Q_{k,i}\right)\sin\alpha_i\right\}}{\sum_i \left\{\dfrac{\left[c'_{k,i}b_i + \left(\dfrac{\gamma_G \gamma_c}{\gamma_\phi}\right)\left(W_{k,i} + \left(\gamma_Q/\gamma_G\right)Q_{k,i} - u_{k,i}b_i\right)\tan\phi_{k,i}\right]\sec\alpha_i}{1 + \tan\alpha_i\left(\dfrac{\tan\phi_{k,i}}{\gamma_\phi}\right)\Lambda_{GEO}}\right\}}$$

The values of the partial factor grouping in this equation are given on page 269 for each of Eurocode 7's Design Approaches.

Design Approach 1 applies partial factors to actions in Combination 1 and to unfavourable variable actions and material properties in Combination 2.

Design Approach 2 applies partial factors to the effects of actions and to resistances, but not to material properties. Different factors are applied to favourable and unfavourable actions.   *[EN 1997-1 §2.4.7.3.4.2 NOTE 2]*

Design Approach 3 applies partial factors to material properties and a small factor to variable actions, but not to other actions or resistances. In Design Approach 3, all actions are treated as 'geotechnical' when designing slopes and embankments.   *[EN 1997-1 §2.4.7.3.4.4 NOTE 2]*

In slip circle analysis, a search is performed to find the critical slip surface and hence the slope's minimum factor of safety. A different critical failure mechanism may be found if partial factors are applied to actions, material properties, or resistances. To avoid this, a traditional slope stability analysis could be carried out first, to establish the critical mechanism, and then a further set of calculations performed on this mechanism, with partial factors from Eurocode 7, to test whether it is stable. However, this complication is probably unnecessary in practice, since the effects of applying partial factors are unlikely to affect the critical mechanism greatly.

The above discussion has focussed on circular mechanisms, whereas it may be more appropriate to consider non-circular or compound mechanisms. For these cases, the same principles apply and the partial factors should be applied in a similar manner to the circular analysis discussed above.

In a traditional analysis it is usual to modify the required factor of safety depending on the perceived level of risk. For example, a very high slope may pose a significantly greater risk than a low height slope, so the required factor of safety would be increased accordingly. In Eurocode 7, one set of partial factors are recommended for all slopes and this may lead engineers to ignore the effects of scale. We therefore recommend that the partial factors should only be used for 'normal' slopes and risk levels. Where the levels of risk to life and property are considered to be large, consideration should be given to using higher partial factors.

9.5.5 Design charts for finite slopes

Based on the preceding analysis, we have developed a series of design charts that allow finite slopes to be designed according to Eurocode 7. The charts cover the verification of limit state GEO using Design Approach 1 and are presented in full in Appendix A. An example of one of these charts is shown in **Figure 9.8**. The slip circle analyses on which the charts are based were performed using commonly available computer software.[9]

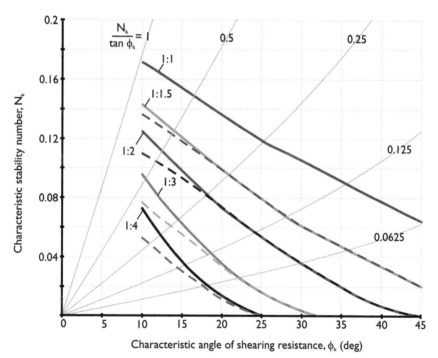

Figure 9.8. Stability number required to verify limit state GEO for a finite slope with $r_u = 0.3$ according to Eurocode 7's Design Approaches

The chart in **Figure 9.8** shows the characteristic stability number N_k for different slope gradients from 1:1 to 1:4. The solid lines assume that the ratio D/H (where D is the depth to a strong layer and H is the slope height) is equal to 2. The dashed lines assume D/H = 1, i.e. the strong layer is coincident with the toe of the slope.

The charts have been derived by analysing a slope geometry for a given design angle of friction φ_d and then determining the required design value of effective cohesion c'_d that gives a global factor of safety = 1.0. Due to the limited number of φ angles used, the curves are not totally smooth.

An example illustrates how to use this chart. Imagine a 2.5m high slope which is required to stand at a gradient of 1:2. The slope comprises soil with characteristic weight density γ_k = 20 kN/m³ and characteristic angle of shearing resistance φ_k = 30°. From **Figure 9.8**, the characteristic stability number needed to verify a 1:2 slope is approximately N_k = 0.036. Hence the characteristic effective cohesion c'_k that the soil must possess is:

$$c'_k = N_k \gamma_k H \geq 0.036 \times 20kN / m^3 \times 2.5m = 1.8kPa$$

9.6 Supervision, monitoring, and maintenance

Slopes must be monitored when the occurrence of limit states cannot be proven sufficiently unlikely by calculation or prescriptive measures, or when the assumptions made in calculations are not based on reliable data.

Monitoring should provide ground-water levels or pore-water pressures for effective stress analysis, lateral and vertical ground movements, the depth and shape of any existing slide for remedial work, and rates of movement. For embankments, Eurocode 7 Part 1 lists additional situations needing monitoring and specifies the contents of a suitable monitoring programme.

9.7 Summary of key points

The design of slopes and embankments is not as simple as it would first appear from a superficial assessment of the partial factor approach. Unless the 'Single Source Principle' is invoked, Eurocode 7 requires different partial factors to be applied to unfavourable and favourable actions. However, this is not possible with typical limit equilibrium slope stability calculations, which cannot easily separate favourable and unfavourable actions.

For those Design Approaches that require permanent actions to be factored, one solution is to factor the soil's weight density. This amounts to treating both favourable and unfavourable actions as coming from a single source. Alternatively for these approaches an enhanced resistance factor is used.

It is evident from the analysis carried out in preparing the examples for this book that Design Approach 2 does not provide sufficient reliability in the design and is therefore unsuitable for slope stability analysis. As Chapter 6 revealed, most countries within Europe who have selected Design Approach 2 for general foundation design switch to Design Approach 3 for the design of slopes and embankments.

9.8   Worked examples

The worked examples in this chapter involve verification of strength of an infinite soil slope overlying permeable rock (Example 9.1); an infinite soil slope overlying impermeable rock (Example 9.2); a road cutting designed using charts and slope stability software (Examples 9.3 and 9.4); and a road embankment over an alluvial flood plain (Example 9.5).

Specific parts of the calculations are marked ❶, ❷, ❸, etc., where the numbers refer to the notes that accompany each example.

9.8.1   Infinite soil slope overlying permeable rock

Example 9.1 considers an infinite soil slope overlying permeable rock, with no groundwater in the slope owing to vertical seepage into the underlying rock, as shown in **Figure 9.9**.

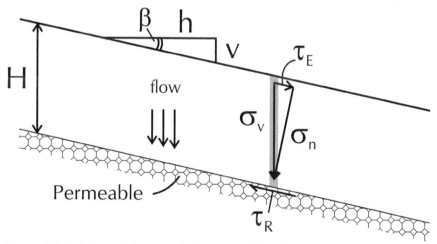

Figure 9.9. Infinite soil slope overlying permeable rock

This example highlights the effects of applying partial factors to material properties and the ambiguities that arise when partial factors are applied to actions. In particular, there are difficulties where an action can have both favourable and unfavourable effects. For slope stability problems, this is a particular issue as the stresses due to the weight of the ground are both the source of the instability and a key contributor to the resistance.

Notes on Example 9.1

❶ The contact area is slightly larger than the horizontal area of the slice, owing to the slope angle.

❷ The calculation of normal stress takes into account both the resolution of the vertical stress normal to the slope and the increased contact area of the slice discussed in ❶.

❸ Here the normal stress is considered to be an unfavourable action, even though its subsequent effect on shear resistance is favourable. Both the unfavourable and favourable effects are considered as coming from a 'single source' and hence a single partial factor (γ_G) is applied to σ_{nk} and τ_{Ek}.

❹ This design is governed by Combination 2, which – with a utilization factor less than 100% – means that the strength of the slope has been verified successfully for Design Approach 1.

❺ In Design Approach 2, partial factors are more commonly applied to effects of actions rather than to actions. Therefore, in this calculation, the normal stress is considered a favourable action effect, which subsequently minimizes the shear resistance. The single source principle is not invoked and different partial factors ($\gamma_{G,fav}$ and γ_G) are applied to σ_{nk} and τ_{Ek} respectively.

❻ Design Approach 2 is the only one which employs a partial factor on resistance > 1.0.

❼ The design just works for Design Approach 2 (degree of utilization is close enough to 100% not to worry). If the single source principle had been applied (see ❺ above), then the utilization would have dropped to 76%.

❽ In Design Approach 3, partial factors from Set A2 are applied to all geotechnical actions (there are no structural actions in this design situation).

❾ The design works for Design Approach 3, giving the same degree of utilization as Design Approach 1.

❿ The traditional global factor of safety for this situation is ≈1.5 which would be considered more than adequate for long-term conditions.

Example 9.1
Infinite soil slope overlying permeable rock
Verification of strength (GEO)

Design situation

Consider a dry hill slope comprising $H = 5m$ of sandy CLAY overlying rock. The clay's characteristic dry weight density is $\gamma_k = 18\dfrac{kN}{m^3}$ and its characteristic drained strength parameters are $c'_k = 2kPa$ and $\varphi_k = 25°$. The hill slopes at $v = 1m$ vertically to every $h = 3m$ horizontally, i.e. $\beta = \tan^{-1}\left(\dfrac{v}{h}\right) = 18.4°$

Design Approach 1

Actions and effects

Contact area per unit width x unit run $A_n = \dfrac{1}{\cos(\beta)} = 1.054\dfrac{m^2}{m^2}$ ❶

Characteristic values at soil/rock interface:

Vertical stress $\sigma_{vk} = \gamma_k \times H = 90\ kPa$

Normal stress $\sigma_{nk} = \dfrac{\sigma_{vk} \times \cos(\beta)}{A_n} = 81\ kPa$ ❷

Shear stress $\tau_{Ek} = \dfrac{\sigma_{vk} \times \sin(\beta)}{A_n} = 27\ kPa$

Partial factors from Sets $\begin{pmatrix} A1 \\ A2 \end{pmatrix} : \gamma_G = \begin{pmatrix} 1.35 \\ 1 \end{pmatrix}$

Design normal stress $\sigma_{nd} = \gamma_G \times \sigma_{nk} = \begin{pmatrix} 109.3 \\ 81 \end{pmatrix} kPa$ ❸

Design shear stress $\tau_{Ed} = \gamma_G \times \tau_{Ek} = \begin{pmatrix} 36.5 \\ 27 \end{pmatrix} kPa$

Material properties and resistance

Characteristic resistance acting parallel to the rock surface, per unit area of slope is $\tau_{Rk} = c'_k + \sigma_{nk} \tan(\varphi_k) = 39.8\ kPa$

Partial factors from Sets $\begin{pmatrix} M1 \\ M2 \end{pmatrix} : \gamma_\varphi = \begin{pmatrix} 1 \\ 1.25 \end{pmatrix}$ and $\gamma_c = \begin{pmatrix} 1 \\ 1.25 \end{pmatrix}$

Design angle of shearing resistance $\varphi_d = \tan^{-1}\left(\dfrac{\tan(\varphi_k)}{\gamma_\varphi}\right) = \left(\dfrac{25}{20.5}\right)^\circ$

Design effective cohesion $c'_d = \dfrac{c'_k}{\gamma_c} = \left(\dfrac{2}{1.6}\right)$ kPa

Partial factors from Set $\begin{pmatrix} R1 \\ R1 \end{pmatrix}$; $\gamma_{Re} = \begin{pmatrix} 1 \\ 1 \end{pmatrix}$

Design shear resistance is $\tau_{Rd} = \dfrac{\overrightarrow{c'_d + \sigma_{nd} \times \tan(\varphi_d)}}{\gamma_{Re}} = \left(\dfrac{53}{31.8}\right)$ kPa

Verification of strength against sliding

Degree of utilization is $\boxed{\Lambda_{GEO,1} = \dfrac{\tau_{Ed}}{\tau_{Rd}} = \left(\dfrac{69}{85}\right)\%}$ ❹

The design is unacceptable if degree of utilization is > 100%

Design Approach 2

Actions and effects

Characteristic values = same as for Design Approach 1

Partial factor from Set A1: $\gamma_G = 1.35$ and $\gamma_{G,fav} = 1$

Design normal stress $\sigma_{nd} = \gamma_{G.fav} \times \sigma_{nk} = 81$ kPa ❺

Design shear stress $\tau_{Ed} = \gamma_G \times \tau_{Ek} = 36.5$ kPa

Material properties and resistance

Characteristic resistance = same as for Design Approach 1

Partial factors from Set M1: $\gamma_\varphi = 1$ and $\gamma_c = 1$

Design angle of shearing resistance $\varphi_d = \tan^{-1}\left(\dfrac{\tan(\varphi_k)}{\gamma_\varphi}\right) = 25^\circ$

Design effective cohesion $c'_d = \dfrac{c'_k}{\gamma_c} = 2$ kPa

Partial factor on sliding resistance from Set R2: $\gamma_{Re} = 1.1$ ❻

Shear resistance $\tau_{Rd} = \dfrac{c'_d + \sigma_{nd} \times \tan(\varphi_d)}{\gamma_{Re}} = 36.2$ kPa

Verification of strength against sliding

Degree of utilization is $\boxed{\Lambda_{GEO,2} = \dfrac{\tau_{Ed}}{\tau_{Rd}} = 101\%}$ ❼

The design is unacceptable if degree of utilization is > 100%

Design Approach 3

Actions and effects

Characteristic values = same as for Design Approach 1

Partial factor from Set A2: $\gamma_G = 1$ ❽

Design normal stress $\sigma_{nd} = \gamma_G \times \sigma_{nk} = 81\ kPa$

Design shear stress $\tau_{Ed} = \gamma_G \times \tau_{Ek} = 27\ kPa$

Material properties and resistance

Characteristic resistance = same as for Design Approach 1

Partial factors from Set M2: $\gamma_\varphi = 1.25$ and $\gamma_c = 1.25$

Design angle of shearing resistance $\varphi_d = \tan^{-1}\left(\dfrac{\tan(\varphi_k)}{\gamma_\varphi}\right) = 20.5°$

Design effective cohesion $c'_d = \dfrac{c'_k}{\gamma_c} = 1.6\ kPa$

Partial factor from Set R3: $\gamma_{Re} = 1$

Shear resistance $\tau_{Rd} = \dfrac{c'_d + \sigma_{nd} \times \tan(\varphi_d)}{\gamma_{Re}} = 31.8\ kPa$

Verification of strength against sliding

Degree of utilization is $\boxed{\Lambda_{GEO,3} = \dfrac{\tau_{Ed}}{\tau_{Rd}} = 85\%}$ ❾

The design is unacceptable if degree of utilization is > 100%

Traditional design

Traditional global factor of safety is $F = \dfrac{\tau_{Rk}}{\tau_{Ek}} = 1.47$ ❿

9.8.2   Infinite soil slope overlying impermeable rock

Example 9.2 considers an infinite soil slope overlying impermeable rock, with groundwater present in the slope and seepage parallel to the underlying rock, as shown in **Figure 9.10**.

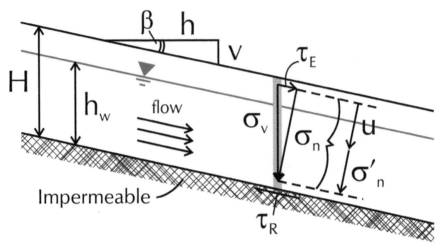

Figure 9.10. *Infinite soil slope overlying impermeable rock*

Notes on Example 9.2

❶ The clay's characteristic weight density (which was 18 kN/m³ in Example 9.1) has been increased here to 20 kN/m³ to account for the water table saturating the soil.

❷ The pore pressure ratio is calculated on the basis of the anticipated maximum water level during the slope's design working life (i.e. the persistent design situation). This is assumed to be 4m above the soil/rock boundary.

❸ As for Example 9.1, both the unfavourable and favourable effects of the soil's weight density are assumed to come from a 'single source' and hence a single partial factor (γ_G) is applied to σ'_{nk} and τ_{Ek}.

❹ This design is governed by Combination 2, which – with a utilization factor greater than 100% – means that the slope does *not* satisfy Design Approach 1. Measures to rectify this include lowering the water table to 3.8m or flattening the slope to 1 in 4.1.

Example 9.2
Infinite soil slope overlying impermeable rock
Verification of strength (GEO)

Design situation

Consider a saturated hill slope comprising $H = 5m$ of sandy CLAY overlying an impermeable rock. The clay's characteristic saturated weight density is

$\gamma_k = 20 \dfrac{kN}{m^3}$ ❶ and its characteristic drained strength parameters are

$c'_k = 2kPa$ and $\varphi_k = 25°$. The hill slopes at $v = 1m$ vertically to $h = 4m$ horizontally. Groundwater, which is at height $h_w = 4m$ above the soil/rock

interface, has characteristic weight density $\gamma_w = 9.81 \dfrac{kN}{m^3}$ and flows parallel

to the interface. The slope angle $\beta = atan\left(\dfrac{v}{h}\right) = 14°$.

<u>Design Approach 1</u>

Actions and effects

Contact area per unit width x unit run $A_n = \dfrac{1}{\cos(\beta)} = 1.031 \dfrac{m^2}{m^2}$

Pore pressure ratio $r_u = \dfrac{\gamma_w \times h_w}{\gamma_k \times H} = 0.39$ ❷

Characteristic values at soil/rock interface:

Vertical total stress $\sigma_{vk} = \gamma_k \times H = 100 \, kPa$

Normal total stress $\sigma_{nk} = \dfrac{\sigma_{vk} \times \cos(\beta)}{A_n} = 94.1 \, kPa$

Pore water pressure $u_k = \dfrac{\gamma_w \times h_w \times \cos(\beta)}{A_n} = 36.9 \, kPa$

Normal effective stress $\sigma'_{nk} = \sigma_{nk} - u_k = 57.2 \, kPa$

Shear stress from soil $\tau_{Ek} = \sigma_{vk} \times \sin(\beta) \times \cos(\beta) = 23.5 \, kPa$

Partial factors from Sets $\begin{pmatrix} A1 \\ A2 \end{pmatrix}$: $\gamma_G = \begin{pmatrix} 1.35 \\ 1 \end{pmatrix}$

Design normal effective stress $\sigma'_{nd} = \gamma_G \times \sigma'_{nk} = \begin{pmatrix} 77.2 \\ 57.2 \end{pmatrix}$ kPa ❸

Design shear stress $\tau_{Ed} = \gamma_G \times \tau_{Ek} = \begin{pmatrix} 31.8 \\ 23.5 \end{pmatrix}$ kPa

Material properties and resistance

Characteristic resistance parallel to rock surface, per unit area of slope

$$\tau_{Rk} = c'_k + \sigma'_{nk} \times \tan(\varphi_k) = 28.7 \text{ kPa}$$

Partial factors from Sets $\begin{pmatrix} M1 \\ M2 \end{pmatrix}$: $\gamma_\varphi = \begin{pmatrix} 1 \\ 1.25 \end{pmatrix}$ and $\gamma_c = \begin{pmatrix} 1 \\ 1.25 \end{pmatrix}$

Design angle of shearing resistance $\varphi_d = \tan^{-1}\left(\dfrac{\tan(\varphi_k)}{\gamma_\varphi} \right) = \begin{pmatrix} 25 \\ 20.5 \end{pmatrix}^\circ$

Design effective cohesion $c'_d = \dfrac{c'_k}{\gamma_c} = \begin{pmatrix} 2 \\ 1.6 \end{pmatrix}$ kPa

Partial factors from Sets $\begin{pmatrix} R1 \\ R1 \end{pmatrix}$: $\gamma_{Re} = \begin{pmatrix} 1 \\ 1 \end{pmatrix}$

Design shear resistance $\tau_{Rd} = \dfrac{c'_d + \sigma'_{nd} \times \tan(\varphi_d)}{\gamma_{Re}} = \begin{pmatrix} 38 \\ 22.9 \end{pmatrix}$ kPa

Verification of strength against sliding

Degree of utilization is $\boxed{\Lambda_{GEO,1} = \dfrac{\tau_{Ed}}{\tau_{Rd}} = \begin{pmatrix} 84 \\ 103 \end{pmatrix}}$ % ❹

The design is unacceptable if the degree of utilization is > 100%

Design Approach 2

Actions and effects

Characteristic values = same as for Design Approach 1

Partial factor from Set A1: $\gamma_G = 1.35$

Design normal stress $\sigma'_{nd} = \gamma_{G.fav} \times \sigma'_{nk} = 57.2$ kPa ❺

Design shear thrust $\tau_{Ed} = \gamma_G \times \tau_{Ek} = 31.8$ kPa

Material properties and resistance

Characteristic resistance = same as for Design Approach 1

Partial factors from Set M1: $\gamma_\varphi = 1$ and $\gamma_c = 1$

Design angle of shearing resistance $\varphi_d = \tan^{-1}\left(\dfrac{\tan(\varphi_k)}{\gamma_\varphi}\right) = 25°$

Design effective cohesion $c'_d = \dfrac{c'_k}{\gamma_c} = 2\,kPa$

Partial factor from Set R2: $\gamma_{Re} = 1.1$

Design shear resistance $\tau_{Rd} = \dfrac{c'_d + \sigma'_{nd} \times \tan(\varphi_d)}{\gamma_{Re}} = 26.1\,kPa$

Verification of strength against sliding

Degree of utilization is $\boxed{\Lambda_{GEO,2} = \dfrac{\tau_{Ed}}{\tau_{Rd}} = 122\,\%}$ **⑥**

The design is unacceptable if the degree of utilization is > 100%

Design Approach 3

Actions and effects
Characteristic values = same as for Design Approach 1
Partial factor from Set A2: $\gamma_G = 1$

Design normal stress $\sigma'_{nd} = \gamma_G \times \sigma'_{nk} = 57.2\,kPa$

Design shear thrust $\tau_{Ed} = \gamma_G \times \tau_{Ek} = 23.5\,kPa$

Material properties and resistance
Characteristic resistance = same as for Design Approach 1
Partial factors from Set M2: $\gamma_\varphi = 1.25$ and $\gamma_c = 1.25$

Design angle of shearing resistance $\varphi_d = \tan^{-1}\left(\dfrac{\tan(\varphi_k)}{\gamma_\varphi}\right) = 20.5°$

Design effective cohesion $c'_d = \dfrac{c'_k}{\gamma_c} = 1.6\,kPa$

Partial factor from Set R3: $\gamma_{Re} = 1$

Design shear resistance $\tau_{Rd} = \dfrac{c'_d + \sigma'_{nd} \times \tan\left(\varphi_d\right)}{\gamma_{Re}} = 22.9\ kPa$

Verification of strength against sliding

Degree of utilization is $\boxed{\Lambda_{GEO,3} = \dfrac{\tau_{Ed}}{\tau_{Rd}} = 103\ \%}$ **7**

The design is unacceptable if the degree of utilization is > 100%

Traditional design

Traditional global factor of safety $F = \dfrac{\tau_{Rk}}{\tau_{Ek}} = 1.22$ **8**

Accidental Design Situation - all Design Approaches

Design situation

Re-consider the slope from above. Owing to extensive flooding, the groundwater level rises to a height $h_w = H = 5\,m$ above the soil/rock

interface. This is regarded as an Accidental Design Situation. **9**

Actions and effects

Pore pressure ratio $r_u = \dfrac{\gamma_w \times h_w}{\gamma_k \times H} = 0.49$

Characteristic values at soil/rock interface:

$\quad$ Pore water pressure $u_k = \dfrac{\gamma_w \times h_w \times \cos(\beta)}{A_n} = 46.2\ kPa$

$\quad$ Normal effective stress $\sigma'_{nk} = \sigma_{nk} - u_k = 48\ kPa$

$\quad$ Shear stress from soil $\tau_{Ek} = \sigma_{vk} \times \sin(\beta) \times \cos(\beta) = 23.5\ kPa$

Partial factor for accidental situation $\gamma_G = 1$

Design normal effective stress $\sigma'_{nd} = \gamma_G \times \sigma'_{nk} = 48\ kPa$

Design shear stress $\tau_{Ed} = \gamma_G \times \tau_{Ek} = 23.5\ kPa$

Material properties and resistance

Characteristic resistance parallel to rock surface, per unit area of slope

$$\tau_{Rk} = c'_k + \sigma'_{nk} \times \tan\left(\varphi_k\right) = 24.4 \text{ kPa}$$

Partial factors for accidental situation $\gamma_\varphi = 1$, $\gamma_c = 1$, and $\gamma_{Re} = 1$

Design angle of shearing resistance $\varphi_d = \tan^{-1}\left(\dfrac{\tan\left(\varphi_k\right)}{\gamma_\varphi}\right) = 25°$

Design effective cohesion $c'_d = \dfrac{c'_k}{\gamma_c} = 2 \text{ kPa}$

Partial factor for accidental situation

Design shear resistance $\tau_{Rd} = \dfrac{\overrightarrow{c'_d + \sigma'_{nd} \times \tan\left(\varphi_d\right)}}{\gamma_{Re}} = 24.4 \text{ kPa}$

<u>*Verification of strength against sliding*</u>

Degree of utilization is $\boxed{\Lambda_{GEO} = \dfrac{\tau_{Ed}}{\tau_{Rd}} = 97\%}$ ⑩

The design is unacceptable if the degree of utilization is > 100%

❺As for Example 9.1, different partial factors ($\gamma_{G,fav}$ and γ_G) are applied to favourable and unfavourable effects of the soil's weight density.

❻ The slope does *not* satisfy Design Approach 2, but can be made to do so by lowering the water table to 2.5m or flattening the slope to 1 in 5.

❼ The result of verification to Design Approach 3 is identical to that obtained using Design Approach 1.

❽ The traditional global factor of safety for this situation is ≈1.2 which would be considered less than adequate for long-term conditions (for which F = 1.3 might be required).

❾ The calculations so far have considered the slope in its persistent state, but we should also check more extreme conditions, such as flooding to ground surface. Eurocode 7 allows 'extreme water pressures ... to be treated as accidental actions', for which Eurocode 7 specifies partial factors equal to 1.0 (i.e. lower than in the persistent case) – independent of Design Approach.

❿ The soil is just strong enough to withstand accidental flooding to ground surface.

9.8.3   Road cutting (using design charts)

Example 9.3 considers the design of a cutting for a new bypass, as shown in **Figure 9.11**. As there is no surcharge, we will use Taylor's charts to assess the safe slope angle for short and long term conditions.

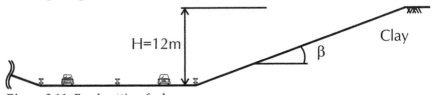

Figure 9.11. *Road cutting for bypass*

Taylor's charts (see **Figure 9.12**) relate a stability number, the angle of friction of the soil and the angle of the slope. To use the chart in a conventional manner requires the same factor of safety to be applied both to c (i.e. effective cohesion c' or undrained strength c_u) and to tan φ to assess the factor of safety for the slope. However, a Eurocode 7 design requires partial factors to be applied to different parts of the equations (to actions, material properties, and resistances) and therefore the use of simple charts is made more complex. In the following example, we have highlighted how the partial factors may be applied for the three Design Approaches when using Taylor's charts.

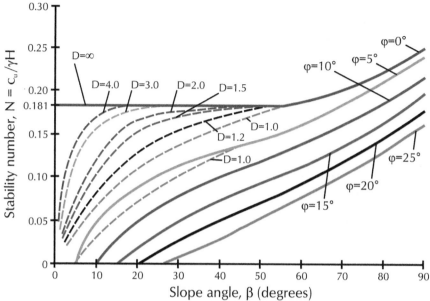

Figure 9.12. *Taylor's chart for determining the stability of undrained slopes*

Notes on Example 9.3

❶ It has been assumed that $\gamma_k \times H$ represents the unfavourable action and therefore γ_G has been applied to this part of the stability number. For DA1-1, the stability number has been calculated using $\gamma_G = 1.35$ and $\gamma_M = 1.0$; and for DA1-2, $\gamma_G = 1.0$ and $\gamma_M = 1.4$. The net effect is to give a lower stability number for DA1-2 than for DA1-1.

❷ For the undrained case, the resultant utilization factor is low, indicating that in the short term the slope more than adequately satisfies DA1.

❸ For the drained case, the stability number for DA1-1 is higher than for DA1-2 as $\gamma_M \times \gamma_G$ is smaller for DA1-2 than for DA1-1.

❹ As would be expected, the drained situation is more critical and DA1-2 governs the design.

Example 9.3
Road cutting (using design charts)
Verification of strength (GEO)

Design situation

Consider a road cutting for a new bypass. Ground conditions in the cutting consist of a uniform CLAY with the following characteristic parameters:

weight density $\gamma_k = 20\dfrac{kN}{m^3}$, undrained strength $c_{uk} = 75kPa$, effective

cohesion $c'_k = 5kPa$, and effective angle of shearing resistance $\varphi_k = 20°$.

The slope is dry. The slope stands at an angle $\beta_k = 19°$ and, at its maximum

depth, the cutting is $H = 12m$ deep. The slope passes through farmland and hence surcharge loading at its crest may be ignored.

Design Approach 1

Actions and effects

Partial factors from Sets $\begin{pmatrix} A1 \\ A2 \end{pmatrix}$: $\gamma_G = \begin{pmatrix} 1.35 \\ 1 \end{pmatrix}$

Material properties and resistance

Partial factors from Sets $\begin{pmatrix} M1 \\ M2 \end{pmatrix}$: $\gamma_\varphi = \begin{pmatrix} 1 \\ 1.25 \end{pmatrix}$, $\gamma_c = \begin{pmatrix} 1 \\ 1.25 \end{pmatrix}$, $\gamma_{cu} = \begin{pmatrix} 1 \\ 1.4 \end{pmatrix}$

Design undrained strength $c_{ud} = \dfrac{c_{uk}}{\gamma_{cu}} = \begin{pmatrix} 75 \\ 53.6 \end{pmatrix} kPa$

Design angle of shearing resistance $\varphi_d = \tan^{-1}\left(\dfrac{\tan(\varphi_k)}{\gamma_\varphi}\right) = \begin{pmatrix} 20 \\ 16.2 \end{pmatrix}°$

Design effective cohesion $c'_d = \dfrac{c'_k}{\gamma_c} = \begin{pmatrix} 5 \\ 4 \end{pmatrix} kPa$

Verification of strength for short-term (undrained) conditions

Design stability number (undrained) $N_{ud} = \dfrac{c_{ud}}{\gamma_G \times \gamma_k \times H} = \begin{pmatrix} 0.231 \\ 0.223 \end{pmatrix}$ **❶**

From Taylor's chart, the maximum design slope angle is $\beta_d = \begin{pmatrix} 84 \\ 79 \end{pmatrix}°$

Degree of utilization is $\boxed{\Lambda_{GEO,1} = \dfrac{\beta_k}{\beta_d} = \begin{pmatrix} 23 \\ 24 \end{pmatrix}\%}$ **❷**

The design is unacceptable if the degree of utilization is > 100%

Verification of strength for long-term (drained) conditions

Design stability number (drained) $N_d = \dfrac{c'_d}{\gamma_G \times \gamma_k \times H} = \begin{pmatrix} 0.015 \\ 0.017 \end{pmatrix}$ **❸**

Design angle of shearing resistance $\varphi_d = \begin{pmatrix} 20 \\ 16.2 \end{pmatrix}°$

From Taylor's chart, the maximum design slope angle is $\beta_d = \begin{pmatrix} 26 \\ 23 \end{pmatrix}°$

Degree of utilization is $\boxed{\Lambda_{GEO,1} = \dfrac{\beta_k}{\beta_d} = \begin{pmatrix} 73 \\ 83 \end{pmatrix}\%}$ **❹**

The design is unacceptable if the degree of utilization is > 100%

Design Approach 2

Actions and effects
Partial factor from Set A1: $\gamma_G = 1.35$

Material properties and resistance
Partial factors from Set M1: $\gamma_\varphi = 1$, $\gamma_c = 1$, $\gamma_{cu} = 1$, $\gamma_{Re} = 1.1$ **⑤**

Design undrained strength $c_{ud} = \dfrac{c_{uk}}{\gamma_{cu}} = 75\ kPa$

Design angle of shearing resistance $\varphi_d = \tan^{-1}\left(\dfrac{\tan\left(\varphi_k\right)}{\gamma_\varphi}\right) = 20\,°$

Design effective cohesion $c'_d = \dfrac{c'_k}{\gamma_c} = 5\ kPa$

Verification of strength for short-term (undrained) conditions

Design stability number (undrained) $N_{ud} = \dfrac{c_{ud}}{\left(\gamma_G \times \gamma_k \times H\right) \times \gamma_{Re}} = 0.21$ **⑥**

From Taylor's chart, the maximum design slope angle is $\beta_d = 74.5°$

Degree of utilization is $\boxed{\Lambda_{GEO,2} = \dfrac{\beta_k}{\beta_d} = 26\ \%}$

The design is unacceptable if the degree of utilization is > 100%

Verification of strength for long-term (drained) conditions

Design stability number (drained) $N_d = \dfrac{c'_d}{\left(\gamma_G \times \gamma_k \times H\right) \times \gamma_{Re}} = 0.014$ **⑥**

Angle of shearing resistance for chart $\varphi_{chart} = \tan^{-1}\left(\dfrac{\tan\left(\varphi_d\right)}{\gamma_{Re}}\right) = 18.3\,°$**⑥**

From Taylor's chart, the maximum design slope angle is $\beta_d = 23.5°$

Degree of utilization is $\boxed{\Lambda_{GEO,2} = \dfrac{\beta_k}{\beta_d} = 81\ \%}$**⑦**

The design is unacceptable if the degree of utilization is > 100%

Design Approach 3

Actions and effects

Partial factor from Set A1: $\gamma_G = 1$ Ⓢ

Material properties and resistance

Partial factors from Set M2: $\gamma_\varphi = 1.25$, $\gamma_c = 1.25$, $\gamma_{cu} = 1.4$

Design undrained strength $c_{ud} = \dfrac{c_{uk}}{\gamma_{cu}} = 53.6 \text{ kPa}$

Design angle of shearing resistance $\varphi_d = \tan^{-1}\left(\dfrac{\tan(\varphi_k)}{\gamma_\varphi}\right) = 16.2\,°$

Design effective cohesion $c'_d = \dfrac{c'_k}{\gamma_c} = 4 \text{ kPa}$

Verification of strength for short-term (undrained) conditions

Design stability number (undrained) $N_{ud} = \dfrac{c_{ud}}{\gamma_G \times \gamma_k \times H} = 0.223$

From Taylor's chart, the maximum design slope angle is $\beta_d = 79°$

Degree of utilization is $\boxed{\Lambda_{GEO,3} = \dfrac{\beta_k}{\beta_d} = 24\,\%}$ Ⓣ

The design is unacceptable if the degree of utilization is > 100%

Verification of strength for long-term (drained) conditions

Design stability number (drained) $N_d = \dfrac{c'_d}{\gamma_G \times \gamma_k \times H} = 0.017$

Design angle of shearing resistance $\varphi_d = 16.2\,°$

From Taylor's chart, the maximum design slope angle is $\beta_d = 23°$

Degree of utilization is $\boxed{\Lambda_{GEO,3} = \dfrac{\beta_k}{\beta_d} = 83\,\%}$ Ⓣ

The design is unacceptable if the degree of utilization is > 100%

Traditional design

Factor of safety
Use lumped factors $F_{undrained} = 1.5$ and $F_{drained} = 1.3$

Verification of strength for short-term (undrained) conditions

Stability number (undrained) $N_u = \dfrac{c_{uk}}{F_{undrained} \times \gamma_k \times H} = 0.208$

From Taylor's chart, the maximum design slope angle is $\beta_d = 72.5°$

Degree of utilization is $\boxed{\Lambda_{trad} = \dfrac{\beta_k}{\beta_d} = 26\,\%}$ ⑩

The design is unacceptable if the degree of utilization is > 100%

Verification of strength for long-term (drained) conditions

Stability number (drained) $N = \dfrac{c'_k}{F_{drained} \times \gamma_k \times H} = 0.016$

Angle of shearing resistance $\varphi_d = \tan^{-1}\left(\dfrac{\tan(\varphi_k)}{F_{drained}}\right) = 15.6°$

From Taylor's chart, the maximum design slope angle is $\beta_d = 21°$

Degree of utilization is $\boxed{\Lambda_{trad} = \dfrac{\beta_k}{\beta_d} = 90\,\%}$ ⑩

The design is unacceptable if the degree of utilization is > 100%

❺ In DA2, the partial factor $\gamma_G = 1.35$ increases the actions but the material properties are not changed, since $\gamma_M = 1.0$. An additional partial factor $\gamma_{Re} = 1.1$ is applied to resistance.

❻ It has been assumed that the resistance factor should be applied to both c (i.e. c' or c_u) and to tan φ.

❼ As for DA1, the drained analysis is critical. DA2 has a marginally lower degree of utilization than DA1.

❽ In DA3, partial factors greater than one are only applied to material properties.

❾ DA3 is identical to DA1-2, since for slopes the actions are treated as 'geotechnical' and are therefore not modified.

❿ Traditional calculations apply different factors of safety in undrained and drained analyses (e.g. 1.5 and 1.3, respectively). Using these traditional factors of safety produces a marginally more conservative design.

Charts do not enable the application of different partial factors to permanent actions (e.g. self-weight) and variable actions (e.g. surcharges). This makes their use for anything but the simplest of cases difficult. Although the ready availability of slope stability software largely precludes the need to use charts, they do provide a simple and quick means of assessing the likely safe angle for a slope and to check the results obtained from software.

9.8.4   Road cutting (using slope stability software)

Example 9.4 considers the 8m deep motorway cutting shown in **Figure 9.13**, which passes through 1.5m of granular head deposits overlying firm to stiff, medium strength CLAY. The groundwater table is just below the head deposits and is kept below the carriageway level by drainage. **Figure 9.13** indicates the steady state position of the phreatic surface. The cutting is to be widened and land ownership constraints mean that it would be preferable to steepen the slope to a 1:2.25 gradient.

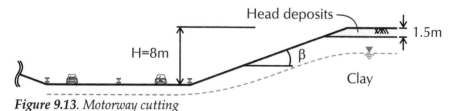

Figure 9.13. Motorway cutting

Such problems are not amenable to simple analysis using charts so the problem has been analysed using Bishop's Simplified Method.[10] This example demonstrates the issues of applying standard slope stability software in the context of design to EN 1997-1 and highlights the difficulties involved in applying DA2 to slope stability problems.

Notes on Example 9.4

❶ In Design Approach 1 Combination 1 (DA1-1), partial factors $\gamma_G = 1.35$ and $\gamma_Q = 1.5$ are applied to permanent and variable actions, respectively, and partial factors on material properties and resistances are 1.0. In slope stability software, this can be achieved by factoring the soils' weight densities by 1.35 and any applied surcharges by 1.5. A search is then made for the critical slip circle with a 'target' factor of safety of 1.0.

❷ In Design Approach 1 Combination 2 (DA1-2), partial factors $\gamma_G = 1.0$ and $\gamma_Q = 1.3$ are applied to permanent and variable actions, respectively; partial factors on resistances are 1.0; and partial factors on material properties are $\gamma_\varphi = \gamma_c = 1.25$. This can be achieved by factoring the soils' effective strengths down by 1.25 and any applied surcharges up by 1.3. A search is then made for the critical slip circle with a target factor of safety of 1.0.

❸ In Design Approach 2 (DA2), partial factors $\gamma_G = 1.35$ and $\gamma_Q = 1.5$ are applied to permanent and variable actions, respectively; partial factors on material properties are $\gamma_M = 1.0$; and on sliding resistance $\gamma_{Re} = 1.1$. If the 'single source' principle is applied, then the same value of γ_G is applied to both favourable and unfavourable actions. This can be achieved by factoring the soils' weight densities by 1.35 and any applied surcharges by 1.5. A search is then made for the critical slip circle with a target factor of safety of 1.1.

❹ In the variation of Design Approach 2 known as DA2*, favourable and unfavourable actions are meant to be treated separately, with $\gamma_{G,fav} = 1.0$ applied to the former and $\gamma_G = 1.35$ to the latter.[11] However, this cannot easily be achieved with existing software, so instead we apply $\gamma_{G,fav} = 1.0$ to *all* permanent actions and an 'intermediate' factor $\gamma_{Q/G} = \gamma_Q/\gamma_G = 1.5/1.35 = 1.11$ to any applied surcharges, and then search for the critical slip circle with a target factor of safety of $\gamma_R \times \gamma_G = 1.1 \times 1.35 = 1.485$.

Example 9.4. Road cutting (using slope stability software)

Input parameters	Design Approaches 1-1 and 2				Design Approaches 1-2† and 3‡			
	γ (kN/m³)	c′ (kPa)	φ (°)	q (kPa)	γ (kN/m³)	c′ (kPa)	φ (°)	q (kPa)
Partial factor	1.35 ❶❸	1.0 ❶❸	1.0 ❶❸	1.5 ❶❸/1.11 ❹	1.0 ❷	1.25 ❷	1.25 ❷	1.3†❷/1.0‡❺
Head deposits	24.3	0	35	-	18	0	29.27	-
Medium strength clay	27	5	23	-	20	4	18.76	-
Surcharge	-	-	-	15/11.1	-	-	-	13†/10‡

Results of analysis	Target FoS	FoS obtained	Utilization	Required φ for c′ = 5kPa	Required c′ for φ = 23°
DA 1-1	1.0 ❶	1.248	(80%)	(18°)	(0 kPa)
DA 1-2	1.0 ❷	0.996	100% ❼	23°	5 kPa
DA 2	1.1 ❸	1.243	88%	20°	1.7 kPa
DA2*	1.485 ❹	1.248	119%	27.9°	10.2 kPa
DA 3	1.0	1.001	100% ❼	23°	5 kPa
Traditional	1.3 ❻	1.251	104%	24°	6 kPa

Note that the results obtained from a Design Approach 2 analysis depend critically on how the partial factors are applied. Neither of the two methods described in ❸ and ❹ above is commensurate with traditional practice, which is to search for a required (material) factor of safety of 1.3. Owing to these uncertainties, Design Approach 2 is not recommended for slope stability assessments.

❺ The approach for Design Approach 3 (DA3) is identical to that for DA1-1, except that the imposed surcharge is treated as a geotechnical – rather than structural – action (and hence is less onerous).

❻ The required factor of safety for a traditional analysis is not prescribed, but a value of 1.3 is often considered acceptable for long-term analysis.

❼ Both DA1-2 and DA3 imply similar degrees of utilization to the traditional analysis. This gives confidence that the manner of applying the partial factors in DA1-2 and DA3 is sensible and likely to yield reliable designs.

Where the perceived level of risk cannot be accounted for by using the partial factors prescribed in EN 1997-1, greater conservatism would be required in the assessment of the characteristic values for the material properties.

9.8.5   Road embankment over an alluvial flood plain

Example 9.5 considers a road embankment that is going to be constructed over an alluvial flood plain, as shown in **Figure 9.14**, where ground conditions comprise 2m of soft, very low strength CLAY, overlying 3m of loose SAND, overlying firm, medium strength CLAY.

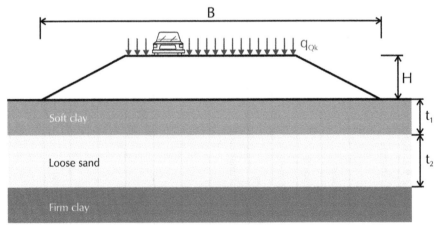

Figure 9.14. *Road embankment over alluvial flood plain*

The soils' characteristic parameters are given in the table below.

Soil	Characteristic parameters				
	γ_k (kN/m³)	c'_k (kPa)	φ_k (°)	c_{uk} (kPa)	q_k (kPa)
Fill	18	0	35	-	-
Soft clay	17	0	27	16	-
Loose sand	19	0	32	-	-
Firm clay	19	5	22	60	-
Surcharge	-	-	-	-	10

As before, this example uses Bishop's Simplified Method and highlights the issues of using software for a Eurocode 7 design. It also demonstrates the requirement for a bearing capacity check for embankments on soft clays.

The box below summarizes the analyses undertaken for this worked example.

Notes on Example 9.5

❶ The calculated factor of safety in the undrained case for DA1-1 is satisfactory as the calculated factor of safety is greater than the target factor.

❷ As for other slope stability examples, even in the short term DA1-2 gives a lower factor of safety than DA1-1 and is therefore the governing case. This example implies that the design does not meet the requirements of Eurocode 7.

❸ For DA2 the same factor of safety as for DA1-1 is calculated.

❹ The utilization factors for DA1-1 and DA2 are different, since for DA2 a target factor of safety of 1.1 is required. Thus DA2 would indicate that the design does not meet the requirements of Eurocode 7.

❺ The degree of utilization for the long-term case is less than for the short-term, which is as would be expected for an embankment on soft clay.

Eurocode 7 requires that embankments on soft clays are checked for bearing capacity. As for the stability analysis above, the undrained (short-term) situation should be the most critical. The thickness of the soft clay is relatively small and therefore a full bearing failure surface cannot develop.

[EN 1997-1 §12.4 (3) and (4)P]

For a simple verification of bearing resistance, an equivalent strip footing 15m wide due to an embankment 3.5m high will be considered with a 10kPa surcharge. A modified bearing capacity formula[12] developed for a shallow depth of soft clay overlying stiffer strata will be used:

$$q_{ult} = c'N_c^* + qN_q^* + \frac{\gamma B N_\gamma^*}{2}$$

where N_c^*, N_q^*, and N_γ^* are (modified) bearing capacity factors, c' is the soil's effective cohesion, q the overburden pressure at the base of the embankment, γ the weight density of the foundation soil, and B the width of the embankment.

In Design Approach 1, Combination 1, partial factors on actions: $\gamma_G = 1.35$ and $\gamma_Q = 1.5$ and on material properties: $\gamma_\varphi = \gamma_c = 1.0$ and $\gamma_{cu} = 1.0$.
  For B/H = 15/2 = 7.5, $N_c^* = 8.1$ ❻, and q = 0
  R_d = 16 x 8.1 = 129.6 kPa
  E_d = 3.5 x 18 x 1.35 + 10 x 1.5 = 100.1 kPa
  $R_d > E_d$ therefore OK, degree of utilization = 77% ❼

In Design Approach 1, Combination 2, partial factors on actions: $\gamma_G = 1.0$ and $\gamma_Q = 1.3$ and on material properties: $\gamma_\varphi = \gamma_c = 1.25$ and $\gamma_{cu} = 1.4$.
  For B/H = 15/2 = 7.5, $N_c^* = 8.1$ ❻, and q = 0
  R_d = 16/1.4 x 8.1 = 92.6 kPa
  E_d = 3.5 x 18 x 1.0 + 10 x 1.3 = 76 kPa
  $R_d > E_d$ therefore OK Degree of utilization = 82% ❼

❻ For the short-term case $\varphi = 0$ and therefore the N_q^* and N_γ^* terms do not need to be considered.

❼ In this example, the slope stability analysis based on DA1-2 or DA3 is critical having higher degrees of utilization than other design approaches or the bearing capacity assessment.

For a complete analysis of the problem a serviceability limit state set of calculations or assessments would need to be made to ensure that the total expected deformations were within tolerable levels.

Example 9.5. Road embankment over an alluvial flood plain

Input parameters	Design Approaches 1-1 and 2					Design Approaches 1-2† and 3‡				
	γ (kN/m³)	c' (kPa)	φ (°)	c_u (kPa)	q (kPa)	γ (kN/m³)	c' (kPa)	φ (°)	c_u (kPa)	q (kPa)
Partial factor	1.35	1.0	1.0	1.0	1.5	1.0	1.25	1.25	1.4	1.3†/1.0‡
Fill	24.3	0	35	-	-	18	0	29.3	-	-
Soft clay	23.0	0	27	16	-	17	0	22.2	11.4	-
Loose sand	25.7	0	32	-	-	19	0	26.6	-	-
Firm clay	25.7	5	22	60	-	19	4	17.9	42.9	-
Surcharge	-	-	-	-	15	-	-	-	-	13†/10‡

Results of analysis	Undrained analysis			Drained analysis		
	Target FoS	FoS obtained	Utilization	Target FoS	FoS obtained	Utilization
DA 1-1	1.0	1.074 ❶	(93%) ❹❼	1.0	1.428	(69%) ❺
DA 1-2	1.0	0.983 ❷	102% ❼	1.0	1.082	92% ❺
DA 2	1.1	1.074 ❸	102% ❹	1.1	1.428	77% ❺
DA 3	1.0	1.011	99%	1.0	1.092	92% ❺
Traditional	1.4	1.388	101%	1.3	1.365	95%

9.9 Notes and references

1. Examples taken from Bromhead, E. N. (1992) *The stability of slopes* (2nd edition), Glasgow, Blackie Academic & Professional, 411pp.

2. See, for example, Bromhead (ibid.) or Simons, N., Menzies, B., and Matthews, M. (2001) *A short course in soil and rock slope engineering*, Thomas Telford, 448pp.

3. Skempton, A. W., and Delory, F. A. (1957) 'Stability of natural slopes in London Clay', *Proc. 4th Int. Conf. on Soil Mechanics & Foundation Engng*, London, 2, pp. 378–381.

4. BS 6031: 1981, Code of practice for earthworks, British Standards Institution.

5. Bromhead, ibid.

6. Bishop, A. W. (1955) 'The use of the slip circle in the stability analysis of slopes', *Géotechnique*, 5, pp. 7–17.

7. Bishop, ibid.

8. See, for example: Janbu, N. (1973) *Slope stability computations*. In Hirschfield, E., and Poulos, S. (eds.) *Embankment dam engineering*, Casagrande Memorial Volume, New York: John Wiley; Morgenstern, N. R., and Price, V. E. (1965) 'The analysis of the stability of general slip surfaces', *Géotechnique*, 15, pp. 79-93; and Sarma, S. K. (1973) 'Stability analysis of embankments and slopes', *Géotechnique*, 23, pp. 423-433.

9. The analysis was performed using the computer program Slide v5.0, available from Rocscience (www.rocscience.com).

10. The analysis was performed using Rocscience's computer program Slide, ibid.

11. Frank, R., Bauduin, C., Kavvadas, M., Krebs Ovesen, N., Orr, T., and Schuppener, B. (2004), *Designers' guide to EN 1997-1: Eurocode 7: Geotechnical design – General rules*, London: Thomas Telford.

12. Mandel, J., and Salencon, J. (1972), 'Force portante d'un sol sur une assise rigide (étude theoretique)', *Géotechnique*, 22(1), pp. 79-93.

Chapter 10

Design of footings

The design of footings is covered by Section 6 of Eurocode 7 Part 1, 'Spread foundations', whose contents are as follows:
- §6.1   General (2 paragraphs)
- §6.2   Limit states (1)
- §6.3   Actions and design situations (3)
- §6.4   Design and construction considerations (6)
- §6.5   Ultimate limit state design (32)
- §6.6   Serviceability limit state design (30)
- §6.7   Foundations on rock; additional design considerations (3)
- §6.8   Structural design of foundations (6)
- §6.9   Preparation of the subsoil (2)

Section 6 of EN 1997-1 applies to pad, strip, and raft foundations and some provisions may be applied to deep foundations, such as caissons.

[EN 1997-1 §6.1(1)P and (2)]

10.1   Ground investigation for footings

Annex B.3 of Eurocode 7 Part 2 provides outline guidance on the depth of investigation points for spread foundations, as illustrated in **Figure 10.1**. (See Chapter 4 for guidance on the spacing of investigation points.)

The recommended minimum depth of investigation, z_a, for spread foundations supporting high-rise structures and civil engineering projects is the greater of:

$z_a \geq 3b_F$ and $z_a \geq 6m$

where b_F is the breadth of the foundation. For raft foundations:

$z_a \geq 1.5b_B$

where b_B is the breadth of the raft.

The depth z_a may be reduced to 2m if the

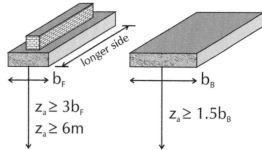

Figure 10.1. Recommended depth of investigation for spread foundations

foundation is built on competent strata[†] with 'distinct' (i.e. known) geology. With 'indistinct' geology, at least one borehole should go to at least 5m. If bedrock is encountered, it becomes the reference level for z_a.

[EN 1997-2 §B.3(4)]

Greater depths of investigation may be needed for very large or highly complex projects or where unfavourable geological conditions are encountered.   [EN 1997-2 §B.3(2)NOTE and B.3(3)]

10.2   Design situations and limit states

Figure 10.2 shows some of the ultimate limit states that spread foundations must be designed to withstand. From left to right, these include: (top) loss of stability owing to an applied moment, bearing failure, and sliding owing to an applied horizontal action; and (bottom) structural failure of the foundation base and combined failure in the structure and the ground.

Loss of stability   Bearing failure   Sliding

Structural failure   Combined failure in ground & structure

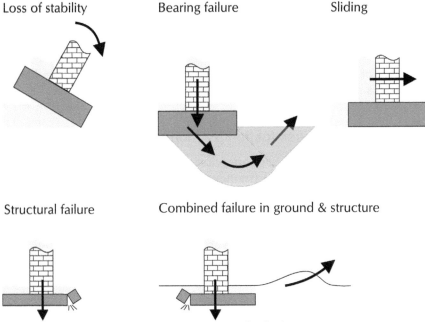

Figure 10.2. Examples of ultimate limit states for footings

[†]i.e. weaker strata are unlikely to occur at depth, structural weaknesses such as faults are absent, and solution features and other voids are not expected.

Eurocode 7 lists a number of things that must be considered when choosing the depth of a spread foundation, some of which are illustrated in **Figure 10.3**.   *[EN 1997-1 §6.4(1)P]*

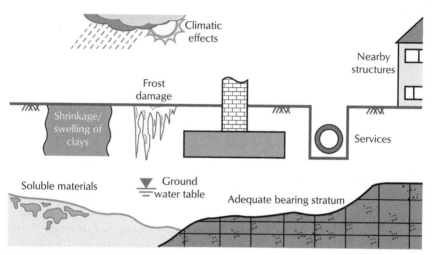

Figure 10.3. Design considerations for footings

10.3   Basis of design

Eurocode 7 requires spread foundations to be designed using one of the following methods: *[EN 1997-1 §6.4(5)P]*

Method	Description	Constraints
Direct	Carry out separate analyses for each limit state, both ultimate (ULS) and serviceability (SLS)	(ULS) Model envisaged failure mechanism
		(SLS) Use a serviceability calculation
Indirect	Use comparable experience with results of field and laboratory measurements and observations	Choose SLS loads to satisfy requirements of all limit states
Prescriptive	Use conventional and conservative design rules and specify control of construction	Use presumed bearing resistance

The *indirect method* is used predominantly for Geotechnical Category 1 structures, where there is good local experience, ground conditions are well known and uncomplicated, and the risks associated with potential failure or excessive deformation of the structure are low. Indirect methods may also be applied to higher risk structures where it is difficult to predict the structural behaviour with sufficient accuracy from analytical solutions. In these cases, reliance is placed on the observational method and identification of a range of potential behaviour. Depending on the observed behaviour, the final design of the foundation can be decided. This approach ensures that the serviceability condition is met but does not explicitly provide sufficient reserve against ultimate conditions. It is therefore important that the limiting design criteria for serviceability are suitably conservative.

The *prescriptive method* may be used for Geotechnical Category 1 structures, where ground conditions are well known. Unlike British standard BS 8004 – which gives allowable bearing pressures for rocks, non-cohesive soils, cohesive soils, peat and organic soils, made ground, fill, high porosity chalk, and Keuper Marl (now called the Mercia Mudstone)[1] – Eurocode 7 only provides values of presumed bearing resistance for rock (via a series of charts[†] in Annex G).

The *direct method* is discussed in some detail in the remainder of this chapter.

This book does not attempt to provide complete guidance on the design of spread foundations, for which the reader should refer to any well-established text on the subject.[2]

10.4   Footings subject to vertical actions

For a spread foundation subject to vertical actions, Eurocode 7 requires the design vertical action V_d acting on the foundation to be less than or equal to the design bearing resistance R_d of the ground beneath it:

$$V_d \leq R_d \qquad\qquad \text{[EN 1997-1 exp (6.1)]}$$

V_d should include the self-weight of the foundation and any backfill on it.

This equation is merely a re-statement of the inequality:

$$E_d \leq R_d$$

discussed at length in Chapter 6. Rather than work in terms of forces, engineers more commonly consider pressures and stresses, so we will rewrite this equation as:

[†]which also appear in BS 8004.

$$q_{Ed} \leq q_{Rd}$$

where q_{Ed} is the design bearing pressure on the ground (an action effect), and q_{Rd} is the corresponding design resistance.

Figure 10.4 shows a footing carrying characteristic vertical actions V_{Gk} (permanent) and V_{Qk} (variable) imposed on it by the super-structure. The characteristic self-weights of the footing and of the backfill upon it are both permanent actions (W_{Gk}). The following sub-sections explain how q_{Ed} and q_{Rd} are obtained from V_{Gk}, V_{Qk}, W_{Gk}, and ground properties.

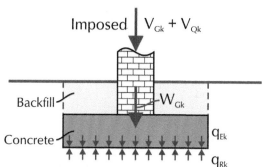

Figure 10.4. Vertical actions on a spread foundation

10.4.1 Effects of actions

The characteristic bearing pressure q_{Ek} shown in **Figure 10.4** is given by:

$$q_{Ek} = \frac{\sum V_{rep}}{A'} = \frac{(V_{Gk} + \sum_i \psi_i V_{Qk,i}) + W_{Gk}}{A'}$$

where V_{rep} is a representative vertical action; V_{Gk}, V_{Qk}, and W_{Gk} are as defined above; A' is the footing's effective area (defined in Section 10.4.2); and ψ_i is the combination factor applicable to the ith variable action (see Chapter 2).

If we assume that only one variable action is applied to the footing, this equation simplifies to:

$$q_{Ek} = \frac{(V_{Gk} + V_{Qk,1}) + W_{Gk}}{A'}$$

since $\psi = 1.0$ for the leading variable action ($i = 1$).

The design bearing pressure q_{Ed} beneath the footing is then:

$$q_{Ed} = \frac{\sum V_d}{A'} = \frac{\gamma_G(V_{Gk} + W_{Gk}) + \gamma_Q V_{Qk,1}}{A'}$$

where γ_G and γ_Q are partial factors on permanent and variable actions, respectively.

10.4.2   Eccentric loading and effective foundation area

The ability of a spread foundation to carry forces reduces dramatically when those forces are applied eccentrically from the centre of the foundation.

To prevent contact with the ground being lost at the footing's edges, it is customary to keep the total action within the foundation's 'middle-third'. In other words, the eccentricity of the action from the centre of the footing is kept within the following limits:

$$e_B \le \frac{B}{6} \text{ and } e_L \le \frac{L}{6}$$

where B and L are the footing's breadth and length, respectively; and e_B and e_L are eccentricities in the direction of B and L (see **Figure 10.5**).

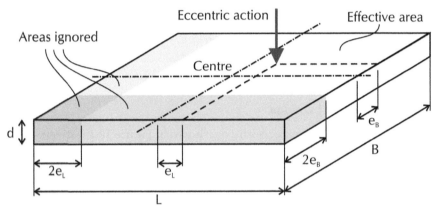

Figure 10.5. *Effective area of spread foundation*

Eurocode 7 Part 1 requires 'special precautions' to be taken where:

> *... the eccentricity of loading exceeds 1/3 of the width of a rectangular footing or [60%] of the radius of a circular footing.*   *[EN 1997-1 §6.5.4(1)P]*

Note that this is *not* the middle-third rule, but rather a 'middle-two-thirds' rule. We recommend that foundations continue to be designed using the middle-third rule until the implications of Eurocode 7's more relaxed Principle have been thoroughly tested in practice.

Bearing capacity calculations take account of eccentric loading by assuming that the load acts at the centre of a smaller foundation, as shown in **Figure 10.5**. The shaded parts of the foundation are therefore ignored. The actual foundation area is therefore reduced to an 'effective area' A', which can be calculated from:[3]

$$A' = B' \times L' = (B - 2e_B) \times (L - 2e_L)$$

where B' and L' are the footing's effective breadth and length, respectively; and the other symbols are as defined above.

10.4.3   Drained bearing resistance

The drained ultimate bearing capacity of a spread foundation q_{ult} has traditionally been calculated from the so-called 'triple-N' formula, which in its original form[4] is given by:

$$q_{ult} = c'N_c + q'N_q + \frac{\gamma'BN_\gamma}{2}$$

where c' is the soil's effective cohesion; q' the effective overburden pressure at the foundation base; γ' the effective weight density of the soil below the foundation; and N_c, N_q and N_γ are bearing capacity factors.

The overburden and cohesion factors N_q and N_c were established in the 1920s by Reissner[5] and Prandtl,[6] respectively, in terms of the soil's angle of shearing resistance φ:

$$N_q = e^{\pi \tan\varphi} \tan^2\left(45° + \frac{\varphi}{2}\right)$$

$$N_c = (N_q - 1)\cot\varphi$$

and these equations are used almost universally in geotechnical practice. However, there is no consensus regarding the value of the factor N_γ.

Design practice in many parts of Europe[7] has traditionally used Brinch-Hansen's[8] equation for N_γ:

$$N_\gamma = 1.5(N_q - 1)\tan\varphi$$

while in America designers typically employ Meyerhof's[9] equation:

$$N_\gamma = (N_q - 1)\tan(1.4\varphi)$$

and offshore structures engineers[10] use Vesic's[11] equation:

$$N_\gamma = 2(N_q + 1)\tan\varphi$$

which recent research[12] suggests may over-predict N_γ. Chen's[13] equation:

$$N_\gamma = 2(N_q - 1)\tan\varphi$$

is also popular and appears in Eurocode 7 Annex D. Note that Chen's equation assumes a rough base with interface friction ≥ 0.5 times the soil's angle of shearing resistance.

Values of these bearing capacity factors for different angles of shearing resistance are illustrated in **Figure 10.6**. The curves for Meyerhof's and Brinch-Hansen's N_γ are virtually co-incident for φ < 30° and diverge only marginally as φ approaches 60°. Chen's formulation for N_γ is slightly more conservative than Vesic's but significantly more optimistic than Brinch-Hansen's, particularly at large angles of shearing resistance.

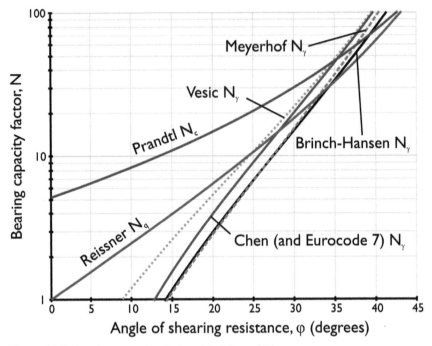

Figure 10.6. *Bearing capacity factors N_q, N_c, and N_γ*

10.4.4   Non-dimensional factors applied to the triple-N formula

Various modifications to the triple-N equation have been proposed, usually involving the introduction of factors to cater for the effects of the foundation's shape, depth, and, base inclination; inclination of the applied load; and inclination of the ground surface.

The 'complete' expression for q_{ult} is:

$$q_{ult} = c'N_c s_c d_c i_c g_c b_c + q'N_q s_q d_q i_q g_q b_q + \frac{\gamma'BN_\gamma s_\gamma d_\gamma i_\gamma g_\gamma b_\gamma}{2}$$

where s_c, s_q, and s_γ are shape factors; d_c, d_q, and d_γ are depth factors; i_c, i_q, and i_γ are load inclination factors; g_c, g_q, and g_γ are ground inclination factors; and b_c, b_q, and b_γ are base inclination factors.

In Europe, this equation is credited to Brinch-Hansen,[14] but in America more usually to Meyerhof[15] (who uses just the shape, depth, and load inclination factors).

Annex D of EN 1997-1 gives an equation for the drained bearing resistance of a spread foundation that omits the depth and ground inclination factors, which are commonly found in bearing capacity formulations. The omission of the former is uneconomic, but neglecting the latter is unsafe. The UK's National Annex emphasizes this point and suggests that an alternative method, including depth and ground inclination factors, may be used.

The equations given in Eurocode 7 Annex D for the shape factors s_c, s_q, and s_γ are summarized in the table below. The equations for s_c and s_q are the ones recommended by Brinch-Hansen and Vesic.

Factor		Cohesion c	Overburden q	Body-weight γ
Shape	s_x	$1+\dfrac{N_q}{N_c}\dfrac{B}{L}$	$1+\dfrac{B}{L}\sin\varphi$	$1-k\dfrac{B}{L}\dagger$
Depth	d_x	No equations given in Eurocode 7		
Load inclination	i_x	Refer to EN 1997-1 Annex D for details		
Ground inclination	g_x	No equations given in Eurocode 7		
Base inclination	b_x	Refer to EN 1997-1 Annex D for details		

$\dagger k = 0.3$ in Eurocode 7 (applies for $B/H \leq 1$); Brinch-Hansen and Vesic recommend $k = 0.4$; a survey[16] of European practice showed five countries using $k = 0.4$, three using $k = 0.3$; and one using $k = 0.2$

10.4.5   Undrained bearing resistance

Annex D of EN 1997-1 gives the following equation for the undrained bearing resistance R of a spread foundation:

$$\frac{R}{A'} = (\pi + 2)c_u b_c s_c i_c + q$$

where c_u is the soil's undrained shear strength; q the total overburden pressure at the foundation base; and the other symbols are as defined for the drained equation (see Section 10.4.3). The value $(\pi + 2)$ is obtained by setting

$\varphi = 0$ in Prandtl's[17] expression for N_c (see Section 10.4.3). The shape factor s_c is given by:[18]

$$s_c = 1 + 0.2\frac{B}{L}$$

As with the drained equation discussed in Section 10.4.3, R/A' omits the depth factor d_c and ground inclination factor g_c. The omission of the former is uneconomic, but neglecting the latter is unsafe. The full equation is:

$$\frac{R}{A'} = (\pi + 2)c_u b_c s_c i_c d_c g_c + q$$

Recent finite element limit analysis[19] of the bearing capacity of foundations in clay has found that the following expression provides a good approximation for the depth factor d_c for clays:

$$d_c = 1 + 0.27\sqrt{\frac{D}{B}}$$

where B and D are the foundation's breadth and depth respectively. This is a better fit than Meyerhof's and Brinch-Hansen's equations for d_c. For consistency, this improved depth factor should be used with the following expression for the shape factor s_c (for $D/B \le 1$):

$$s_c = 1 + 0.12\frac{B}{L} + 0.17\sqrt{\frac{D}{B}}$$

where L is the foundation's length and B and D are as defined above. Note that this equation implies that s_c is not equal to 1.0 even for a strip footing.

10.4.6   Gross or net bearing resistance?

In traditional calculations, the allowable bearing capacity[†] q_a is normally written in terms of net pressures as:

$$q_{a,net} = \left(\frac{q_{ult,net}}{F}\right) \Rightarrow q_a = \left(\frac{q_{ult} - q_0}{F}\right) + q_0$$

where $q_{a,net}$ = net allowable, $q_{ult,net}$ = net ultimate, q_a = gross allowable, and q_{ult} = gross ultimate bearing capacities; q_0 = overburden pressure; and F = a factor of safety.

The question arises, when using the Design Approach 2, whether the resistance factor γ_{Rv} should be applied to the ground's gross vertical

---

[†]Sometimes, on soft soil sites, large settlements may occur under loaded foundations without actual shear failure occurring; in such cases, the allowable bearing capacity is based on the maximum allowable settlement.

resistance R_v or to its nett resistance $R_{v,net}$ – in other words, should the traditional practice of factoring net resistance be followed?

If net resistance is factored in Design Approach 2, the design bearing resistance q_{Rd} is given by:

$$q_{Rd} = \left(\frac{R_{v,k}/A' - \sigma_v}{\gamma_{Rv}} \right) + \sigma_v$$

where σ_v is the total overburden at the base of the foundation and A' is the effective area of the footing. If gross resistance is factored, the design bearing resistance is given by:

$$q_{Rd} = \frac{R_{v,d}}{A'} = \frac{1}{A'} \left(\frac{R_{v,k}}{\gamma_{Rv}} \right)$$

Eurocode 7 is silent on this issue (which only applies to Design Approach 2, since in the other Design Approaches $\gamma_{Rv} = 1.0$). To be consistent with calculations for other geotechnical structures, we recommend that the resistance factor be applied to gross rather than to net resistance.

10.5 Footings subject to horizontal actions

For a spread foundation subject to horizontal actions, Eurocode 7 requires the design horizontal action H_d acting on the foundation to be less than or equal to the sum of the design resistance R_d from the ground beneath the footing and any design passive thrust R_{pd} on the side of the foundation:

$$H_d \leq R_d + R_{pd} \qquad \qquad \textit{[EN 1997-1 exp (6.21)]}$$

which is merely a re-statement of the inequality:

$$E_d \leq R_d$$

discussed at length in Chapter 6. Rather than work in terms of forces, engineers prefer to use shear stresses, so we will re-write this inequality as:

$$\tau_{Ed} \leq \tau_{Rd}$$

where τ_{Ed} is the design shear stress acting across the base of the footing (an action effect) and τ_{Rd} is the design resistance to that shear stress.

Figure 10.7 shows the footing from **Figure 10.4** subject to characteristic horizontal actions H_{Gk} (permanent) and H_{Qk} (variable), in addition to characteristic vertical actions V_{Gk} (permanent), V_{Qk} (variable), and W_{Gk} (permanent).

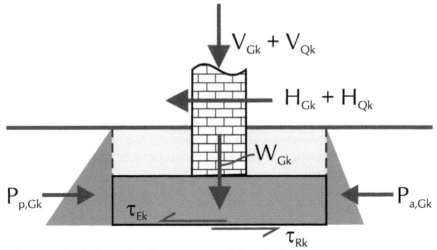

Figure 10.7. *Horizontal actions on a spread foundation*

10.5.1   Effects of actions

The characteristic shear stress τ_{Ek} shown in **Figure 10.7** is given by:

$$\tau_{Ek} = \frac{\sum H_{rep}}{A'} = \frac{(H_{Gk} + \sum_i \psi_i H_{Qk,i}) + P_{a,Gk}}{A'}$$

where H_{rep} is a representative horizontal action; H_{Gk} and H_{Qk} are defined above; $P_{a,Gk}$ is the characteristic thrust due to active earth pressures on the side of the footing (a permanent action), A' is the footing's effective area (defined in Section 10.4.2), and ψ_i is the combination factor applicable to the ith variable action.

If we assume that only one variable horizontal action is applied to the footing, this equation simplifies to:

$$\tau_{Ek} = \frac{(H_{Gk} + H_{Qk,1}) + P_{a,Gk}}{A'}$$

since $\psi = 1.0$ for the leading variable action (i = 1).

The design shear stress is then:

$$\tau_{Ed} = \frac{\sum H_d}{A'} = \frac{\gamma_G(H_{Gk} + P_{a,Gk}) + \gamma_Q H_{Qk,1}}{A'}$$

where γ_G and γ_Q are partial factors on permanent and variable actions, respectively.

10.5.2   Drained sliding resistance

Under drained conditions, the characteristic shear resistance τ_{Rk} shown in **Figure 10.7** (ignoring passive pressures for the time being) is given by:

$$\tau_{Rk} = \frac{V'_{Gk} \tan \delta_k}{A'} = \frac{(V_{Gk} - U_{Gk}) \tan \delta_k}{A'}$$

where V_{Gk} and V'_{Gk} represent the characteristic total and effective permanent vertical actions on the footing, respectively; U_{Gk} is the characteristic uplift owing to pore water pressures acting on the underside of the base (also a permanent action); δ_k is the characteristic angle of interface friction between the base and the ground; and the other symbols are as defined above. Variable actions have been excluded from this equation, since they are favourable.

This expression conservatively ignores any effective adhesion between the footing and the ground, as suggested by Eurocode 7.   *[EN 1997-1 §6.5.3(10)]*

The design shear resistance τ_{Rd} (ignoring passive pressures) is then given by:

$$\tau_{Rd} = \frac{V'_{Gd} \tan \delta_d}{\gamma_{Rh} A'} = \frac{(V_{Gd} - U_{Gd}) \tan \delta_d}{\gamma_{Rh} A'}$$

where γ_{Rh} is a partial factor on horizontal sliding resistance and the subscripts 'd' denote design values.

The vertical action V_{Gd} is *favourable*, since an increase in its value increases the shear resistance; whereas U_{Gd} is *unfavourable* action, since an increase in its value decreases the resistance. Introducing into this equation partial factors on favourable and unfavourable permanent actions ($\gamma_{G,fav}$ and γ_G) results in:

$$\tau_{Rd} = \frac{(\gamma_{G,fav} V_{Gk} - \gamma_G U_{Gk}) \tan \delta_k}{\gamma_{Rh} \gamma_\varphi A'} = \left[\left(\frac{\gamma_{G,fav}}{\gamma_{Rh} \times \gamma_\varphi} \right) V_{Gk} - \left(\frac{\gamma_G}{\gamma_{Rh} \times \gamma_\varphi} \right) U_{Gk} \right] \times \left(\frac{\tan \delta_k}{A'} \right)$$

where γ_φ is the partial factor on shearing resistance.

If, however, partial factors are applied to the effects of actions rather than to the actions themselves, then the previous equation becomes:

$$\tau_{Rd} = \frac{\gamma_{G,fav} (V_{Gk} - U_{Gk}) \tan \delta_k}{\gamma_{Rh} \gamma_\varphi A'} = \left(\frac{\gamma_{G,fav}}{\gamma_{Rh} \times \gamma_\varphi} \right) (V_{Gk} - U_{Gk}) \times \left(\frac{\tan \delta_k}{A'} \right)$$

where γ_φ is the partial factor on shearing resistance.

The table below summarizes the values of these partial factors for each of Eurocode 7's three Design Approaches (see Chapter 6).

Individual partial factor or partial factor 'grouping'	Design Approach			
	1		2	3
	Combination 1	Combination 2		
γ_G	1.35	1.0	1.35	1.35/1.0*
$\gamma_{G,fav}$	1.0	1.0	1.0	1.0
γ_φ	1.0	1.25	1.0	1.25
γ_{cu}	1.0	1.4	1.0	1.4
γ_{Rh}	1.0	1.0	1.1	1.0
$\gamma_{G,fav} / (\gamma_{Rh} \times \gamma_\varphi)$	1.0	0.8	0.91	0.8
$\gamma_G / (\gamma_{Rh} \times \gamma_\varphi)$	1.35	0.8	1.23	1.08/0.8
$1 / (\gamma_{Rh} \times \gamma_{cu})$	1.0	0.71	0.91	0.71

*Factor from Set A2 on geotechnical actions

Eurocode 7 allows δ_d to be determined from the following equation:

$$\delta_d = k\varphi_{cv,d} = k \tan^{-1}\left(\frac{\tan \varphi_{cv,k}}{\gamma_\varphi}\right)$$

where $\varphi_{cv,d}$ is the ground's design *constant-volume* (aka 'critical state') angle of shearing resistance; γ_φ is defined above; and $k = 1$ for concrete cast in situ or $k = \frac{2}{3}$ for precast concrete. Values of $\varphi_{cv,d}$ are rarely measured, but are usually assessed from simple rules-of-thumb.[20]

Figure 10.8 illustrates key differences between a soil's characteristic *peak* angle of shearing resistance $\varphi_{p,k}$ and its constant-volume counterpart $\varphi_{cv,k}$. The former is inherently more variable than the latter and is only significantly larger in dense (dilatant) soils while strains in the soil are small. By applying a partial factor $\gamma_\varphi = 1.25$ to the cautious estimate of $\tan \varphi_{p,k}$, Eurocode 7 ensures that the calculation of sliding resistance is sufficiently reliable for design. However, applying the same partial factor $\gamma_\varphi = 1.25$ to a cautious estimate of $\tan \varphi_{cv,k}$ may be overly conservative.

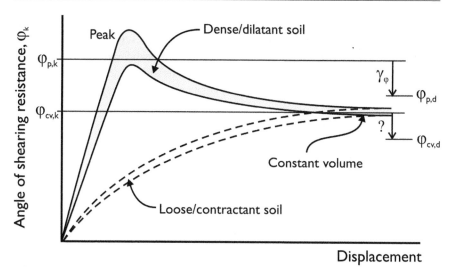

Figure 10.8. Variability of peak and constant volume angles of shearing resistance

According to the theory of critical state soil mechanics,[21] φ_{cv} represents the smallest angle of shearing resistance mobilized at large displacement (assuming residual shear surfaces do not form in the soil). Hence the UK National Annex to BS EN 1997-1 notes that:

'it might be more appropriate to select the design value of φ'_{cv} directly.'

One way of determining $\varphi_{cv,d}$ would be to replace the partial factor γ_φ in the previous equation with a partial factor $\gamma_{\varphi,cv} < \gamma_\varphi$, i.e.:

$$\delta_d = k\varphi_{cv,d} = k\tan^{-1}\left(\frac{\tan\varphi_{cv,k}}{\gamma_{\varphi,cv}}\right)$$

where $\gamma_{\varphi,cv}$ might be as low as 1.0 (depending on how cautious your selection of φ_{cv} is). If this approach is taken, then an additional check that $\varphi_{cv,d}$ is smaller than φ_d should also be made.

It is often the case that a portion of the horizontal load is derived from an inclined load on the footing. In this case, both the horizontal and vertical components of load come from the same source. However, the vertical component is favourable as far as an ultimate limit state of sliding is concerned, whereas the horizontal component is unfavourable. Except in Design Approach 1 Combination 2 (see Chapter 6), a favourable permanent action attracts a partial factor $\gamma_{G,fav} = 1.0$ and an unfavourable permanent action a partial factor $\gamma_G = 1.35$. If the components of the inclined action are treated differently, then the inclination of that action will change. This is a prime example of where the Single Source Principle discussed in Chapter 3 should be invoked and the whole action should be treated either as unfavourable or as favourable (whichever is more onerous).

10.5.3   Undrained sliding resistance

Under undrained conditions, the characteristic shear resistance τ_{Rk} shown in **Figure 10.7** (ignoring passive pressures) is given by:

$$\tau_{Rk} = c_{uk}$$

where c_{uk} represents the characteristic undrained shear strength of the soil.

The design shear resistance τ_{Rd} (ignoring passive pressures) is then given by:

$$\tau_{Rd} = \frac{c_{ud}}{\gamma_{Rh}} = \frac{c_{uk}}{\gamma_{cu} \times \gamma_{Rh}}$$

where γ_{Rh} is the partial factor on horizontal sliding resistance and γ_{cu} is the partial factor on undrained shear strength.

The table on page 318 summarizes the values of these partial factors for each of Eurocode 7's three Design Approaches (see Chapter 6).

10.5.4   Passive earth pressure – favourable action or resistance?

The characteristic shear resistance τ_{Rk} shown in **Figure 10.7** is given by:

$$\tau_{Rk} = \frac{R_{hk} + P_{p,Gk}}{A'}$$

where R_{hk} represents the characteristic horizontal resistance over the base of the footing (i.e. the interface between the footing and the ground); $P_{p,Gk}$ represents the characteristic thrust due to passive earth pressures which helps to restrain the footing; and A' is the footing's effective area (defined in Section 10.4.2).

The design shear resistance τ_{Rd} is then either (in Design Approaches 1 and 3):

$$\tau_{Rd} = \frac{1}{A'}\left(R_{hd} + \gamma_{G,fav}P_{p,Gk}\right)$$

or (in Design Approach 2):

$$\tau_{Rd} = \frac{1}{A'}\left(\frac{R_{hk}}{\gamma_{Rh}} + \gamma_{G,fav}P_{p,Gk}\right)$$

where γ_{Rh} and $\gamma_{G,fav}$ are partial factors on horizontal resistance and *favourable* permanent actions, respectively.

In the first expression, partial factors are applied to actions and material properties, while in the second expression they are applied to effects of actions and resistance (see Chapter 6).

In both these equations, the passive earth pressure $P_{p,Gk}$ has been treated as a favourable action (and hence multiplied by $\gamma_{G,fav} = 1.0$). If it had been treated as a resistance, the equations would have been written:

$$\tau_{Rd} = \frac{1}{A'}\left(R_{hd} + P_{p,Gd}\right)$$

and:

$$\tau_{Rd} = \frac{1}{A'}\left(\frac{R_{hk} + P_{p,Gk}}{\gamma_{Rh}}\right)$$

Eurocode 7 does not explicitly state which of these assumptions should be adopted (although the phrase 'passive resistance' suggests the latter). In practice, the thrusts due to active and passive earth pressures (P_{ak} and P_{pk}) are often ignored – a simplification which errs on the side of safety, since $P_{p,Gk} > P_{a,Gk}$ in most situations.

10.6   Design for serviceability

As discussed in Chapter 8, Eurocode 7 requires the design movements E_d of a foundation to be less than or equal to the limiting movement C_d specified for the project:

$$E_d \leq C_d$$

The components of settlement that should be considered are:
- immediate settlement (s_0) due to shear at constant volume in saturated soils (or with volume reduction in partially-saturated soils)
- settlement caused by consolidation (s_1)
- settlement caused by creep (s_2)   *[EN 1997-1 §6.6.2(2)]*

Hence, the previous inequality can be re-written for footings as follows:

$$s_{Ed} = s_0 + s_1 + s_2 \leq s_{Cd}$$

where s_{Ed} is the total settlement (an action effect) and s_{Cd} is the limiting value of that settlement.

In verifications of serviceability limit states (SLSs), the combination factors ψ applied to accompanying variable actions are those specified for the characteristic, frequent, or quasi-permanent combinations (see Chapter 2), i.e. $\psi = \psi_2$.

In ultimate limit state (ULS) verifications, combinations of actions for permanent and transient design situations employ combination factors $\psi = \psi_0$. Since ψ_0 is numerically greater than ψ_2 for most actions, representative actions are usually larger for ultimate than for serviceability limit states.

Partial factors for serviceability limit states are normally taken as 1.0.

[EN 1997-1 §2.4.8(2)]

Eurocode 7 states that calculations of settlement must always be carried out for footings on soft clays – and should be carried out for footings on firm to stiff clays[†] when the risk is anything other than negligible (i.e. the footing is not in Geotechnical Category 1).   *[EN 1997-1 §6.6.1(3)P and (4)]*

Calculations must consider settlement of the entire foundation and differential movement between parts of the foundation must include both immediate and delayed components.   *[EN 1997-1 §6.6.1(7)P and 6.6.2(1)P]*

Annex F of EN 1997-1 presents two methods to evaluate settlement.

In the stress-strain method, the total settlement of a foundation may be evaluated by, first, computing the stress distribution in the ground due to the foundation loading (using elasticity theory for homogeneous, isotropic soil); second, computing the strain in the ground from those stresses using an appropriate stress-strain model (and appropriate stiffness); and, finally, integrating the vertical strains to find the settlements.   *[EN 1997-1 §F.1(1)]*

In the adjusted elasticity method, the foundation's total settlement 's' may be evaluated using elasticity theory using the following equation:

$$s = \frac{p \times b \times f}{E_m}$$

where p is the bearing pressure (linearly distributed) on the base of the foundation; b is the foundation's width; f is a settlement coefficient; and E_m is the design value of the modulus of elasticity.   *[EN 1997-1 §F.2(1)]*

Other methods for calculating settlement (from in situ tests) are given in the Annexes to EN 1997-2 (see the list in Chapter 4). Eurocode 7 emphasizes the fact that settlement calculations 'should not be regarded as accurate. They merely provide an approximate indication'.   *[EN 1997-1 §6.6.1(6)]*

[†]Strictly, 'soft' clay refers to 'low strength' clay, 'firm' to 'medium strength' and 'stiff' to 'high strength' – see Chapter 4.

Where the depth of compressible layers is large, it is normal to limit the analysis to depths where the increase in effective vertical stress is greater than 20% of the in situ effective stress.   *[EN 1997-1 §§6.6.2(6)]*

As discussed in Chapter 8, for conventional structures founded on clays, Eurocode 7 requires settlements to be calculated explicitly when the ratio of the characteristic bearing resistance R_k to the applied serviceability loads E_k is less than three. If this ratio is less than two, those calculations should take account of the ground's non-linear stiffness.   *[EN 1997-1 §6.6.2(16)]*

The serviceability limit state may be deemed to have been verified if:

$$E_k \leq \frac{R_k}{\gamma_{R,SLS}}$$

where E_k = characteristic effects of actions, R_k = characteristic resistance to those actions, and $\gamma_{R,SLS}$ = a partial resistance factor ≥ 3.

10.7   Structural design

EN 1997-1 provides a short section on the structural design of spread foundations. It provides no guidance on the procedures for assessing the required amount or detailing of reinforcement in the concrete – this is dealt with by Eurocode 2.[22]

For stiff footings, Eurocode 7 recommends a linear distribution of ground stresses may be used to calculate bending moments and shear stresses in the foundation. For flexible rafts and strip foundations, an analysis based on a deforming continuum or an equivalent spring model is recommended.

When soil structure interaction is significant, numerical methods are likely to be required to assess total and differential settlements.

10.8   Supervision, monitoring, and maintenance

Apart from two paragraphs on the preparation of the ground, §6 of EN 1997-1 provides no additional rules regarding supervision, monitoring, and maintenance to those given in §4.

§4 of EN 1997-1 requires the construction processes and workmanship to be supervised, the structure's performance to be monitored both during and after construction, and the structure to be maintained. As discussed in Chapter 16, these requirements must be stated in the Geotechnical Design Report (GDR) so that responsibilities are clearly articulated and the Client is informed about what to do if monitoring indicates that the structure is not performing adequately. The aims are to ensure the structure is adequately constructed and will perform within the project's acceptance criteria.

10.9   Summary of key points

The design of footings to Eurocode 7 involves checking that the ground has sufficient bearing resistance to withstand vertical actions, sufficient sliding resistance to withstand horizontal and inclined actions, and sufficient stiffness to prevent unacceptable settlement. The first two of these guard against ultimate limit states and the last against a serviceability limit state.

Verification of ultimate limit states is demonstrated by satisfying the inequalities:

$$V_d \leq R_d \text{ and } H_d \leq R_d + R_{pd}$$

(where the symbols are defined in Section 10.3). These equations are merely specific forms of:

$$E_d \leq R_d$$

which is discussed at length in Chapter 6.

Verification of serviceability limit states (SLSs) is demonstrated by satisfying the inequality:

$$s_{Ed} = s_0 + s_1 + s_2 \leq s_{Cd}$$

(where the symbols are defined in Section 10.6). This equation is merely a specific form of:

$$E_d \leq C_d$$

which is discussed at length in Chapter 8. Alternatively, SLSs may be verified by satisfying:

$$E_k \leq \frac{R_k}{\gamma_{R,SLS}}$$

where the partial factor $\gamma_{R,SLS} \geq 3$.

10.10   Worked examples

The worked examples in this chapter consider the design of a pad footing on dry sand (Example 10.1); the same footing but eccentrically loaded (Example 10.2); a strip footing on clay (Example 10.3); and, for the same footing, verification of the serviceability limit state (Example 10.4).

Specific parts of the calculations are marked ❶, ❷, ❸, etc., where the numbers refer to the notes that accompany each example.

10.10.1   Pad footing on dry sand

Example 10.1 considers the design of a simple rectangular spread footing on dry sand, as shown in **Figure 10.9**. It adopts the calculation method given in Annex D of EN 1997-1.

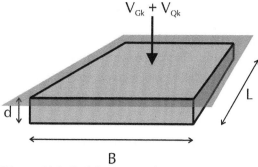

$V_{Gk} + V_{Qk}$

L

d

B

Figure 10.9. Pad footing on dry sand

In this example it is assumed that ground surface is at the top of the footing, i.e. the base of the footing is 0.5m below ground level.

The loading is applied centrally to the footing and therefore eccentricity can be ignored. Ground water is also not considered. The example concentrates on the application of the partial factors under the simplest of conditions. In reality, the assessment of a footing would need to consider a number of other situations before a design may be finalized.

Notes on Example 10.1

❶ In order to concentrate on the EC7 rather than the geotechnical related issues a relatively simple problem has been selected which excludes the effects of groundwater.

❷ The formulas for bearing capacity factors and shape factors are those given in Annex D. Other formulas could be used where they are thought to give a better theoretical/practical model for the design situation being considered.

❸ The suggested method in Annex D does not include depth factors which are present in other formulations of the extended bearing capacity formula (e.g. Brinch-Hansen or Vesic). There has been concern in using these depth factors as their influence can be significant and the reliance on the additional capacity provided by its inclusion is not conservative.

❹ For Design Approach 1, DA1-2 is critical with a utilization factor of 97% implying that the requirements of the code are only just met.

❺ For Design Approach 2 the uncertainty in the calculation is covered through partial factors on the actions and an overall factor on the calculated resistance.

Example 10.1
Pad footing on dry sand
Verification of strength (limit state GEO)

Design situation

Consider a rectangular pad footing of length L = 2.5m, breadth B = 1.5m, and depth d = 0.5m, which is required to carry an imposed permanent action V_{Gk} = 800kN and an imposed variable action V_{Qk} = 450kN, both of which are applied at the centre of the foundation. The footing is founded on dry sand ➊ with characteristic angle of shearing resistance φ_k = 35°, effective cohesion c'_k = 0kPa, and weight density $\gamma_k = 18\dfrac{kN}{m^3}$. The weight density of the reinforced concrete is $\gamma_{ck} = 25\dfrac{kN}{m^3}$ (as per EN 1991-1-1 Table A.1).

Design Approach 1

Actions and effects

Characteristic self-weight of footing is $W_{Gk} = \gamma_{ck} \times L \times B \times d = 46.9\,kN$

Partial factors from Sets $\begin{pmatrix} A1 \\ A2 \end{pmatrix}$: $\gamma_G = \begin{pmatrix} 1.35 \\ 1 \end{pmatrix}$ and $\gamma_Q = \begin{pmatrix} 1.5 \\ 1.3 \end{pmatrix}$

Design vertical action: $V_d = \gamma_G \times \left(W_{Gk} + V_{Gk}\right) + \gamma_Q \times V_{Qk} = \begin{pmatrix} 1818.3 \\ 1431.9 \end{pmatrix} kN$

Area of base: $A_b = L \times B = 3.75\,m^2$

Design bearing pressure: $q_{Ed} = \dfrac{V_d}{A_b} = \begin{pmatrix} 484.9 \\ 381.8 \end{pmatrix} kPa$

Material properties and resistance

Partial factors from Sets $\begin{pmatrix} M1 \\ M2 \end{pmatrix}$: $\gamma_\varphi = \begin{pmatrix} 1 \\ 1.25 \end{pmatrix}$ and $\gamma_c = \begin{pmatrix} 1 \\ 1.25 \end{pmatrix}$

Design angle of shearing resistance is $\varphi_d = \tan^{-1}\left(\dfrac{\tan\left(\varphi_k\right)}{\gamma_\varphi}\right) = \begin{pmatrix} 35 \\ 29.3 \end{pmatrix}°$

Design cohesion is $c'_d = \dfrac{c'_k}{\gamma_c} = \begin{pmatrix} 0 \\ 0 \end{pmatrix} kPa$

Bearing capacity factors

For overburden: $N_q = \left[e^{(\pi \times \tan(\varphi_d))} \times \left(\tan\left(45° + \dfrac{\varphi_d}{2} \right) \right)^2 \right] = \begin{pmatrix} 33.3 \\ 16.9 \end{pmatrix}$

For cohesion: $N_c = \overrightarrow{\left[(N_q - 1) \times \cot(\varphi_d) \right]} = \begin{pmatrix} 46.1 \\ 28.4 \end{pmatrix}$

For self-weight: $N_\gamma = \overrightarrow{\left[2(N_q - 1) \times \tan(\varphi_d) \right]} = \begin{pmatrix} 45.2 \\ 17.8 \end{pmatrix}$ ❷

Shape factors

For overburden: $s_q = \overrightarrow{\left[1 + \left(\dfrac{B}{L} \right) \times \sin(\varphi_d) \right]} = \begin{pmatrix} 1.34 \\ 1.29 \end{pmatrix}$

For cohesion: $s_c = \dfrac{\overrightarrow{(s_q \times N_q - 1)}}{N_q - 1} = \begin{pmatrix} 1.35 \\ 1.31 \end{pmatrix}$

For self-weight: $s_\gamma = 1 - 0.3 \times \left(\dfrac{B}{L} \right) = 0.82$ ❸

Bearing resistance

Overburden at foundation base is $\sigma'_{vk,b} = \gamma_k \times d = 9 \, kPa$

Partial factors from Set $\begin{pmatrix} R1 \\ R1 \end{pmatrix}$: $\gamma_{Rv} = \begin{pmatrix} 1.0 \\ 1.0 \end{pmatrix}$

From overburden $q_{ult_1} = \overrightarrow{(N_q \times s_q \times \sigma'_{vk,b})} = \begin{pmatrix} 402.8 \\ 196.9 \end{pmatrix} kPa$

From cohesion $q_{ult_2} = \overrightarrow{(N_c \times s_c \times c'_d)} = \begin{pmatrix} 0 \\ 0 \end{pmatrix} kPa$

From self-weight $q_{ult_3} = \overrightarrow{\left(N_\gamma \times s_\gamma \times \gamma_k \times \dfrac{B}{2} \right)} = \begin{pmatrix} 500.7 \\ 197.5 \end{pmatrix} kPa$

Total resistance $q_{ult} = \sum_{i=1}^{3} \overrightarrow{q_{ult_i}} = \begin{pmatrix} 903.5 \\ 394.4 \end{pmatrix} kPa$

Design resistance is $q_{Rd} = \dfrac{q_{ult}}{\gamma_{Rv}} = \begin{pmatrix} 903.5 \\ 394.4 \end{pmatrix} kPa$

Utilization factor $\Lambda_{GEO,1} = \dfrac{q_{Ed}}{q_{Rd}} = \begin{pmatrix} 54 \\ 97 \end{pmatrix} \%$ ④

Design is unacceptable if utilization factor is > 100%

Design Approach 2

Actions and effects

Partial factors from Set A1: $\gamma_G = 1.35$ and $\gamma_Q = 1.5$

Design action is $V_d = \gamma_G \times \left(W_{Gk} + V_{Gk} \right) + \gamma_Q \times V_{Qk} = 1818.3 \text{ kN}$

Design bearing pressure is $q_{Ed} = \dfrac{V_d}{A_b} = 484.9 \text{ kPa}$

Material properties and resistance

Partial factors from Set M1: $\gamma_\varphi = 1.0$ and $\gamma_c = 1.0$

Design angle of shearing resistance is $\varphi_d = \tan^{-1}\left(\dfrac{\tan(\varphi_k)}{\gamma_\varphi} \right) = 35°$

Design cohesion is $c'_d = \dfrac{c'_k}{\gamma_c} = 0 \text{ kPa}$

Bearing capacity factors

For overburden: $N_q = e^{\left(\pi \times \tan(\varphi_d) \right)} \left(\tan\left(45° + \dfrac{\varphi_d}{2} \right) \right)^2 = 33.3$

For cohesion: $N_c = \left(N_q - 1 \right) \times \cot(\varphi_d) = 46.1$

For self-weight: $N_\gamma = 2\left(N_q - 1 \right) \times \tan(\varphi_d) = 45.2$

Shape factors

For overburden: $s_q = 1 + \left(\dfrac{B}{L} \right) \times \sin(\varphi_d) = 1.34$

For cohesion: $s_c = \dfrac{s_q \times N_q - 1}{N_q - 1} = 1.35$

For self-weight: $s_\gamma = 1 - 0.3 \times \left(\dfrac{B}{L} \right) = 0.82$

Bearing resistance

Partial factor from Set R2: $\gamma_{Rv} = 1.4$ ⑤

From overburden $q_{ult_1} = N_q \times s_q \times \sigma'_{vk,b} = 402.8 \, \text{kPa}$

From cohesion $q_{ult_2} = N_c \times s_c \times c'_d = 0 \, \text{kPa}$

From self-weight $q_{ult_3} = N_\gamma \times s_\gamma \times \gamma_k \times \dfrac{B}{2} = 500.7 \, \text{kPa}$

Total resistance $q_{ult} = \sum q_{ult} = 903.5 \, \text{kPa}$

Design resistance is $q_{Rd} = \dfrac{q_{ult}}{\gamma_{Rv}} = 645.3 \, \text{kPa}$

Verification of bearing resistance

Utilization factor $\boxed{\Lambda_{GEO,2} = \dfrac{q_{Ed}}{q_{Rd}} = 75\,\%}$ ⑥

Design is unacceptable if utilization factor is > 100%

Design Approach 3

Actions and effects

Partial factors on structural actions, Set A1: $\gamma_G = 1.35$ and $\gamma_Q = 1.5$

Design vertical action $V_d = \gamma_G \times \left(W_{Gk} + V_{Gk} \right) + \gamma_Q \times V_{Qk} = 1818.3 \, \text{kN}$

Design bearing pressure $q_{Ed} = \dfrac{V_d}{A_b} = 484.9 \, \text{kPa}$

Material properties and resistance

Partial factors from Set M2: $\gamma_\varphi = 1.25$ and $\gamma_c = 1.25$ ⑦

Design angle of shearing resistance is $\varphi_d = \tan^{-1}\left(\dfrac{\tan(\varphi_k)}{\gamma_\varphi} \right) = 29.3\,°$

Design cohesion is $c'_d = \dfrac{c'_k}{\gamma_c} = 0 \, \text{kPa}$

Bearing capacity factors

For overburden: $N_q = e^{(\pi \times \tan(\varphi_d))} \times \left(\tan\left(45° + \dfrac{\varphi_d}{2} \right) \right)^2 = 16.9$

For cohesion: $N_c = \left(N_q - 1 \right) \times \cot\left(\varphi_d \right) = 28.4$

For self-weight: $N_\gamma = 2\left(N_q - 1 \right) \times \tan\left(\varphi_d \right) = 17.8$

Shape factors

For overburden: $s_q = 1 + \left(\dfrac{B}{L} \right) \times \sin\left(\varphi_d \right) = 1.29$

For cohesion: $s_c = \dfrac{s_q \times N_q - 1}{N_q - 1} = 1.31$

For self-weight: $s_\gamma = 1 - 0.3 \times \left(\dfrac{B}{L} \right) = 0.82$

Bearing resistance

Partial factor from Set R3: $\gamma_{Rv} = 1$

From overburden $q_{ult_1} = N_q \times s_q \times \sigma'_{vk,b} = 196.9 \, kPa$

From cohesion $q_{ult_2} = N_c \times s_c \times c'_d = 0 \, kPa$

From self-weight $q_{ult_3} = N_\gamma \times s_\gamma \times \gamma_k \times \dfrac{B}{2} = 197.5 \, kPa$

Total resistance $q_{ult} = \sum q_{ult} = 394.4 \, kPa$

Design resistance $q_{Rd} = \dfrac{q_{ult}}{\gamma_{Rv}} = 394.4 \, kPa$

Verification of bearing resistance

Utilization factor $\boxed{\Lambda_{GEO,3} = \dfrac{q_{Ed}}{q_{Rd}} = 123 \, \%}$ **⑧**

Design is unacceptable if utilization factor is > 100%

❻ The calculated utilization factor is 75% which would indicate that according to DA2 the footing is potentially over-designed.

❼ Design Approach 3 applies partial factors to both actions and material properties at the same time.

❽ The resultant utilization factor is 123% thus the DA3 calculation suggests the design is unsafe and re-design would be required.

The three Design Approaches give different assessments of the suitability of the proposed foundation for the design loading. Of the three approaches, DA1 suggests the footing is only just satisfactory whilst DA3 suggests redesign would be required and DA2 may indicate that the footing is overdesigned!

Which approach is the most appropriate cannot be determined although DA3 would appear unnecessarily conservative by providing significant partial factors on both actions and material properties.

10.10.2   Eccentric pad footing on dry sand

Example 10.2 considers the design of a pad footing on dry sand, in which the imposed vertical load from the superstructure is eccentric to the centre of the foundation, as shown in **Figure 10.10**.

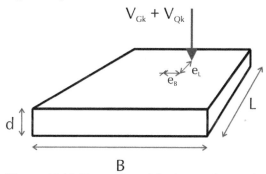

Figure 10.10. Eccentric pad footing on dry sand

Because the load is eccentric, the foundation's design is based on its effective area. The foundation's self weight (which acts through the centre of the footing) helps to reduce the eccentricity of the total load. Eccentric loads should be avoided whenever possible since they make the footing inefficient.

Example 10.2
Eccentric pad footing on dry sand
Verification of strength (limit state GEO)

Design situation

Owing to an error on site, the pad footing from the previous design example is out-of-position on plan, such that the imposed actions act at distances e_B = 75mm and e_L = 100mm from the centre of the footing.

Design Approach 1

Geometry

Eccentricity of total vertical action:

$$e'_B = \frac{\left(\gamma_G V_{Gk} + \gamma_Q V_{Qk}\right) \times e_B}{\gamma_G \times \left(W_{Gk} + V_{Gk}\right) + \gamma_Q V_{Qk}} = \binom{72.4}{72.5} \, mm \; ❶$$

Load is within middle-third of base if $e'_B \leq \dfrac{B}{6} = 250 \, mm$

Effective breadth is $B' = B - 2e'_B = \binom{1.36}{1.35} \, m \; ❷$

Eccentricity of total vertical action:

$$e'_L = \frac{\left(\gamma_G V_{Gk} + \gamma_Q V_{Qk}\right) \times e_L}{\gamma_G \times \left(W_{Gk} + V_{Gk}\right) + \gamma_Q V_{Qk}} = \binom{96.5}{96.7} \, mm$$

Load is within middle-third of base if $e'_L \leq \dfrac{L}{6} = 417 \, mm$

Effective length is $L' = L - 2e'_L = \binom{2.31}{2.31} \, m \; ❷$

Effective area of base is therefore $A'_b = \overrightarrow{(L' \times B')} = \binom{3.13}{3.13} \, m^2$

Actions and effects

From previous calculation: $V_d = \binom{1818.3}{1431.9} \, kN$

Design bearing pressure is $q_{Ed} = \dfrac{V_d}{A'_b} = \binom{581.6}{458.2} \, kPa$

Material properties and resistance

From previous calculation: $\varphi_d = \begin{pmatrix} 35 \\ 29.3 \end{pmatrix}°$ and $c'_d = \begin{pmatrix} 0 \\ 0 \end{pmatrix}$ kPa

Bearing capacity factors: $N_q = \begin{pmatrix} 33.3 \\ 16.9 \end{pmatrix}$, $N_c = \begin{pmatrix} 46.1 \\ 28.4 \end{pmatrix}$, and $N_\gamma = \begin{pmatrix} 45.2 \\ 17.8 \end{pmatrix}$

Shape factors

For overburden: $\overrightarrow{s_q = \left[1 + \left(\dfrac{B'}{L'}\right) \times \sin(\varphi_d)\right]} = \begin{pmatrix} 1.34 \\ 1.29 \end{pmatrix}$

For cohesion: $\overrightarrow{s_c = \dfrac{(s_q N_q - 1)}{N_q - 1}} = \begin{pmatrix} 1.35 \\ 1.31 \end{pmatrix}$

For self-weight: $s_\gamma = 1 - 0.3 \times \left(\dfrac{B'}{L'}\right) = \begin{pmatrix} 0.82 \\ 0.82 \end{pmatrix}$

Bearing resistance

From overburden $q_{ult_1} = \overrightarrow{\left(N_q \times s_q \times \sigma'_{vk,b}\right)} = \begin{pmatrix} 400.6 \\ 196 \end{pmatrix}$ kPa

From cohesion $q_{ult_2} = \overrightarrow{\left(N_c \times s_c \times c'_d\right)} = \begin{pmatrix} 0 \\ 0 \end{pmatrix}$ kPa

From self-weight $q_{ult_3} = \overrightarrow{\left(N_\gamma \times s_\gamma \times \gamma_k \times \dfrac{B'}{2}\right)} = \begin{pmatrix} 454.4 \\ 179.2 \end{pmatrix}$ kPa

Total resistance $q_{ult} = \displaystyle\sum_{i=1}^{3} q_{ult_i} = \begin{pmatrix} 855.1 \\ 375.2 \end{pmatrix}$ kPa

Design resistance is then $q_{Rd} = \dfrac{q_{ult}}{\gamma_{Rv}} = \begin{pmatrix} 855.1 \\ 375.2 \end{pmatrix}$ kPa

Verification of bearing resistance

Utilization factor is $\Lambda_{GEO,1} = \dfrac{q_{Ed}}{q_{Rd}} = \begin{pmatrix} 68 \\ 122 \end{pmatrix}$ % ❸

Design is not acceptable if utilization factor is > 100%

Design Approach 2

Geometry

Eccentricity of total vertical action:

$$e'_B = \frac{\left(\gamma_G V_{Gk} + \gamma_Q V_{Qk}\right) \times e_B}{\gamma_G \times \left(W_{Gk} + V_{Gk}\right) + \gamma_Q V_{Qk}} = 72.4\,\text{mm}$$

Load is within middle-third of base if $e'_B \leq \dfrac{B}{6} = 250\,\text{mm}$

Effective breadth is $B' = B - 2e'_B = 1.36\,\text{m}$

Eccentricity of total vertical action:

$$e'_L = \frac{\left(\gamma_G V_{Gk} + \gamma_Q V_{Qk}\right) \times e_L}{\gamma_G \times \left(W_{Gk} + V_{Gk}\right) + \gamma_Q V_{Qk}} = 96.5\,\text{mm}$$

Load is within middle-third of base if $e'_L \leq \dfrac{L}{6} = 417\,\text{mm}$

Effective length is $L' = L - 2e'_L = 2.31\,\text{m}$

Effective area of base is then $A'_b = L' \times B' = 3.13\,\text{m}^2$

Actions and effects

From previous calculation: $V_d = 1818.3\,\text{kN}$

Design bearing pressure $q_{Ed} = \dfrac{V_d}{A'_b} = 581.6\,\text{kPa}$

Material properties and resistance

From previous calculation: $\varphi_d = 35°$ and $c'_d = 0\,\text{kPa}$

Bearing capacity factors: $N_q = 33.3$, $N_c = 46.1$, and $N_\gamma = 45.2$

Shape factors

For overburden: $s_q = 1 + \left(\dfrac{B'}{L'}\right) \times \sin\left(\varphi_d\right) = 1.34$

For cohesion: $s_c = \dfrac{s_q N_q - 1}{N_q - 1} = 1.35$

For self-weight: $s_\gamma = 1 - 0.3 \times \left(\dfrac{B'}{L'}\right) = 0.82$

Bearing resistance

From overburden $q_{ult_1} = N_q \times s_q \times \sigma'_{vk,b} = 400.6\,kPa$

From cohesion $q_{ult_2} = N_c \times s_c \times c'_d = 0\,kPa$

From self-weight $q_{ult_3} = N_\gamma \times s_\gamma \times \gamma_k \dfrac{B'}{2} = 454.4\,kPa$

Total resistance $q_{ult} = \displaystyle\sum q_{ult} = 8.6 \times 10^5\,Pa$

Design resistance is then $q_{Rd} = \dfrac{q_{ult}}{\gamma_{Rv}} = 610.8\,kPa$

Verification of bearing resistance

Utilization factor $\boxed{\Lambda_{GEO,2} = \dfrac{q_{Ed}}{q_{Rd}} = 95\,\%}$ **4**

Design is unacceptable if utilization factor is > 100%

Design Approach 3

Geometry

Eccentricity of total vertical action:

$$e'_B = \frac{\left(\gamma_G V_{Gk} + \gamma_Q V_{Qk}\right) \times e_B}{\gamma_G \times \left(W_{Gk} + V_{Gk}\right) + \gamma_Q V_{Qk}} = 72.4\,mm$$

Load is within middle-third of base if $e'_B \leq \dfrac{B}{6} = 250\,mm$

Effective breadth is $B' = B - 2e'_B = 1.36\,m$

Eccentricity of total vertical action:

$$e'_L = \frac{\left(\gamma_G V_{Gk} + \gamma_Q V_{Qk}\right) \times e_L}{\gamma_G \times \left(W_{Gk} + V_{Gk}\right) + \gamma_Q V_{Qk}} = 96.5\,mm$$

Load is within middle-third of base if $e'_L \leq \dfrac{L}{6} = 417\,mm$

Effective length is $L' = L - 2e'_L = 2.31\,m$

Effective area of base is then $A'_b = L' \times B' = 3.13\,m^2$

Actions and effects

From previous calculation: $V_d = 1818.3$ kN

Design bearing pressure $q_{Ed} = \dfrac{V_d}{A'_b} = 581.6$ kPa

Material properties and resistance

From previous calculation: $\varphi_d = 29.3°$ and $c'_d = 0$ kPa

Bearing capacity factors: $N_q = 16.9$, $N_c = 28.4$, and $N_\gamma = 17.8$

Shape factors

For overburden: $s_q = 1 + \left(\dfrac{B'}{L'}\right) \times \sin\left(\varphi_d\right) = 1.29$

For cohesion: $s_c = \dfrac{s_q N_q - 1}{N_q - 1} = 1.31$

For self-weight: $s_\gamma = 1 - 0.3 \times \left(\dfrac{B'}{L'}\right) = 0.82$

Bearing resistance

From overburden $q_{ult_1} = N_q \times s_q \times \sigma'_{vk,b} = 196$ kPa

From cohesion $q_{ult_2} = N_c \times s_c \times c'_d = 0$ kPa

From self-weight $q_{ult_3} = N_\gamma \times s_\gamma \times \gamma_k \times \dfrac{B'}{2} = 179.2$ kPa

Total resistance $q_{ult} = \sum q_{ult} = 375.2$ kPa

Design resistance $q_{Rd} = \dfrac{q_{ult}}{\gamma_{Rv}} = 375.2$ kPa

Verification of bearing resistance

Utilization factor $\Lambda_{GEO,3} = \dfrac{q_{Ed}}{q_{Rd}} = 155\%$ ⑤

Design is not acceptable if utilization factor is > 100%

Notes on Example 10.2

❶ The introduction of eccentricity of the applied loads due to construction error needs to recognize that the total action comprises both applied loads and the dead weight of the footing. The dead weight of the footing still acts through the centre of the footing, thus eccentricity of the total action is less than the eccentricity of the applied loads. There is no guidance in the code as to whether the eccentricity should be calculated for the characteristic or design values of the actions, but we consider that it is best to base the calculation on the design actions as shown. For the example given the difference is minimal but as the relative proportions of permanent to variable loads change so the effects may become more apparent.

❷ The effect of eccentricity is to reduce the effective dimensions of the footing. These reduced dimensions are then used throughout the rest of the calculation to check the adequacy of the footing.

❸ The introduction of eccentricity into this example results in the foundation being inadequate for DA1. Thus the footing would need to be re-designed in order to satisfy EC7 requirements. This might entail making the footing larger or repositioning the source of the applied loads.

❹ The footing is just adequate for DA2.

❺ The footing does not satisfy DA3 and needs to be re-designed.

10.10.3   Strip footing on clay

Example 10.3 considers the design of a strip footing on clay, as shown in **Figure 10.11**. Groundwater is at a depth d_w below ground surface.

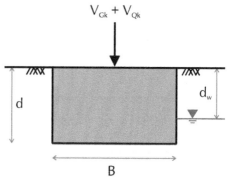

This example demonstrates the use of partial factors for undrained and drained parameters. The inclusion of groundwater above the base of the footing illustrates the complications in applying partial factors to water pressures.

Figure 10.11. Strip footing on clay

Example 10.3
Example 10.3
Strip footing on clay
Verification of strength (limit state GEO)

Design situation

Consider an infinitely long footing of breadth $B = 2.5m$ and depth $d = 1.5m$, which is required to carry an imposed permanent action $V_{Gk} = 250\dfrac{kN}{m}$ and an imposed variable action $V_{Qk} = 110\dfrac{kN}{m}$. The footing is founded on a medium strength clay ❶ with characteristic undrained strength $c_{uk} = 45kPa$, angle of shearing resistance $\varphi_k = 25°$, effective cohesion $c'_k = 5kPa$, and weight density $\gamma_k = 21\dfrac{kN}{m^3}$. The water table is currently at a depth $d_w = 1m$. The weight density of groundwater is $\gamma_w = 9.81\dfrac{kN}{m^3}$ and of reinforced concrete

$$\gamma_{ck} = 25\frac{kN}{m^3} \text{ (EN 1991-1-1 Table A.1)}.$$

Design Approach 1

Geometrical parameters

Design depth of water table $d_{w,d} = 0m$ ❷

Actions and effects

Characteristic self-weight of footing is $W_{Gk} = \gamma_{ck} \times B \times d = 93.8\dfrac{kN}{m}$

Characteristic pore pressure underneath base
$$u_{k,b} = \gamma_w \times (d - d_{w,d}) = 14.7 kPa$$

Partial factors, Set $\begin{pmatrix} A1 \\ A2 \end{pmatrix}$: $\gamma_G = \begin{pmatrix} 1.35 \\ 1 \end{pmatrix}$, $\gamma_{G,fav} = \begin{pmatrix} 1 \\ 1 \end{pmatrix}$, and $\gamma_Q = \begin{pmatrix} 1.5 \\ 1.3 \end{pmatrix}$

Design vertical action: $V_d = \gamma_G \times (W_{Gk} + V_{Gk}) + \gamma_Q \times V_{Qk} = \begin{pmatrix} 629.1 \\ 486.8 \end{pmatrix}\dfrac{kN}{m}$

Design bearing pressure (total stress): $q_{Ed} = \dfrac{V_d}{B} = \begin{pmatrix} 251.6 \\ 194.7 \end{pmatrix} kPa$

Design upthrust (favourable): $u_d = \gamma_{G,fav} \times u_{k,b} = \begin{pmatrix} 14.7 \\ 14.7 \end{pmatrix}$ kPa ③

Design bearing pressure (effective stress): $q'_{Ed} = q_{Ed} - u_d = \begin{pmatrix} 236.9 \\ 180 \end{pmatrix}$ kPa

Material properties and resistance

Partial factors, Set $\begin{pmatrix} M1 \\ M2 \end{pmatrix}$: $\gamma_{cu} = \begin{pmatrix} 1 \\ 1.4 \end{pmatrix}$, $\gamma_\varphi = \begin{pmatrix} 1 \\ 1.25 \end{pmatrix}$ and $\gamma_c = \begin{pmatrix} 1 \\ 1.25 \end{pmatrix}$

Design undrained strength is $c_{ud} = \dfrac{c_{uk}}{\gamma_{cu}} = \begin{pmatrix} 45 \\ 32.1 \end{pmatrix}$ kPa

Design shearing resistance is $\varphi_d = \tan^{-1}\left(\dfrac{\tan(\varphi_k)}{\gamma_\varphi}\right) = \begin{pmatrix} 25 \\ 20.5 \end{pmatrix}$ °

Design cohesion is $c'_d = \dfrac{c'_k}{\gamma_c} = \begin{pmatrix} 5 \\ 4 \end{pmatrix}$ kPa

Drained bearing capacity factors

For overburden: $N_q = \overline{\left[e^{(\pi \times \tan(\varphi_d))} \times \left(\tan\left(45° + \dfrac{\varphi_d}{2}\right) \right)^2 \right]} = \begin{pmatrix} 10.7 \\ 6.7 \end{pmatrix}$

For cohesion: $N_c = \overline{\left[(N_q - 1) \times \cot(\varphi_d) \right]} = \begin{pmatrix} 20.7 \\ 15.3 \end{pmatrix}$

For self-weight: $N_\gamma = \overline{\left[2(N_q - 1) \times \tan(\varphi_d) \right]} = \begin{pmatrix} 9 \\ 4.3 \end{pmatrix}$

Depth and shape factors

Salgado's depth factor for undrained loading: $d_c = 1 + 0.27\sqrt{\dfrac{d}{B}} = 1.21$ ④

Ignore depth factors for drained loading

Salgado's shape factor for undrained loading: $s_c = 1 + 0.17\sqrt{\dfrac{d}{B}} = 1.13$ ④

Depth factors are all 1.0 for drained loading and so can be ignored

Undrained bearing resistance

Total overburden at foundation base is $\sigma_{vk,b} = \gamma_k \times d = 31.5$ kPa

Partial factors from Sets $\begin{pmatrix} R1 \\ R1 \end{pmatrix}$: $\gamma_{Rv} = \begin{pmatrix} 1.0 \\ 1.0 \end{pmatrix}$

Ultimate resistance is $q_{ult} = (\pi + 2) \times c_{ud} \times d_c \times s_c + \sigma_{vk,b} = \begin{pmatrix} 348.1 \\ 257.6 \end{pmatrix}$ kPa

Design resistance is $q_{Rd} = \dfrac{q_{ult}}{\gamma_{Rv}} = \begin{pmatrix} 348.1 \\ 257.6 \end{pmatrix}$ kPa

Drained bearing resistance

Effective overburden at foundation base is $\sigma'_{vk,b} = \sigma_{vk,b} - u_{k,b} = 16.8$ kPa

From overburden $q'_{ult_1} = \overrightarrow{\left(N_q \times \sigma'_{vk,b}\right)} = \begin{pmatrix} 179 \\ 112.5 \end{pmatrix}$ kPa

From cohesion $q'_{ult_2} = \overrightarrow{\left(N_c \times c'_d\right)} = \begin{pmatrix} 103.6 \\ 61.1 \end{pmatrix}$ kPa

From self-weight $q'_{ult_3} = \overrightarrow{\left[N_\gamma \times \left(\gamma_k - \gamma_w\right) \times \dfrac{B}{2}\right]} = \begin{pmatrix} 126.1 \\ 59.5 \end{pmatrix}$ kPa

Total resistance $q'_{ult} = \sum\limits_{i=1}^{3} q'_{ult_i} = \begin{pmatrix} 408.7 \\ 233 \end{pmatrix}$ kPa

Design resistance is $q'_{Rd} = \dfrac{q'_{ult}}{\gamma_{Rv}} = \begin{pmatrix} 408.7 \\ 233 \end{pmatrix}$ kPa

Verification of undrained bearing resistance

Degree of utilization $\Lambda_{GEO,1} = \dfrac{q_{Ed}}{q_{Rd}} = \begin{pmatrix} 72 \\ 76 \end{pmatrix}$ % ⑤

Design is unacceptable if degree of utilization is > 100%

Verification of drained bearing resistance

Degree of utilization $\Lambda'_{GEO,1} = \dfrac{q'_{Ed}}{q'_{Rd}} = \begin{pmatrix} 58 \\ 77 \end{pmatrix}$ % ⑤

Design is unacceptable if degree of utilization is > 100%

Design Approach 2

Actions and effects

Partial factors from set A1: $\gamma_G = 1.35$ and $\gamma_Q = 1.5$ **⑥**

Design action is $V_d = \gamma_G \times \left(W_{Gk} + V_{Gk} \right) + \gamma_Q \times V_{Qk} = 629.1 \dfrac{kN}{m}$

Design bearing pressure (total stress) is $q_{Ed} = \dfrac{V_d}{B} = 251.6\,kPa$

Design upthrust (favourable): $u_d = \gamma_{G,fav} \times u_{k,b} = 14.7\,kPa$

Design bearing pressure (effective stress) is $q'_{Ed} = q_{Ed} - u_d = 236.9\,kPa$

Material properties and resistance

Partial factors from set M1: $\gamma_{cu} = 1.0$, $\gamma_\varphi = 1.0$, and $\gamma_c = 1.0$ **⑥**

Design undrained strength is $c_{ud} = \dfrac{c_{uk}}{\gamma_c} = 45\,kPa$

Design angle of shearing resistance is $\varphi_d = \tan^{-1}\left(\dfrac{\tan\left(\varphi_k\right)}{\gamma_\varphi} \right) = 25\,deg$

Design cohesion is $c'_d = \dfrac{c'_k}{\gamma_c} = 5\,kPa$

Drained bearing capacity factors

For overburden: $N_q = e^{\left(\pi \times \tan\left(\varphi_d\right)\right)} \left(\tan\left(45° + \dfrac{\varphi_d}{2} \right) \right)^2 = 10.7$

For cohesion: $N_c = \left(N_q - 1 \right) \times \cot\left(\varphi_d\right) = 20.7$

For self-weight: $N_\gamma = 2\left(N_q - 1 \right) \times \tan\left(\varphi_d\right) = 9$

Depth and shape factors
Are the same as for Design Approach 1

Undrained bearing resistance
Total overburden at foundation base is $\sigma_{vk,b} = \gamma_k \times d = 31.5\,kPa$

Partial factors from set R2: $\gamma_{Rv} = 1.4$ **⑦**

Ultimate resistance is $q_{ult} = (\pi + 2) \times c_{ud} \times d_c \times s_c + \sigma_{vk,b} = 348.1\,kPa$

Design resistance is $q_{Rd} = \dfrac{q_{ult}}{\gamma_{Rv}} = 248.6\,kPa$

Drained bearing resistance

Effective overburden at foundation base is $\sigma'_{vk,b} = \sigma_{vk,b} - u_{k,b} = 16.8\,kPa$

From overburden $q'_{ult_1} = \overrightarrow{\left(N_q \times \sigma'_{vk,b} \right)} = 179\,kPa$

From cohesion $q'_{ult_2} = \overrightarrow{\left(N_c \times c'_d \right)} = 103.6\,kPa$

From self-weight $q'_{ult_3} = \overrightarrow{\left[N_\gamma \times \left(\gamma_k - \gamma_w \right) \times \dfrac{B}{2} \right]} = 126.1\,kPa$

Total resistance $q'_{ult} = \sum q'_{ult} = 408.7\,kPa$

Design resistance is $q'_{Rd} = \dfrac{q'_{ult}}{\gamma_{Rv}} = 291.9\,kPa$

Verification of undrained bearing resistance

Degree of utilization $\Lambda_{GEO,2} = \boxed{\dfrac{q_{Ed}}{q_{Rd}} = 101\,\%}$ ❽

Design is unacceptable if degree of utilization factor is > 100%

Verification of drained bearing resistance

Degree of utilization $\Lambda'_{GEO,2} = \boxed{\dfrac{q'_{Ed}}{q'_{Rd}} = 81\,\%}$ ❽

Design is unacceptable if degree of utilization is > 100%

Design Approach 3

Actions and effects

Partial factors on actions from set A1: $\gamma_G = 1.35$ and $\gamma_Q = 1.5$ **⑨**

Design vertical action $V_d = \gamma_G \times \left(W_{Gk} + V_{Gk}\right) + \gamma_Q \times V_{Qk} = 629.1\dfrac{kN}{m}$

Design bearing pressure (total stress) $q_{Ed} = \dfrac{V_d}{B} = 251.6\,kPa$

Design upthrust (favourable): $u_d = \gamma_{G,fav} \times u_{k,b} = 14.7\,kPa$

Design bearing pressure (effective stress) $q'_{Ed} = q_{Ed} - u_d = 236.9\,kPa$

Material properties and resistance

Partial factors from set M2: $\gamma_{cu} = 1.4$, $\gamma_\varphi = 1.25$, and $\gamma_c = 1.25$ **⑨**

Design undrained strength is $c_{ud} = \dfrac{c_{uk}}{\gamma_{cu}} = 32.1\,kPa$

Design angle of shearing resistance is $\varphi_d = \tan^{-1}\left(\dfrac{\tan\left(\varphi_k\right)}{\gamma_\varphi}\right) = 20.5\,°$

Design cohesion is $c'_d = \dfrac{c'_k}{\gamma_c} = 4\,kPa$

Drained bearing capacity factors

For overburden: $N_q = e^{\left(\pi \times \tan\left(\varphi_d\right)\right)} \times \left(\tan\left(45° + \dfrac{\varphi_d}{2}\right)\right)^2 = 6.7$

For cohesion: $N_c = \left(N_q - 1\right) \times \cot\left(\varphi_d\right) = 15.3$

For self-weight: $N_\gamma = 2\left(N_q - 1\right) \times \tan\left(\varphi_d\right) = 4.3$

Depth and shape factors
Are the same as for Design Approach 1

Undrained bearing resistance

Total overburden at foundation base is $\sigma_{vk,b} = \gamma_k \times d = 31.5 \, \text{kPa}$

Partial factors from set R3: $\gamma_{Rv} = 1.0$

Ultimate resistance is $q_{ult} = (\pi + 2) \times c_{ud} \times d_c \times s_c + \sigma_{vk,b} = 257.6 \, \text{kPa}$

Design resistance is $q_{Rd} = \dfrac{q_{ult}}{\gamma_{Rv}} = 257.6 \, \text{kPa}$

Drained bearing resistance

Effective overburden at foundation base is $\sigma'_{vk,b} = \sigma_{vk,b} - u_{k,b} = 16.8 \, \text{kPa}$

From overburden $q'_{ult_1} = \overrightarrow{\left(N_q \times \sigma'_{vk,b} \right)} = 112.5 \, \text{kPa}$

From cohesion $q'_{ult_2} = \overrightarrow{\left(N_c \times c'_d \right)} = 61.1 \, \text{kPa}$

From self-weight $q'_{ult_3} = \overrightarrow{\left[N_\gamma \times \left(\gamma_k - \gamma_w \right) \times \dfrac{B}{2} \right]} = 59.5 \, \text{kPa}$

Total resistance $q'_{ult} = \sum q'_{ult} = 233 \, \text{kPa}$

Design resistance is $q'_{Rd} = \dfrac{q'_{ult}}{\gamma_{Rv}} = 233 \, \text{kPa}$

Verification of undrained bearing resistance

Degree of utilization $\Lambda_{GEO,3} = \dfrac{q_{Ed}}{q_{Rd}} = 98 \, \%$ ⑩

Design is unacceptable if degree of utilization is > 100%

Verification of drained bearing resistance

Degree of utilization $\Lambda'_{GEO,3} = \dfrac{q'_{Ed}}{q'_{Rd}} = 102 \, \%$ ⑩

Design is unacceptable if degree of utilization is > 100%

Notes on Example 10.3

❶ The term 'medium strength' is defined as an undrained strength between 40 and 75 kPa in EN ISO 14688-2.[23]

❷ In an ultimate limit state, the design water level should represent the most onerous that could occur during the design lifetime of the structure. Hence, here it is taken at ground surface.

❸ The water pressure acting beneath the footing is a *favourable* action, since it resists the weight of the foundation.

❹ For the undrained case, d_c and s_c are based on formulae developed from finite element studies (see main text). Note that EN 1997-1 does not include depth factors in its recommendations in Annex D. Shape factors for strip footing are normally taken as 1.0.

❺ The calculation indicates that the drained (long-term) situation is slightly more critical than the undrained (short-term). Combination 2 governs in both cases and is verified, since the utilization factors are less than 100%.

❻ Partial factors for Design Approach 2 are applied principally to actions.

❼ A resistance factor of 1.4 is applied to the resistance, in combination with the factors on actions.

❽ The calculation indicates that the undrained (short-term) situation is more critical than the drained (long-term) situation and is marginally unsatisfactory.

❾ For Design Approach 3, the partial factors result in a simultaneous increase in actions and decrease in soil strength. Design Approach 3 will therefore always be more conservative than Design Approach 1.

❿ Both for the drained and undrained case, Design Approach 3 suggests the footing is just satisfactory (degree of utilization ≈ 100%).

10.10.4 Settlement of strip footing on clay

Example 10.4 looks at verification of serviceability of the strip footing from Example 10.3 (see **Figure 10.11** on page 337). It is assumed that a rigid layer exists below the foundation, limiting settlement to within that layer.

This example attempts to verify serviceability implicitly (through an ultimate limit state calculation), and then repeats the exercise explicitly (using a serviceability limit state calculation).

Notes on Example 10.4

❶ For serviceability limit states, the design depth of the water table is the most adverse level that could occur 'in normal circumstances'. Hence, we have decided *not* to raise the water table to ground level. This is a less severe requirement than for ultimate limit states (see Example 10.3).

❷ The pore pressure beneath the base is lower for the serviceability limit state than for ultimate (see Example 10.3).

❸ Partial factors for serviceability limit state are normally taken as 1.0.

❹ The water pressure acting beneath the footing is a *favourable* action, since it resists the weight of the foundation.

❺ For footings on clays (excluding soft clays), serviceability limit states may be verified without an explicit settlement calculation provided a minimum resistance factor of 3.0 is applied.

❻ The calculation based on a resistance factor of 3.0 does not work for either the undrained or drained situations and therefore an explicit settlement calculation is required.

❼ The calculation model used[24] is one of many that are available and follows the guidance given in Annex F to EN 1997-1.

❽ The calculation model chosen is commonly used in UK practice but is only consistent with the model given in EN 1997-1 Annex F if $E = 1/m_v$.

❾ Only immediate and consolidation settlement have been considered. The creep component is considered negligible in this example. The analysis has ignored any correction that may be applied to adjust for consolidation settlements based on one dimensional analysis. A depth correction factor is not normally applied to infinitely long footings.

❿ The limiting value depends on the structure's specific requirements. In this example, serviceability is satisfied by the explicit calculation (degree of utilization = 92%).

Example 10.4
Settlement of strip footing on clay
Verification of serviceability

Design situation

Consider the infinitely long strip footing from the previous example. There is a rigid layer underlying the footing at a depth of $d_R = 4.5m$ The clay's undrained Young's modulus is assumed to be $E_{uk} = 600c_{uk} = 27$ MPa and its characteristic coefficient of compressibility $m_{vk} = 0.12\dfrac{m^2}{MN}$

Implicit verification of serviceability (based on ULS check)

Geometrical parameters

Design depth of water table $d_{w,d} = d_w = 1m$ ❶

Actions and effects

From previous calculation, characteristic actions are:

  imposed permanent $V_{Gk} = 250$ kN/m

  imposed variable action $V_{Qk} = 110$ kN/m

  self-weight of footing $W_{Gk} = 93.8$ kN/m

Characteristic pore pressure under base $u_{k.b} = \gamma_w \times (d - d_{w,d}) = 4.9$ kPa ❷

Partial load factors for SLS: $\gamma_G = 1$, $\gamma_{G.fav} = 1$, and $\gamma_Q = 1$ ❸

Design vertical action: $V_d = \gamma_G \times (W_{Gk} + V_{Gk}) + \gamma_Q \times V_{Qk} = 453.8$ kN/m

Design bearing pressure (total stress): $q_{Ed} = \dfrac{V_d}{B} = 181.5$ kPa

Design upthrust (favourable): $u_d = \gamma_{G.fav} \times u_{k.b} = 4.9$ kPa ❹

Design bearing pressure (effective stress): $q'_{Ed} = q_{Ed} - u_d = 176.6$ kPa

Material properties and resistance

From previous calculation, characteristic material properties are: undrained strength $c_{uk} = 45$ kPa, shearing resistance $\varphi_k = 25°$, cohesion $c'_k = 5$ kPa

Partial material factors for SLS: $\gamma_{cu} = 1$, $\gamma_\varphi = 1$ and $\gamma_c = 1$ ❸

Design undrained strength $c_{ud} = c_{uk} \div \gamma_{cu} = 45$ kPa

Design shearing resistance $\varphi_d = \tan^{-1}\left(\tan\left(\varphi_k\right) \div \gamma_\varphi\right) = 25°$

Design cohesion $c'_d = c'_k \div \gamma_c = 5\,kPa$

Drained bearing capacity factors

For overburden: $N_q = e^{\left(\pi \times \tan\left(\varphi_d\right)\right)} \times \left(\tan\left(45° + \dfrac{\varphi_d}{2}\right)\right)^2 = 10.7$

For cohesion: $N_c = \left(N_q - 1\right) \times \cot\left(\varphi_d\right) = 20.7$

For self-weight: $N_\gamma = 2\left(N_q - 1\right) \times \tan\left(\varphi_d\right) = 9$

Depth and shape factors

Previous calculation: $d_c = 1 + 0.27\sqrt{\dfrac{d}{B}} = 1.21$ & $s_c = 1 + 0.17\sqrt{\dfrac{d}{B}} = 1.13$

Undrained bearing resistance

From previous calculation, total overburden under base is $\sigma_{vk,b} = 31.5\,kPa$

Partial resistance factor for SLS: $\gamma_{Rv.SLS} = 3.0$ **⑤**

Ultimate resistance is $q_{ult} = (\pi + 2) \times c_{ud} \times d_c \times s_c + \sigma_{vk,b} = 348.1\,kPa$

Design resistance is $q_{Rd} = \dfrac{q_{ult}}{\gamma_{Rv,SLS}} = 116\,kPa$

Drained bearing resistance

Effective overburden under base is $\sigma'_{vk,b} = \sigma_{vk,b} - u_d = 26.6\,kPa$

From overburden $q'_{ult_1} = N_q \times \sigma'_{vk,b} = 283.6\,kPa$

From cohesion $q'_{ult_2} = N_c \times c'_d = 103.6\,kPa$

From self-weight $q'_{ult_3} = N_\gamma \times \left(\gamma_k - \gamma_w\right) \times \dfrac{B}{2} = 126.1\,kPa$

Total resistance $q'_{ult} = \sum q'_{ult} = 513.3\,kPa$

Design resistance is $q'_{Rd} = \dfrac{q'_{ult}}{\gamma_{Rv,SLS}} = 171.1\,kPa$

Verification of undrained bearing resistance

Degree of utilization $\Lambda_{SLS} = \boxed{\dfrac{q_{Ed}}{q_{Rd}}} = 156\,\%$ ⑥

Design is unacceptable if the degree of utilization is > 100%

Verification of drained bearing resistance

Degree of utilization $\Lambda'_{SLS} = \boxed{\dfrac{q'_{Ed}}{q'_{Rd}}} = 103\,\%$ ⑥

Design is unacceptable if the degree of utilization is > 100%

Explicit verification of serviceability

Actions and effects

Increase in bearing pressure is $\Delta q_d = q_{Ed} - \sigma_{vk,b} = 150\,kPa$

Immediate settlement (Christian & Carrier)

Settlement factor for D/B is $\dfrac{d}{B} = 0.6$, giving $\mu_0 = 0.93$ (from chart) ⑦

Settlement factor for H/B is $\dfrac{d_R - d}{B} = 1.2$, giving $\mu_1 = 0.4$ (from chart)

Immediate settlement $s_0 = \dfrac{\Delta q_d\,B\,\mu_0\,\mu_1}{E_{uk}} = 5.2\,mm$

Consolidation settlement

Divide clay layer into $N = 5$ sub-layers of thickness $\Delta t = \dfrac{\left(d_R - d\right)}{N} = 0.6\,m$

For each layer $i = 1..N$, the depth below base to the centre of each layer is

given by $z_i = (\Delta t \times i) - \dfrac{\Delta t}{2}$ and the normalized foundation half breadth by

$m_i = \dfrac{B}{2\,z_i}$. The influence factor $I_{q_i} = I_{q.\infty}(m_i)$ can be found from Fadum's

chart. The change in vertical stress in each layer is $\Delta\sigma_{v_i} = 4 I_{q_i} \Delta q_d$ and the settlement in each layer $\rho_{c_i} = m_{vk}\Delta\sigma_{v_i}\Delta t.$ ⑧

Substituting values into the previous expressions gives:

$$z = \begin{pmatrix} 0.3 \\ 0.9 \\ 1.5 \\ 2.1 \\ 2.7 \end{pmatrix} m \quad m = \begin{pmatrix} 4.17 \\ 1.39 \\ 0.83 \\ 0.6 \\ 0.46 \end{pmatrix} \quad I_q = \begin{pmatrix} 0.25 \\ 0.23 \\ 0.19 \\ 0.16 \\ 0.13 \end{pmatrix} \quad \Delta\sigma_v = \begin{pmatrix} 149.2 \\ 135.7 \\ 113.3 \\ 93.2 \\ 77.8 \end{pmatrix} kPa \quad \rho_c = \begin{pmatrix} 10.7 \\ 9.8 \\ 8.2 \\ 6.7 \\ 5.6 \end{pmatrix} mm$$

The total consolidation settlement is $s_1 = \displaystyle\sum_{i=1}^{N} \rho_{c_i} = 41\,mm$

Total settlement

Sum of settlements is $s = s_0 + s_1 = 46\,mm$ ⑨

Design effect of actions is $s_{Ed} = s = 46\,mm$

Verification of settlement

Limiting values of foundation movement for isolated foundation is

$s_{Cd} = 50mm$ ⑩

Degree of utilization is $\boxed{\Lambda_{SLS} = \dfrac{s_{Ed}}{s_{Cd}} = 92\,\%}$

Design is unacceptable if the degree of utilization is > 100%

10.11   Notes and references

1. See Tables 1–3 of BS 8004: 1986, Code of practice for foundations, British Standards Institution.

2. See, for example, Tomlinson, M. J. (2000) *Foundation design and construction* (7th edition), Prentice Hall or Bowles, J. E. (1997) *Foundation analysis and design* (5th edition), McGraw-Hill.

3. Meyerhof, G. G. (1963) 'Some recent research on the bearing capacity of foundations', *Can. Geotech. J.*, 1(1), pp. 16–26.

4. Buisman, A. S. K. (1940) *Grondmechanica*, Delft, The Netherlands: Waltman; Terzaghi, K. (1943) *Theoretical soil mechanics*, New York: Wiley.

5. Reissner, H. (1924) 'Zum Erddruckproblem', *1st Int. Conf. on Applied Mechanics*, Delft, pp. 295–311.

6. Prandtl, L. (1921), 'Uber die Eindringungsfestigkeit plastischer Baustoffe und die Festigkeit von Schneiden', *Zeitsch. Angew. Mathematik und Mechanik*, 1, 15–20.

7. Sieffert, J. G., and Bay-gress, C. (2000) 'Comparison of European bearing capacity calculation methods for shallow foundations', *Geotechnical Engineering*, 143, pp. 65–75.

8. Brinch-Hansen, J. (1970) *A revised and extended formula for bearing capacity*, Danish Geotechnical Institute, Bulletin No. 28, 6pp.

9. Meyerhof, ibid.

10. American Petroleum Institute (2000), *Recommended practice for planning, designing and constructing fixed offshore platforms — Working Stress Design.* 226pp.

11. Vesic, A. S. (1973) 'Analysis of ultimate loads of shallow foundations', *J. Soil Mech. Found. Div.*, Am. Soc. Civ. Engrs, 99(1), pp. 45–73.

12. Ukritchon, B., Whittle, A., and Klangvijit, C. (2003) 'Calculations of bearing capacity factor N_γ using numerical limit analyses', *Journal of Geotechnical and Geoenvironmental Engineering*, Am. Soc. Civ. Engrs.

13. Chen, W.F. (1975) *Limit analysis and soil plasticity*, Elsevier.

14. Brinch-Hansen, ibid.

15. Meyerhof, ibid.

16. Sieffert et al., ibid.

17. Prandtl, ibid.

18. Meyerhof, ibid.

19. Salgado, R., Lyamin, A. V., Sloan, S. W., and Yu, H. S. (2004) 'Two- and three-dimensional bearing capacity of foundations in clay', *Géotechnique*, 54, pp. 297–306.

20. Bolton, M.D. (1986) 'The strength and dilatancy of sands', *Géotechnique*, 36(1), pp. 65–78.

21. See, for example, Schofield, A. N., and C. P. Wroth (1968) *Critical State Soil Mechanics*, McGraw-Hill; or Bolton, M. D. (1991) *A guide to soil mechanics* (3rd edition), M.D. & K. Bolton.

22. EN 1992, Eurocode 2 — Design of concrete structures, European Committee for Standardization, Brussels.

23. BS EN ISO 14688, Geotechnical investigation and testing — Identification and classification of soil, Part 2: Principles for a classification, British Standards Institution.

24. Christian, J.T., and W.D. Carrier (1978) 'Janbu, Bjerrum and Kjaernsli's chart reinterpreted', *Can. Geo. J.*, 15, p. 5.

Chapter 11

Design of gravity walls

The design of gravity walls is covered by Section 9 of Eurocode 7 Part 1, 'Retaining structures', whose contents are as follows:

§9.1   General (6 paragraphs)
§9.2   Limit states (4)
§9.3   Actions, geometrical data and design situations (26)
§9.4   Design and construction considerations (10)
§9.5   Determination of earth pressures (23)
§9.6   Water pressures (5)
§9.7   Ultimate limit state design (26)
§9.8   Serviceability limit state design (14)

A gravity wall is a structure whose self-weight (including any backfill on its base) plays a significant role in the support of the retained material. Such walls are usually made of stone or concrete and have a base footing (with or without heel), a ledge, and – when necessary – a buttress.   *[EN 1997-1 §9.1.2.1]*

Structures composed of elements of both gravity and embedded walls are called 'composite' in Eurocode 7. These include earth structures reinforced by tendons, geotextiles, or grouting and structures with multiple rows of ground anchorages or soil nails. Composite walls should be designed according to the rules discussed here and in Chapter 12.   *[EN 1997-1 §9.1.2.3]*

Section 9 of EN 1997-1 applies to structures which retain ground (soil, rock, or backfill) and water, where 'retained' means 'kept at a slope steeper than it would eventually adopt if no structure were present'.   *[EN 1997-1 §9.1.1(1)P]*

11.1   Ground investigation for gravity walls

Annex B.3 of Eurocode 7 Part 2 provides outline guidance on the depth of investigation points for retaining structures, as illustrated in **Figure 11.1**. (See Chapter 4 for guidance on the spacing of investigation points.)

The recommended minimum depth of investigation, z_a, for excavations where the groundwater table is below formation level is the greater of:
$z_a \geq 0.4h$ and $z_a \geq (t + 2m)$

Where the groundwater is above formation, z_a is the greater of:

$$z_a \geq (H + 2m) \text{ and } z_a \geq (t + 2m)$$

If all the strata encountered are impermeable, the depth of investigation should also satisfy:

$$z_a \geq (t + 5m)$$

The depth z_a may be reduced to 2m if the wall is built on competent strata[†] with 'distinct' (i.e. known) geology. With 'indistinct' geology, at least one borehole should go to at least 5m.

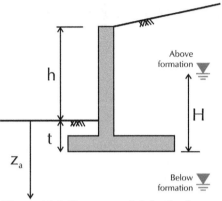

Figure 11.1. *Recommended depth of investigation points for excavations*

If bedrock is encountered, it becomes the reference level for z_a.

[EN 1997-2 §B.3(4)]

Greater depths of investigation may be needed for very large or highly complex projects or where unfavourable geological conditions are encountered.   *[EN 1997-2 §B.3(2)NOTE and B.3(3)]*

11.2   Design situations and limit states

§9 of Eurocode 7 Part 1 includes a series of illustrations showing limit modes for gravity walls, including overall stability, foundation failure, and structural failure. **Figure 11.2** gives examples of ultimate limit states that can affect mass gravity walls. From left to right, these include: toppling, sliding, and bearing failure.

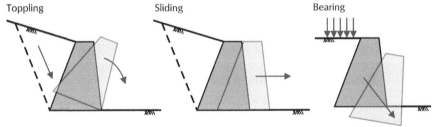

Figure 11.2. *Examples of ultimate limit states for mass gravity walls*

[†]i.e. weaker strata are unlikely to occur at depth, structural weaknesses such as faults are absent, and solution features and other voids are not expected.

Examples of ultimate limit states that can affect L- and T-shaped gravity walls are shown in **Figure 11.3**. From left to right, these include: (top) toppling, sliding, and bearing failure; and (bottom) structural failure of the wall's stem and toe.

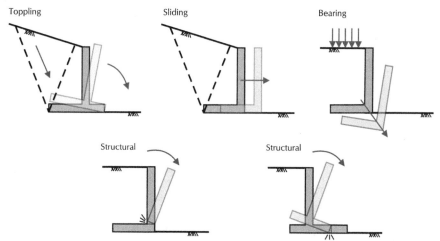

Figure 11.3. Examples of ultimate limit states for L- and T-shaped gravity walls

11.3   Basis of design

Eurocode 7 requires gravity walls to be designed so that ultimate limit states of bearing and sliding ('foundation failures') will not occur. Verification of these limit states must follow the Principles of §6 of EN 1997-1, 'Spread foundations', which are discussed in Chapter 10.   *[EN 1997-1 §9.7.3(1)P]*

Additionally, gravity walls must not fail due to overall instability of the ground in which they are installed.   *[EN 1997-1 §9.7.2(1)P]*

Regrettably, Eurocode 7 gives little detailed guidance on the design of gravity retaining walls. However, it does indicate the level of reliability that is required whichever design method is adopted. Reliability is ensured through a combination of partial factors on parameters (see Chapter 6), tolerances on dimensions (see Section 11.3.1), and selection of suitable water levels (see Section 11.3.2).

This book does not attempt to provide complete guidance on the design of gravity walls, for which the reader should refer to any well-established text on the subject.[1]

11.3.1   Unplanned excavation

EN 1997-1 requires an allowance to be made for the possibility of an unplanned excavation reducing formation level on the restraining side of a retaining wall (see **Figure 11.4**).

When normal levels of site control are employed, verification of ultimate limit states should assume an increase in retained height ΔH given by:

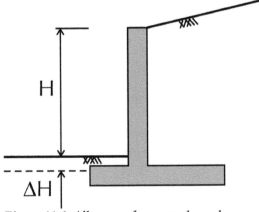

$$\Delta H = \frac{H}{10} \leq 0.5m$$

[EN 1997-1 §9.3.2.2(2)]

Eurocode 7 also warns that, where the surface level is particularly uncertain, larger values of ΔH should be used. However, it also allows smaller values

Figure 11.4. *Allowance for an unplanned excavation*

(including ΔH = 0) to be assumed when measures are put in place throughout the execution period to control the formation reliably.

[EN 1997-1 §9.3.2.2(3) and (4)]

The rules for unplanned excavations apply to ultimate limit states only and not to serviceability limit states. Anticipated excavations in front of the wall should be considered specifically – they are not, by definition, *unplanned*.

These rules provide the designer with considerable flexibility in dealing with the risk of over-digging. A more economical design may be obtained by adopting ΔH = 0, but the risk involved must be controlled during construction – leading to a supervision requirement that must be specified in the Geotechnical Design Report (see Chapter 16). A designer who wants to minimize the need for supervision must guard against the effects of over-digging by adopting ΔH = 10%H.

The rules for unplanned excavations in BS 8002[2] were changed in 2001 to match those of (the then draft) Eurocode 7 Part 1.

11.3.2   Selection of water levels

Selection of appropriate water levels is particularly important in retaining wall design and Eurocode 7 makes specific demands in this regard, as illustrated in **Figure 11.5**.

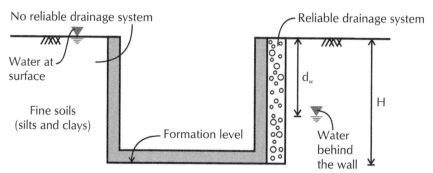

Figure 11.5. Assumed water levels behind retaining walls

When the wall retains medium or low permeability (fine) soils, then the wall must be designed for a water level above formation level. In other words, d_w < H, where d_w is the depth to the water table and H the height of the wall.

Eurocode 7 states that, without reliable drainage, the water level should normally[†] be taken at the surface of the retained material, i.e. d_w = 0 on the left-hand side of **Figure 11.5**. If, however, a reliable drainage system *is* provided, then the water level may be assumed to occur below the top of the wall, i.e. d_w > 0 on the right-hand side of **Figure 11.5**. In addition, a maintenance programme for the drainage system must be specified (in the Geotechnical Design Report – see Chapter 16) or the drains must be capable of operating adequately without maintenance.   *[EN 1997-1 §9.4.2(1)P and 9.6(3)]*

Current British practice[3] takes the distance d_w, when drainage is absent, as:

$$d_w = \frac{H}{4} \text{ for } H \leq 4m \text{ and } d_w = 1m \text{ for } H > 4m$$

which is less onerous than Eurocode 7's requirements.

It is rare for the water table to rise to ground surface and thus Eurocode 7's requirement could be regarded as too conservative. When the natural water

---

[†]Eurocode 7's requirements have been changed from a Principle ('shall') to an Application Rule ('should') in Corrigendum AC:2009

table can be demonstrated to come below formation level, it is advisable to assume less onerous water levels even in fine soils.

11.3.3   Wall friction and adhesion

The design angle of wall friction δ_d may be taken as:

$$\delta_d = k \times \varphi_{cv,d}$$

where $\varphi_{cv,d}$ is the design value of the soil's constant volume (i.e. critical state) angle of shearing resistance. For pre-cast concrete walls supporting sand or gravel, k = 2/3 and, for cast-in-place concrete walls, k = 1.

[EN 1997-1 §9.5.1(6) and 9.5.1(7)]

As discussed in Chapter 10, it may be preferable to select $\varphi_{cv,d}$ directly rather than calculate it from:

$$\varphi_{cv,d} = \tan^{-1}\left(\frac{\tan \varphi_{cv,k}}{\gamma_\varphi}\right)$$

where $\varphi_{cv,k}$ is the soil's characteristic constant volume angle of shearing resistance and γ_φ is the partial factor on shearing resistance.

11.3.4   Passive earth pressure – favourable action or resistance?

A question of particular importance in the design of retaining walls is whether passive earth pressure should be regarded as a resistance or as a favourable action. As a favourable action, the characteristic thrust $P_{p,k}$ produced by the passive earth pressure would be multiplied by the partial factor $\gamma_{G,fav}$, as follows:

$$P_{p,d} = \gamma_{G,fav} P_{p,k}$$

whereas, as a resistance, it would be divided by the partial factor γ_{Re}:

$$P_{p,d} = \frac{P_{p,k}}{\gamma_{Re}}$$

The values of these partial factors for each of Eurocode 7's Design Approaches (see Chapter 6) are summarized in the table below.

Individual partial factor or partial factor 'grouping'	Design Approach			
	1		2	3
	Combination 1	Combination 2		
$\gamma_{G,fav}$	1.0	1.0	1.0	1.0
γ_{Re}	1.0	1.0	1.4	1.0

Thus the answer to this question only has any consequence in verifications using Design Approach 2. From a practical point of view, the requirement to allow for an unplanned excavation in front of the wall (see Section 11.3.1) makes it unlikely that passive earth pressure will have a significant influence on the outcome of the verifications. A simple (and conservative) expedient is to ignore the presence of passive earth pressure.

11.4   Reinforced concrete walls

Figure 11.6 shows the pressures that act on a T-shaped gravity wall, assuming that a surcharge q exists at ground surface and the water table is located above formation level. The assumption is made that the wall's heel is wide enough for a Rankine zone to form within the backfill that sits on top of the wall's heel (see Section 11.4.4 for further discussion of this point).

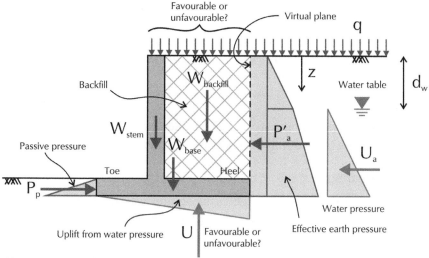

Figure 11.6. Earth pressures acting on a reinforced concrete wall

The horizontal *total* earth pressure σ_a acting on the virtual plane at depth z below ground surface is given by:[†]

$$\sigma_a = \sigma'_a + u = K_a \left(\int_0^z \gamma \, dz + q - u \right) - 2c'\sqrt{K_a} + u$$

[†]This equation is missing from Eurocode 7 (which only gives an equation for dry soil), but a version of it appears in the UK National Annex.

where σ'_a = the horizontal *effective* earth pressure and u = the pore water pressure in the ground at z; K_a = the ground's active earth pressure coefficient, γ = its weight density, and c' = its effective cohesion; and q = the surcharge at ground surface behind the virtual plane. (This equation ignores the beneficial effect of any adhesion along the virtual plane.)

The pore water pressure is given by:

$$u = \gamma_w \times (z - d_w)$$

where γ_w = the weight density of water and d_w = the depth of the water table below the top of the wall.

For the special case where the pore pressure is zero (i.e. above the water table), the equation for σ_a reduces to:

$$\sigma_a = K_a \left(\int_0^z \gamma dz + q \right) - 2c' \sqrt{K_a} \qquad \text{[EN 1997-1 exp (C.1 modified)]}$$

Integration of these equations allows the active effective earth and water thrusts (P'_a and U_a on **Figure 11.6**) to be calculated:

$$P'_a = \int_0^H \sigma'_a dz \text{ and } U_a = \int_0^H u dz$$

where H is the height of the virtual plane above the base of the wall. Both of these actions are unfavourable for bearing, sliding, and toppling of the wall.

11.4.1   Bearing

Section 6 of Eurocode 7 Part 1 requires the design vertical action V_d acting on the wall's foundation to be less than or equal to the design bearing resistance R_d of the ground beneath it:

$$V_d \leq R_d \qquad \text{[EN 1997-1 exp (6.1)]}$$

which, as discussed in Chapter 10, may be re-written as:

$$q_{Ed} \leq q_{Rd}$$

where q_{Ed} is the design bearing pressure on the ground (an action effect) and q_{Rd} is the corresponding design resistance.

The design bearing pressure q_{Ed} beneath the base of the wall is given by:

$$q_{Ed} = \frac{\gamma_G W_{Gk} + \sum_i \gamma_{Q,i} \psi_i V_{Qk,i}}{A'} = \gamma_G \left(\frac{W_{Gk}}{A'} \right) + \sum_i \gamma_{Q,i} \psi_i q_{Qk,i}$$

where W_{Gk} is the wall's characteristic permanent self-weight (including backfill); V_{Qk} is a characteristic variable vertical action imposed on the wall (to the left of the virtual plane); A' is the effective area of the base; γ_G and γ_Q

are partial factors on permanent and variable actions, respectively; and ψ_i is the combination factor applicable to the ith variable action (see Chapter 2).

The wall's self-weight is simply the sum of the weights of its stem and base, plus that of the backfill to the left of the virtual plane (W_{stem}, W_{base}, and $W_{backfill}$ on **Figure 11.7**). Since these are *unfavourable* actions in bearing, the characteristic weight densities should be selected as upper (or 'superior') values.

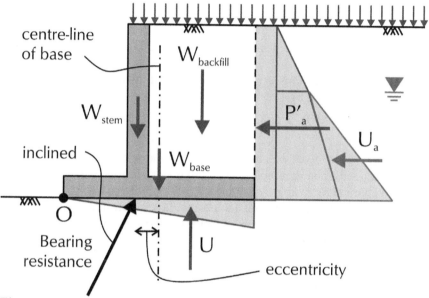

Figure 11.7. Inclined, eccentric resistance to actions on reinforced concrete wall

Bearing resistance is calculated using the methods discussed in Chapter 6 for footings subject to vertical actions. However, whereas most footings are subject to simple vertical loading, the footing of a reinforced concrete wall has to resist an eccentric, inclined action owing to the combination of the wall's self-weight and the horizontal thrust on the virtual plane ($P'_a + U_a$ on **Figure 11.7**). These factors have a strong detrimental influence on the bearing resistance that the ground provides. They also make the calculation of that resistance much more complicated than for a simple footing subject to vertical loading, as the worked examples at the end of this chapter show.

11.4.2   Sliding

Section 6 of Eurocode 7 Part 1 requires the design horizontal action H_d acting on the virtual plane to be less than or equal to the sum of the design resistance R_d from the ground beneath the footing and any design passive resistance R_{pd} on the side of the wall (see **Figure 11.6**):

$$H_d \le R_d + R_{pd}$$ [EN 1997-1 exp (6.2)]

which may be re-written as:

$$H_{Ed} \le H_{Rd}$$

where H_{Ed} is the design horizontal action effect and H_{Rd} is the corresponding total design horizontal resistance.

The design horizontal action effect H_{Ed} is given by:

$$H_{Ed} = H_d = P'_{ad} + U_{ad}$$

where P'_{ad} and U_{ad} are the design values of P'_a and U_a shown on **Figure 11.6**.

When drained conditions apply, the magnitude of the design horizontal resistance R_d is given by:

$$R_d = \frac{(W_{Gd} - U_{Gd}) \times \tan \delta_d}{\gamma_{Rh}} = \left(\frac{\gamma_{G,fav} W_{Gk} - \gamma_G U_{Gk}}{\gamma_{Rh}} \right) \times \left(\frac{\tan \delta_k}{\gamma_\varphi} \right)$$

where W_{Gk} = the wall's characteristic permanent self-weight, including backfill; U_{Gk} = the characteristic permanent water upthrust beneath the base; δ_k = the characteristic angle of interface friction between the base and the ground; $\gamma_{G,fav}$ and γ_G = partial factors on favourable and unfavourable actions, respectively; γ_{Rh} = a partial factor on sliding resistance; and γ_φ = a partial factor on shearing resistance.

In this equation, the wall's self-weight is regarded as a *favourable* action since an increase in its value would increase the sliding resistance, R_d; whereas the water upthrust is treated as an *unfavourable* action since an increase in its value would decrease R_d. Hence a larger factor $\gamma_G > \gamma_{G,fav}$ is applied to U_{Gk} than to W_{Gk}.

Another reason why the water upthrust should be treated as an unfavourable action is the fact that the horizontal water thrust U_a shown on **Figure 11.7** is clearly an unfavourable action. Since both forces arise from the same source, for consistency they should be factored the same way. This argument could be extended to the weight of the backfill, since it contributes both to the effective earth thrust P'_a and the weight of the wall. However, it is not at all certain that the backfill on top of the wall heel will be compacted as much as that outside the virtual plane, and we therefore recommend treating the earth pressure and the self-weight of the backfill as separate actions (this is also a more conservative assumption).

If the previous equation had been written in terms of submerged weight, as:

$$R_d = \frac{W'_{Gd} \times \tan \delta_d}{\gamma_{Rh}}$$

then only a single partial factor (presumably $\gamma_{G,fav}$) could be applied to the term W'_{Gd} (= $W_{Gd} - U_{Gd}$) and the design resistance would have been overestimated. Because of this, we recommend that the traditional use of submerged weights in gravity wall calculations should be discontinued.

The weight of the imposed surcharge should be omitted from the calculation of design resistance, since it is (usually) a variable action and therefore a more critical design situation arises when it is absent.

11.4.3 Toppling

Verification of resistance to toppling requires the design destabilizing moment $M_{Ed,dst}$ acting about the wall's toe (point 'O' on **Figure 11.7**) to be less than or equal to the design stabilizing moment $M_{Ed,stb}$ acting about the same point:

$$M_{Ed,dst} \leq M_{Ed,stb}$$

The forces that contribute to the destabilizing moment are the effective earth thrust (P'_a) and the water thrust (U_a) behind the virtual plane, plus the water uplift (U) – as shown on **Figure 11.7**. The effective earth thrust should include the contribution from the imposed surcharge acting at ground surface behind the virtual plane.

The forces that contribute to the stabilizing moment are the self-weights of the wall stem (W_{stem}) and base (W_{base}), plus that of the backfill on the wall heel ($W_{backfill}$) – again, as shown on **Figure 11.7**. The weight of the imposed surcharge should be omitted from the stabilizing moment, since it is (usually) a variable action and therefore a more critical design situation arises when it is absent.

11.4.4 Reinforced concrete walls with narrow heels

The discussions of bearing, sliding, and toppling in the previous sections have assumed that the heel of the reinforced concrete wall is large enough for a Rankine zone to form within the area bounded by the so-called 'virtual plane' (see **Figure 11.8**, top). This assumption greatly simplifies the calculations that need to be undertaken to verify the wall's strength and stability (because active forces can be resolved vertically and horizontally along the virtual plane and friction ignored).

The assumption is valid when the heel's width 'b' satisfies the inequality:

$$b \geq h \times \tan\left(45° - \frac{\varphi}{2}\right)$$

where h is the retained height of the wall, measured from the underside of its base, and φ is the angle of shearing resistance of the backfill. When this inequality is satisfied, no friction occurs along the virtual plane[4] and the active effective earth thrust P'_a is inclined at the same angle β to the horizontal as is the ground surface behind the wall.

When the inequality is not satisfied, calculations of wall strength and stability are normally based on the mechanism shown at the bottom of **Figure 11.8**. Backfill inside the inclined virtual plane is considered to be part of the wall and the forces acting on that virtual plane are treated as destabilizing actions.

The horizontal and vertical components of P'_a are given by:

$$P'_{ah} = P'_a \cos(\theta + \delta)$$

and

$$P'_{av} = P'_a \sin(\theta + \delta)$$

where δ is the angle of interface friction along the virtual plane and the angle θ (shown on **Figure 11.8**) is given by:

$$\theta = \tan\left(\frac{b}{h}\right)$$

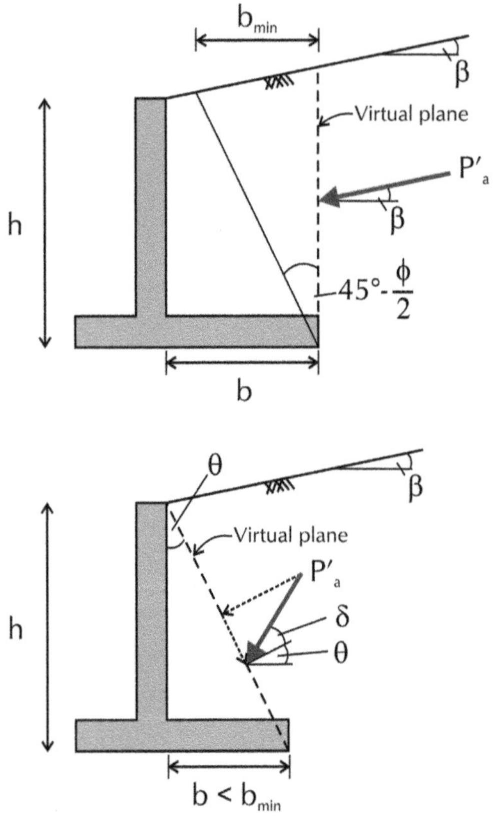

Figure 11.8. (Top) Virtual plane with no friction at back of reinforced concrete wall and (bottom) inclined virtual plane with friction

The procedures used to design such walls are identical to those used for mass gravity walls, as discussed at length in Section 11.5 below.

11.5   Mass gravity walls

Figure 11.9 shows the pressures that act on a mass gravity wall, assuming that a surcharge exists at ground surface and the water table is located above formation level. Because the back face of the wall is inclined at an angle θ to the vertical, the effective earth pressures acting on the wall are inclined. The simplification made for reinforced concrete walls, of a Rankine zone behind the wall (see Section 11.4.4), is not valid for this situation.

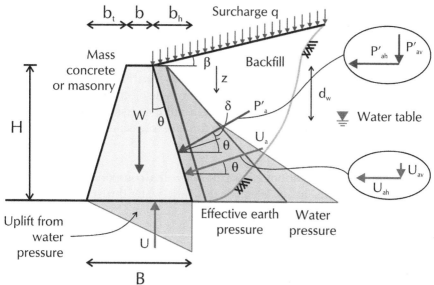

Figure 11.9. Earth pressures acting on a mass gravity wall

The normal component σ'_{an} of the *effective* earth pressure σ'_a acting on the back of the wall at depth z below ground surface is given by:

$$\sigma'_{an} = K_{a\gamma}\left(\int_0^z \gamma dz - u\right) + K_{aq}q - K_{ac}c'$$

where $K_{a\gamma}$, K_{aq}, K_{ac} = active earth pressure coefficients for soil weight, surcharge loading, and effective cohesion, respectively; γ = the ground's weight density and c' = its effective cohesion; and q = the surcharge at ground surface behind the wall.

The pore water pressure u acting on the back of the wall at depth z below ground surface is given by:

$$u = \gamma_w \times (z - d_w) \geq 0$$

where γ_w = the weight density of water and d_w = the depth of the water table below the top of the wall.

The normal component σ_{an} of the *total* earth pressure σ_a acting on the back of the wall at depth z below ground surface is then given by:

$$\sigma_{an} = \sigma'_{an} + u$$

Annex C of EN 1997-1 provides a numerical procedure[5] for determining the active earth pressure coefficients $K_{a\gamma}$, K_{aq}, K_{ac} based on the following expressions:

$$K_{a\gamma} = K_n \times \cos\beta \times \cos(\beta - \theta)$$

$$K_{aq} = K_n \times \cos^2\beta$$

$$K_{ac} = (K_n - 1) \times \cot\varphi$$

where the auxiliary coefficient K_n is a function of the soil's angle of shearing resistance (φ), the angle of interface friction between the soil and the wall (δ), the slope of the ground surface (β), and the angle of inclination of the back face of the wall (θ). Chapter 12 provides a full discussion of the numerical procedure given in Annex C, which includes expressions for passive earth pressure coefficients as well.

Integration of these equations allows the horizontal components (P'_{ah} and U_{ah} on **Figure 11.9**) of the active effective earth and water thrusts (P'_a and U_a) to be calculated:

$$P'_{ah} = \int_0^H \sigma'_a \cos(\theta + \delta) dz = \int_0^H \sigma'_{an} \cos\theta dz \text{ and } U_{ah} = \int_0^H u \cos\theta dz$$

where H is the wall's height above formation level. Both of these actions are unfavourable for bearing and sliding of the wall and destabilizing for toppling.

The vertical components (P'_{av} and U_{av} on **Figure 11.9**) of the active effective earth and water thrusts (P'_a and U_a) can then be determined from the horizontal components (P'_{ah} and U_{ah}), as follows:

$$P'_{av} = P'_{ah} \times \tan(\theta + \delta) \text{ and } U_{av} = U_{ah} \times \tan\theta$$

These actions are unfavourable for bearing, favourable for sliding, and stabilizing for toppling of the wall. However, the Single-Source Principle (discussed in Chapter 3) requires these forces to be treated as *unfavourable* in all these verifications.

Verification of bearing, sliding, and toppling resistance for a mass gravity wall is similar to that for a reinforced concrete wall (discussed in Sections 11.4.1–11.4.3), except that the self-weight of the mass gravity wall does not include the weight of backfill. Instead, any vertical force that backfill imposes on the wall is incorporated in the vertical component P'_{av} of the active effective earth thrust P'_a.

11.5.1   Bearing

Verification of bearing resistance for a mass gravity wall is similar to that for a reinforced concrete wall (discussed in Section 11.4.1).

The total design bearing pressure q_{Ed} beneath the base of the wall is given by:

$$q_{Ed} = \gamma_G \left(\frac{W_{Gk} + P'_{av,Gk} + U_{av,Gk}}{A'} \right) + \sum_i \gamma_{Q,i} \psi_i q_{Qk,i}$$

where W_{Gk} is the wall's characteristic permanent self-weight; $P'_{av,Gk}$ is the vertical component of the characteristic permanent active effective earth thrust; $U_{av,Gk}$ is the vertical component of the characteristic permanent water thrust; q_{Qk} is a characteristic variable vertical surcharge on the ground surface behind the wall; A' is the effective area of the base; γ_G and γ_Q are partial factors on permanent and variable actions, respectively; and ψ_i is the combination factor applicable to the ith variable action (see Chapter 2).

Since the self-weight of the wall and backfill are *unfavourable* actions for bearing, their characteristic weight densities should be selected as upper (or 'superior') values.

The effective design bearing pressure q'_{Ed} beneath the base is given by:

$$q'_{Ed} = q_{Ed} - \gamma_G \left(\frac{U_{Gk}}{A'} \right)$$

where U_{Gk} is the characteristic permanent uplift from water pressure beneath the wall's base. Note that the uplift here is treated as an unfavourable action (and multiplied by γ_G not $\gamma_{G,fav}$) owing to the 'Single-Source Principle' discussed in Chapter 3 (see the section *Distinction between favourable and unfavourable actions*). Since it arises from the same water table as U_{ah} and U_{av} (which are unfavourable actions), so the uplift U is also treated as unfavourable.

Chapter 3 discusses whether water pressures should be factored or not.

11.5.2   Sliding

Verification of the sliding resistance for a mass gravity wall is similar to that for a reinforced concrete wall (as discussed in Section 11.4.2).

The effective earth thrust (P'_a) is an unfavourable action for sliding. Thus both its horizontal and vertical components (P'_{ah} and P'_{av}) should be treated as unfavourable and design values obtained by multiplying by the partial factor γ_G.

The water thrust (U_a) is also an unfavourable action for sliding. Both its horizontal and vertical components $(U_{ah}$ and $U_{av})$ should be treated as unfavourable and design values obtained by multiplying by the partial factor γ_G. Since the 'Single-Source Principle' applies, the uplift beneath the wall (U_v) should also be treated as unfavourable.

11.5.3   Toppling

Verification of toppling resistance for a mass gravity wall is similar to that for a reinforced concrete wall (as discussed in Section 11.4.3).

11.6   Reinforced fill structures

'Eurocode 7 ... does not cover the detailed design of reinforced fill structures. The values of the partial factors ... given in EN 1997-1 have not been calibrated for reinforced fill structures.'[6]

The design of reinforced fill structures is currently carried out according to national standards, such as British Standard BS 8006.[7] Although national standards share many common features, differences in working practices, geology, and climate, etc. have delayed the development of a single design method accepted throughout Europe. European standard EN 14475 (discussed in Chapter 15) provides guidance on the execution of reinforced fill structures; a future European standard will cover their design.[8]

11.7   Design for serviceability

Design values of earth pressures for the verification of serviceability limit states must be derived using characteristic soil parameters, taking account of the initial stress, stiffness, and strength of the ground; the stiffness of structural elements; and the allowable deformation of the structure. These earth pressures may not reach limiting (i.e. fully active or passive) values.

[EN 1997-1 §9.8.1(2)P, (4), and (5)]

Calculations of wall movement are not necessarily required to verify the avoidance of serviceability limit states. As discussed in Chapter 8, Eurocode 7 acknowledges that deformations can be kept within required serviceability limits provided 'a sufficiently low fraction of the ground strength is mobilised'.

[EN 1997-1 §2.4.8(4)]

For conventional structures founded on clays, the serviceability limit state may be deemed to have been verified by an ultimate limit state calculation, provided the ratio of the characteristic bearing resistance R_k to the applied serviceability loads E_k is at least equal to three:

$$E_k \leq \frac{R_k}{\gamma_{R,SLS}}$$

in which $\gamma_{R,SLS}$ = a partial resistance factor ≥ 3. *[EN 1997-1 §6.6.2(16)]*

Since Eurocode 7 fails to provide a method for calculating the displacement of gravity walls, when such calculations are required, simple rules such as those given in CIRIA 516[9] could be used. The settlement 's' (in mm) of a gravity wall in a coarse-grained material can be estimated from:

$$s = \frac{2.5qB^{0.7}}{N^{1.4}}$$

where q = bearing pressure (kPa), B = the breath of the wall (m), and N = standard penetration test blow count. In an overconsolidated clay:

$$s = \frac{15qB}{c_u}$$

where c_u = the foundation soil's undrained strength and the other symbols are defined above.

Limiting values for the allowable wall and ground displacements must take into account the tolerance to those displacements of supported structures and services. *[EN 1997-1 §9.8.2(1)P]*

See Chapter 12 for further discussion of Eurocode 7's serviceability requirements for retaining structures.

11.8   Structural design

Gravity walls must be verified against structural failure according to the provisions of Eurocode 2[10] (for reinforced and mass concrete walls) and Eurocode 6 (for masonry walls).[11] *[EN 1997-1 §9.7.6(1)P]*

The Concrete Centre has published a series of 'How to...' guides[12] that will help engineers in these tasks and the Institution of Structural Engineers is also preparing a series of guides[13] on the subject.

11.9   Supervision, monitoring, and maintenance

There are no paragraphs in EN 1997-1 that specifically cover supervision, monitoring, and maintenance of gravity walls. There are also no execution standards that cover the topic. Practice should therefore continue to be based on existing national standards.[14]

11.10   Summary of key points

The design of gravity walls to Eurocode 7 involves checking that the ground beneath the wall has sufficient bearing resistance to withstand inclined, eccentric actions, sufficient sliding resistance to withstand horizontal and inclined actions, sufficient stability to avoid toppling, and sufficient stiffness to prevent unacceptable settlement or tilt. The first three of these guard against ultimate limit states and the last against a serviceability limit state.

Verification of the ultimate limit states is demonstrated by satisfying the inequalities:

$$V_d \leq R_d \text{, and } H_d \leq R_d + R_{pd} \text{, and } M_{Ed,dst} \leq M_{Ed,stb}$$

(where the symbols are defined in Section 10.3). These equations are merely specific forms of:

$$E_d \leq R_d$$

which is discussed at length in Chapter 6.

11.11   Worked examples

The worked examples in this chapter consider the design of a T-shaped gravity wall retaining dry fill under undrained conditions (Example 11.1); the same wall under drained conditions (Example 11.2); the same wall again, retaining wet fill under drained conditions (Example 11.3); and a mass concrete wall retaining granular fill (Example 11.4).

Specific parts of the calculations are marked ❶, ❷, ❸, etc., where the numbers refer to the notes that accompany each example.

11.11.1 T-shaped gravity wall retaining dry fill (undrained analysis)

Example 11.1 considers the design of a T-shaped gravity wall retaining dry fill, as shown in **Figure 11.10**.

Even though the wall is founded on clay, the backfill has been assumed to be granular (which would be typical for this type of wall). A fully effective drain at the heel of the wall has been included so that water pressures may (for simplicity) be ignored.

Notes on Example 11.1

❶ The effect of the weight of the back fill is favourable for sliding and toppling but unfavourable for bearing, thus for design different characteristic values for the weight density should be used depending on whether the weight of the back fill is favourable or unfavourable. However, the variation in weight density is likely to be small and in order to demonstrate more important features of the calculations a single weight density for the backfill has been used.

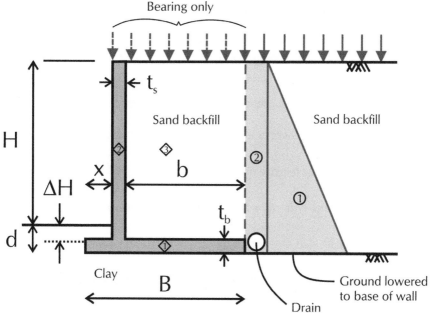

Figure 11.10. T-shaped gravity wall retaining dry fill

❷ Retaining walls are the one geotechnical structure where it is normal to apply an adjustment to the dimensions to allow for the potential for unplanned excavation on the passive side of the wall.

❸ The characteristic surcharge acts across the whole backfill and on top of the wall, where it is considered to be unfavourable (for bearing). Where the surcharge is considered to be favourable, it is excluded from the calculation.

❹ The partial factor sets are those given in Annex A. They are unmodified by the UK National Annex to BS EN 1997-1.

❺ For the assumption of a virtual back with Rankine earth pressures to be true a check needs to be carried out to ensure that it can develop fully. It should be noted that using standard design rules for proportioning T-shaped concrete walls will generally result in the conditions for a virtual back with Rankine earth pressures not being satisfied (as is the case here for Combination 2).

Example 11.1
T-shaped gravity wall retaining dry fill (undrained analysis)
Verification of undrained strength (limit state GEO)

Design situation

Consider a T-shaped gravity retaining wall, t_s = 250mm thick, which retains

H = 3.0m of sand fill and sits upon a medium strength clay. A drain will be installed at the wall heel to keep the fill dry. The base of the wall is B = 2.7m wide and t_b = 300mm thick, and its toe extends x = 0.5m in front of the stem. The underside of the base is d = 0.5m below formation level. The weight

density of reinforced concrete is $\gamma_{ck} = 25\dfrac{kN}{m^3}$ (as per EN 1991-1-1 Table

A.1). The fill has characteristic drained strength $\varphi_k = 36°$ and c'_k = 0kPa

The fill's weight density is $\gamma_k = 18\dfrac{kN}{m^3}$. ❶ The clay below the wall has

characteristic undrained strength $c_{uk,fdn}$ = 45kPa and weight density

$\gamma_{k,fdn} = 22\dfrac{kN}{m^3}$. A variable surcharge q_{Qk} = 10kPa may act at the top of the

wall during persistent and transient situations.

Design Approach 1

Geometrical parameters

Unplanned excavation $\Delta H = \min(10\% H, 0.5m) = 0.3$ m ❷
Design retained height $H_d = H + \Delta H = 3.3$ m
Width of heel $b = B - t_s - x = 1.95$ m

Actions

Characteristic vertical actions and moments (about toe) due to self-weight:

$$\text{Wall base: } W_{Gk_1} = \gamma_{ck} \times B \times t_b = 20.3\frac{kN}{m}$$

$$\text{Moment from base: } M_{k_1} = W_{Gk_1} \times \frac{B}{2} = 27.3\frac{kNm}{m}$$

$$\text{Wall stem: } W_{Gk_2} = \gamma_{ck} \times \left(H + d - t_b\right) \times t_s = 20\frac{kN}{m}$$

Moment from stem: $M_{k_2} = W_{Gk_2} \times \left(\dfrac{t_s}{2} + x\right) = 12.5 \dfrac{kNm}{m}$

Backfill: $W_{Gk_3} = \gamma_k \times b \times \left(H + d - t_b\right) = 112.3 \dfrac{kN}{m}$

Moment from backfill: $M_{k_3} = W_{Gk_3} \times \left(\dfrac{b}{2} + t_s + x\right) = 193.8 \dfrac{kNm}{m}$

Total characteristic self-weight $W_{Gk} = \sum W_{Gk} = 152.6 \dfrac{kN}{m}$

Total characteristic stabilizing moment $M_{Ek,stb} = \sum M_k = 233.6 \dfrac{kNm}{m}$

Characteristic surcharge (variable) $Q_{Qk} = q_{Qk} \times (B - x) = 22 \dfrac{kN}{m}$ ③

Material properties

Partial factors Sets $\begin{pmatrix} M1 \\ M2 \end{pmatrix}$: $\gamma_\varphi = \begin{pmatrix} 1 \\ 1.25 \end{pmatrix}$ $\gamma_c = \begin{pmatrix} 1 \\ 1.25 \end{pmatrix}$ $\gamma_{cu} = \begin{pmatrix} 1 \\ 1.4 \end{pmatrix}$ ④

Design shearing resistance of fill $\varphi_d = \tan^{-1}\left(\dfrac{\tan(\varphi_k)}{\gamma_\varphi}\right) = \begin{pmatrix} 36 \\ 30.2 \end{pmatrix}°$

Design effective cohesion of fill $c'_d = \dfrac{c'_k}{\gamma_c} = \begin{pmatrix} 0 \\ 0 \end{pmatrix} kPa$

Design undrained strength of clay $c_{ud,fdn} = \dfrac{c_{uk,fdn}}{\gamma_{cu}} = \begin{pmatrix} 45 \\ 32.1 \end{pmatrix} kPa$

Minimum width for Rankine $b_{min} = (H + d) \times \tan\left(45° - \dfrac{\varphi_d}{2}\right) = \begin{pmatrix} 1.78 \\ 2.01 \end{pmatrix} m$ ⑤

Effects of actions

Partial factors Sets $\begin{pmatrix} A1 \\ A2 \end{pmatrix}$: $\gamma_G = \begin{pmatrix} 1.35 \\ 1 \end{pmatrix}$ $\gamma_{G,fav} = \begin{pmatrix} 1 \\ 1 \end{pmatrix}$ $\gamma_Q = \begin{pmatrix} 1.5 \\ 1.3 \end{pmatrix}$

Design vertical actions:

unfavourable: $V_d = \gamma_G \times W_{Gk} + \gamma_Q \times Q_{Qk} = \begin{pmatrix} 239 \\ 181.2 \end{pmatrix} \dfrac{kN}{m}$ ⑥

favourable: $V_{d,fav} = \gamma_{G,fav} \times W_{Gk} = \begin{pmatrix} 152.6 \\ 152.6 \end{pmatrix} \dfrac{kN}{m}$ ⑥

Active earth pressure coefficient for fill $K_a = \dfrac{1 - \sin(\varphi_d)}{1 + \sin(\varphi_d)} = \begin{pmatrix} 0.26 \\ 0.331 \end{pmatrix}$ **7**

Design thrust on virtual back and destabilizing moments (about toe)

$$\text{Backfill } \overrightarrow{P_{ad_1}} = \left[\frac{\gamma_G \times K_a \times \gamma_k \times (H+d)^2}{2} \right] = \begin{pmatrix} 38.6 \\ 36.5 \end{pmatrix} \frac{kN}{m}$$

$$\text{Moment from backfill } M_{d_1} = P_{ad_1} \times \left(\frac{H+d}{3} \right) = \begin{pmatrix} 45.1 \\ 42.6 \end{pmatrix} \frac{kNm}{m}$$

$$\text{Surcharge } \overrightarrow{P_{ad_2}} = \left[\gamma_Q \times K_a \times q_{Qk} \times (H+d) \right] = \begin{pmatrix} 13.6 \\ 15.1 \end{pmatrix} \frac{kN}{m}$$

$$\text{Moment from surcharge } M_{d_2} = P_{ad_2} \times \left(\frac{H+d}{2} \right) = \begin{pmatrix} 23.9 \\ 26.4 \end{pmatrix} \frac{kNm}{m}$$

$$\text{Total design horizontal thrust } H_{Ed} = \sum_{i=1}^{2} P_{ad_i} = \begin{pmatrix} 52.3 \\ 51.6 \end{pmatrix} \frac{kN}{m}$$

$$\text{Total design destabilizing moment } M_{Ed,dst} = \sum_{i=1}^{2} M_{d_i} = \begin{pmatrix} 68.9 \\ 69 \end{pmatrix} \frac{kNm}{m}$$

Sliding resistance

$$\text{Partial factors from Sets } \begin{pmatrix} R1 \\ R1 \end{pmatrix} : \gamma_{Rh} = \begin{pmatrix} 1 \\ 1 \end{pmatrix} \text{ and } \gamma_{Rv} = \begin{pmatrix} 1 \\ 1 \end{pmatrix}$$

$$\text{Design undrained sliding resistance } \overrightarrow{H_{Rd}} = \left(\frac{c_{ud,fdn} \times B}{\gamma_{Rh}} \right) = \begin{pmatrix} 121.5 \\ 86.8 \end{pmatrix} \frac{kN}{m}$$

Bearing resistance

Design stabilizing moment due to self-weight and surcharge:

$$M_{Ed,stb} = \gamma_G \times M_{Ek,stb} + \gamma_Q \times Q_{Qk} \times \frac{(B+x)}{2} = \begin{pmatrix} 368.1 \\ 279.3 \end{pmatrix} \frac{kNm}{m}$$

$$\text{Eccentricity of load } e_B = \left(\frac{B}{2} - \frac{M_{Ed,stb} - M_{Ed,dst}}{V_d} \right) = \begin{pmatrix} 0.1 \\ 0.19 \end{pmatrix} m$$

Load is within middle-third of base if e_B is less than or equal to $\dfrac{B}{6} = 0.45 \, m$

Effective breadth and area are $B' = B - 2e_B = \begin{pmatrix} 2.5 \\ 2.32 \end{pmatrix}$ m and $A' = B'$

For cohesion $i_c = \left[\dfrac{1}{2} \left(1 + \sqrt{1 - \dfrac{\overrightarrow{H_{Ed}}}{A' \times c_{ud,fdn}}} \right) \right] = \begin{pmatrix} 0.87 \\ 0.78 \end{pmatrix}$

Total overburden at foundation base is $\sigma_{vk,b} = \gamma_{k,fdn} \times (d - \Delta H) = 4\,\text{kPa}$

Ultimate resistance $q_{ult} = \left[(\pi + 2) \times c_{ud,fdn} \times \overrightarrow{i_c} + \sigma_{vk,b} \right] = \begin{pmatrix} 204.8 \\ 133 \end{pmatrix}$ kPa ⑧

Design resistance $q_{Rd} = \dfrac{q_{ult}}{\gamma_{Rv}} = \begin{pmatrix} 204.8 \\ 133 \end{pmatrix}$ kPa

Toppling resistance
Design stabilizing moment due to self-weight alone:
$$M_{Ed,stb} = \gamma_{G,fav} \times M_{Ek,stb} = \begin{pmatrix} 233.6 \\ 233.6 \end{pmatrix} \begin{matrix} \text{kNm} \\ \text{m} \end{matrix}$$

Verifications
For undrained sliding $H_{Ed} = \begin{pmatrix} 52.3 \\ 51.6 \end{pmatrix} \dfrac{\text{kN}}{\text{m}}$ and $H_{Rd} = \begin{pmatrix} 121.5 \\ 86.8 \end{pmatrix} \dfrac{\text{kN}}{\text{m}}$

Degree of utilization $\left[\Lambda_{GEO,1} = \dfrac{H_{Ed}}{H_{Rd}} = \begin{pmatrix} 43 \\ 59 \end{pmatrix} \% \right]$ ⑨

For undrained bearing $q_{Ed} = \dfrac{V_d}{B'} = \begin{pmatrix} 95.4 \\ 78 \end{pmatrix}$ kPa and $q_{Rd} = \begin{pmatrix} 204.8 \\ 133 \end{pmatrix}$ kPa

Degree of utilization $\left[\Lambda_{GEO,1} = \dfrac{q_{Ed}}{q_{Rd}} = \begin{pmatrix} 47 \\ 59 \end{pmatrix} \% \right]$ ⑨

For toppling $M_{Ed,dst} = \begin{pmatrix} 68.9 \\ 69 \end{pmatrix} \dfrac{\text{kNm}}{\text{m}}$ and $M_{Ed,stb} = \begin{pmatrix} 233.6 \\ 233.6 \end{pmatrix} \dfrac{\text{kNm}}{\text{m}}$

Degree of utilization $\left[\Lambda_{GEO,1} = \dfrac{M_{Ed,dst}}{M_{Ed,stb}} = \begin{pmatrix} 30 \\ 30 \end{pmatrix} \% \right]$ ⑨

Design is unacceptable if degree of utilization is > 100%

Design Approach 2 (summary)

Verifications

For undrained sliding $H_{Ed} = 52.3\dfrac{kN}{m}$ and $H_{Rd} = 110.5\dfrac{kN}{m}$

Degree of utilization $\Lambda_{GEO,2} = \dfrac{H_{Ed}}{H_{Rd}} = 47\%$ ⑩

For undrained bearing $q_{Ed} = \dfrac{V_d}{B'} = 95.4\,kPa$ and $q_{Rd} = 146.3\,kPa$

Degree of utilization $\Lambda_{GEO,2} = \dfrac{q_{Ed}}{q_{Rd}} = 65\%$ ⑩

For toppling $M_{Ed,dst} = 68.9\dfrac{kNm}{m}$ and $M_{Ed,stb} = 233.6\dfrac{kNm}{m}$

Degree of utilization $\Lambda_{GEO,2} = \dfrac{M_{Ed,dst}}{M_{Ed,stb}} = 30\%$ ⑩

Design is unacceptable if degree of utilization is > 100%

Design Approach 3 (summary)

Verifications

For undrained sliding $H_{Ed} = 51.6\dfrac{kN}{m}$ and $H_{Rd} = 86.8\dfrac{kN}{m}$

Degree of utilization $\Lambda_{GEO,3} = \dfrac{H_{Ed}}{H_{Rd}} = 59\%$ ⑩

For undrained bearing $q_{Ed} = \dfrac{V_d}{B'} = 85.5\,kPa$ and $q_{Rd} = 132.8\,kPa$

Degree of utilization $\Lambda_{GEO,3} = \dfrac{q_{Ed}}{q_{Rd}} = 64\%$ ⑩

For toppling $M_{Ed,dst} = 69\dfrac{kNm}{m}$ and $M_{Ed,stb} = 233.6\dfrac{kNm}{m}$

Degree of utilization $\Lambda_{GEO,3} = \dfrac{M_{Ed,dst}}{M_{Ed,stb}} = 30\%$ ⑩

Design is unacceptable if degree of utilization is > 100%

❻ The variable surcharge needs to be considered in different positions depending which situation is being considered. Where the variable surcharge could be regarded as favourable it should not be included in the vertical actions and thus a variable surcharge is normally considered across the whole of the wall width for bearing capacity and overall stability only.

❼ Simple Rankine earth pressure coefficients have been used, ignoring friction along the virtual plane.

❽ The formulae for bearing capacity and inclination factors are those given in Annex D to EN 1997-1. Note: shape and depth factors are not included as the footing is considered to be a strip and there is little depth of embedment. As discussed in Chapter 6, EN 1997-1 does not include depth factors.

❾ The degree of utilization in Design Approach 1 for the undrained case is 59% for Combination 2 (for both sliding and bearing)

❿ The results for Design Approaches 2 and 3 are presented in summary only. The full calculations are available from the book's website at www.decodingeurocode7.com.

Design Approach 2 applies factors greater than 1.0 to actions and resistance. Bearing rather than sliding is critical in DA2 and the degree of utilization (65%) is higher than for DA1.

Design Approach 3 applies factors greater than 1.0 to structural actions (i.e. the self-weight of the concrete) and material properties. Bearing is critical and the degree of utilization (64%) is marginally lower than for DA2.

11.11.2   T-shaped gravity wall retaining dry fill (drained analysis)

Example 11.2 continues the design of the wall from Example 11.1 (as shown in **Figure 11.10**), but this time using a drained analysis to verify the wall's design under long-term conditions.

All dimensions are the same as in Example 11.1 and the drain is considered to be fully effective. However, for this example, effective stress parameters for the clay, which are appropriate for considering the long-term behaviour of the wall, have been considered.

Notes on Example 11.2

❶ Similar comments are applicable to aspects of the calculation to those given for the undrained case (see Example 11.1).

Example 11.2
T-shaped gravity wall retaining dry fill (drained analysis)
Verification of drained strength (limit state GEO)

Design situation

Re-consider the T-shaped gravity wall from the previous worked example, but now under persistent conditions. The foundation clay's characteristic peak angle of shearing resistance is $\varphi_{k,fdn} = 26°$, its effective cohesion $c'_{k,fdn} = 5kPa$, and its constant volume angle of shearing resistance $\varphi_{cv,k,fdn} = 20°$. All other parameters remain unchanged from before.

Groundwater has been taken as coincident with the base of the wall. ❶

Design Approach 1

Actions (from previous calculation)

Characteristic total self-weight of wall is $W_{Gk} = 152.6 \, kN/m$

Characteristic surcharge is $Q_{Qk} = 22 \, kN/m$

Characteristic stabilizing moment (about toe) $M_{Ek,stb} = 233.6 \, kNm/m$

Vertical action (unfavourable) $V_d = \begin{pmatrix} 239 \\ 181.2 \end{pmatrix} kN/m$

Vertical action (favourable) $V_{d,fav} = \begin{pmatrix} 152.6 \\ 152.6 \end{pmatrix} kN/m$

Material properties

Partial factors from Sets $\begin{pmatrix} M1 \\ M2 \end{pmatrix}$: $\gamma_\varphi = \begin{pmatrix} 1 \\ 1.25 \end{pmatrix}$ and $\gamma_c = \begin{pmatrix} 1 \\ 1.25 \end{pmatrix}$

Design shearing resistance of clay $\varphi_{d,fdn} = \tan^{-1}\left(\dfrac{\tan\left(\varphi_{k,fdn}\right)}{\gamma_\varphi} \right) = \begin{pmatrix} 26 \\ 21.3 \end{pmatrix}°$

Design effective cohesion of clay $c'_{d,fdn} = \dfrac{c'_{k,fdn}}{\gamma_c} = \begin{pmatrix} 5 \\ 4 \end{pmatrix} kPa$

UK NA to BS EN 1997-1 allows $\varphi_{cv,d}$ to be selected directly. Here, take the smaller of φ_d and $\varphi_{cv,k}$ i.e. $\varphi_{cv,d,fdn} = \overline{min\left(\varphi_{d,fdn}, \varphi_{cv,k,fdn}\right)} = \begin{pmatrix} 20 \\ 20 \end{pmatrix}°$

For concrete cast-in-place, $k = 1$ and $\delta_{d,fdn} = k \times \varphi_{cv,d,fdn} = \left(\dfrac{20}{20}\right)$ °②

Sliding resistance

Partial factors from Sets $\begin{pmatrix} R1 \\ R1 \end{pmatrix}$: $\gamma_{Rh} = \begin{pmatrix} 1 \\ 1 \end{pmatrix}$ and $\gamma_{Rv} = \begin{pmatrix} 1 \\ 1 \end{pmatrix}$

Design drained sliding resistance (ignoring adhesion, as required by EN 1997-1

exp. 6.3a) $H_{Rd} = \left(\dfrac{\overrightarrow{V_{d,fav} \times \tan\left(\delta_{d,fdn}\right)}}{\gamma_{Rh}} \right) = \begin{pmatrix} 55.5 \\ 55.5 \end{pmatrix} \dfrac{kN}{m}$ ③

Bearing resistance
Drained bearing capacity factors:

$$N_q = \left[\overrightarrow{e^{\left(\pi \tan\left(\varphi_{d,fdn}\right)\right)} \left(\tan\left(45° + \dfrac{\varphi_{d,fdn}}{2} \right) \right)^2} \right] = \begin{pmatrix} 11.9 \\ 7.3 \end{pmatrix} ④$$

$$N_c = \left[\overrightarrow{\left(N_q - 1\right) \times \cot\left(\varphi_{d,fdn}\right)} \right] = \begin{pmatrix} 22.3 \\ 16.1 \end{pmatrix} ④$$

$$N_\gamma = \left[\overrightarrow{2\left(N_q - 1\right) \times \tan\left(\varphi_{d,fdn}\right)} \right] = \begin{pmatrix} 10.59 \\ 4.91 \end{pmatrix} ④$$

Drained inclination factors (for effective length $L' = \infty\,m$)

$$\text{Exponent } m_B = \dfrac{\left(2 + \dfrac{B'}{L'} \right)}{\left(1 + \dfrac{B'}{L'} \right)} = \begin{pmatrix} 2 \\ 2 \end{pmatrix}$$

$$i_q = \left[1 - \left(\dfrac{H_{Ed}}{\overrightarrow{V_d + A' \times c'_{d,fdn} \times \cot\left(\varphi_{d,fdn}\right)}} \right) \right]^{m_B} = \begin{pmatrix} 0.64 \\ 0.56 \end{pmatrix} ④$$

$$i_c = \left[i_q - \left[\dfrac{\left(1 - i_q\right)}{\overrightarrow{N_c \times \tan\left(\varphi_{d,fdn}\right)}} \right] \right] = \begin{pmatrix} 0.61 \\ 0.49 \end{pmatrix} ④$$

$$i_\gamma = \left[1 - \left(\dfrac{H_{Ed}}{\overrightarrow{V_d + A' \times c'_{d,fdn} \times \cot\left(\varphi_{d,fdn}\right)}} \right) \right]^{m_B + 1} = \begin{pmatrix} 0.52 \\ 0.42 \end{pmatrix} ④$$

Drained bearing resistance

Effective overburden at foundation base

$$\sigma'_{vk,b} = \gamma_{k,fdn} \times (d - \Delta H) = 4.4 \text{ kPa}$$

From overburden $q_{ult_1} = \overrightarrow{\left(N_q \times i_q \times \sigma'_{vk,b}\right)} = \begin{pmatrix} 33.6 \\ 18 \end{pmatrix} \text{ kPa}$

From cohesion $q_{ult_2} = \overrightarrow{\left(N_c \times i_c \times c'_{d,fdn}\right)} = \begin{pmatrix} 68 \\ 31.7 \end{pmatrix} \text{ kPa}$

From self-weight $q_{ult_3} = \overrightarrow{\left[N_\gamma \times i_\gamma \times \left(\gamma_{k,fdn} - \gamma_w\right) \times \dfrac{B'}{2}\right]} = \begin{pmatrix} 83.5 \\ 29.2 \end{pmatrix} \text{ kPa}$ **⑤**

Total resistance $q_{ult} = \displaystyle\sum_{i=1}^{3} q_{ult_i} = \begin{pmatrix} 185.1 \\ 78.8 \end{pmatrix} \text{ kPa}$

Design resistance $q_{Rd} = \dfrac{q_{ult}}{\gamma_{Rv}} = \begin{pmatrix} 185.1 \\ 78.8 \end{pmatrix} \text{ kPa}$

Toppling resistance

Design stabilizing moment: $M_{Ed,stb} = \gamma_{G,fav} \times M_{Ek,stb} = \begin{pmatrix} 233.6 \\ 233.6 \end{pmatrix} \dfrac{\text{kNm}}{\text{m}}$

Verification of resistance

For drained sliding $H_{Ed} = \begin{pmatrix} 52.3 \\ 51.6 \end{pmatrix} \dfrac{\text{kN}}{\text{m}}$ and $H_{Rd} = \begin{pmatrix} 55.5 \\ 55.5 \end{pmatrix} \dfrac{\text{kN}}{\text{m}}$

Degree of utilization $\boxed{\Lambda_{GEO,1} = \dfrac{H_{Ed}}{H_{Rd}} = \begin{pmatrix} 94 \\ 93 \end{pmatrix} \%}$

For drained bearing $q_{Ed} = \begin{pmatrix} 95.4 \\ 78 \end{pmatrix} \text{ kPa and } q_{Rd} = \begin{pmatrix} 185.1 \\ 78.8 \end{pmatrix} \text{ kPa}$

Degree of utilization $\boxed{\Lambda_{GEO,1} = \dfrac{q_{Ed}}{q_{Rd}} = \begin{pmatrix} 52 \\ 99 \end{pmatrix} \%}$ **⑥**

For toppling $M_{Ed,dst} = \begin{pmatrix} 68.9 \\ 69 \end{pmatrix} \dfrac{\text{kNm}}{\text{m}}$ and $M_{Ed,stb} = \begin{pmatrix} 233.6 \\ 233.6 \end{pmatrix} \dfrac{\text{kNm}}{\text{m}}$

Degree of utilization $\boxed{\Lambda_{GEO,1} = \dfrac{M_{Ed,dst}}{M_{Ed,stb}} = \begin{pmatrix} 30 \\ 30 \end{pmatrix} \%}$

Design is unacceptable if degree of utilization is > 100%

Design Approach 2 (summary)

Verification of resistance

For drained sliding $H_{Ed} = 52.3 \dfrac{kN}{m}$ and $H_{Rd} = 50.5 \dfrac{kN}{m}$

Degree of utilization $\boxed{\Lambda_{GEO,2} = \dfrac{H_{Ed}}{H_{Rd}} = 104 \%}$ **❼**

For drained bearing $q_{Ed} = 95.4 \, kPa$ and $q_{Rd} = 132.2 \, kPa$

Degree of utilization $\boxed{\Lambda_{GEO,2} = \dfrac{q_{Ed}}{q_{Rd}} = 72 \%}$

For toppling $M_{Ed,dst} = 68.9 \dfrac{kNm}{m}$ and $M_{Ed,stb} = 233.6 \dfrac{kNm}{m}$

Degree of utilization $\boxed{\Lambda_{GEO,2} = \dfrac{M_{Ed,dst}}{M_{Ed,stb}} = 30 \%}$

Design is unacceptable if degree of utilization is > 100%

Design Approach 3 (summary)

Verification of resistance

For drained sliding $H_{Ed} = 51.6 \dfrac{kN}{m}$ and $H_{Rd} = 55.5 \dfrac{kN}{m}$

Degree of utilization $\boxed{\Lambda_{GEO,3} = \dfrac{H_{Ed}}{H_{Rd}} = 93 \%}$

For drained bearing $q_{Ed} = 85.5 \, kPa$ and $q_{Rd} = 83.9 \, kPa$

Degree of utilization $\boxed{\Lambda_{GEO,3} = \dfrac{q_{Ed}}{q_{Rd}} = 102 \%}$ **❼**

For toppling $M_{Ed,dst} = 69 \dfrac{kNm}{m}$ and $M_{Ed,stb} = 233.6 \dfrac{kNm}{m}$

Degree of utilization $\boxed{\Lambda_{GEO,3} = \dfrac{M_{Ed,dst}}{M_{Ed,stb}} = 30 \%}$

Design is unacceptable if degree of utilization is > 100%

❷ The selection of an appropriate angle of friction between the base of the wall and the ground is a matter of engineering judgement. The constant volume angle of friction, φ_{cv}, is the minimum likely value for a soil and thus it is considered that it would be unreasonable to adopt a design value lower than φ_{cv}.

❸ The drained sliding resistance excludes any influence of effective cohesion. This is conservative but reflects the likelihood that adhesion would not be generated along a soil/concrete interface. Furthermore, the vertical action should be taken as favourable since this reduces the sliding resistance.

❹ The formulae for bearing capacity and load inclination factors are those given in Annex D of EN 1997-1. As the wall is effectively a strip footing, shape factors are not required. Depth factors are not given in Annex D.

❺ Since the water level could rise to the base of the wall, the submerged weight density ($\gamma_{k,fdn} - \gamma_w$) should be used in this equation.

❻ For drained conditions, bearing is critical and indicates a higher degree of utilization than for the undrained case (99% vs 59%, both for Combination 2). Under drained conditions, the requirements of Eurocode 7 are just met.

The fact that bearing resistance is more critical in the drained case is contrary to the normal expectation that situations improve with time, i.e. bearing and sliding resistances should be greater for the drained situation. This suggests a lack of compatibility between undrained and drained parameters at low effective stresses. However, engineers are not traditionally comfortable using high values of effective c' and/or φ in clays that would give such compatibility.

❼ The results for Design Approaches 2 and 3 are presented in summary only. The full calculations are available from the book's website at www.decodingeurocode7.com.

Design Approach 2 applies factors greater than 1.0 to actions and resistance. Sliding is critical in DA2 and the degree of utilization (104%) is higher than Eurocode 7 allows.

Design Approach 3 applies factors greater than 1.0 to structural actions (i.e. the self-weight of the concrete) and material properties. Bearing is critical and the degree of utilization (102%) is marginally unacceptable under Eurocode 7.

11.11.3 T-shaped gravity wall retaining wet fill

Example 11.3 re-considers the design of the T-shaped gravity wall from Examples 11.1 and 11.2 (shown in **Figure 11.10**), allowing for the absence of a drain that is fully effective in the long term. (Alternatively, this example could represent the situation where a drain is positioned behind the wall stem, not behind the heel.) This potentially leads to a build-up of water behind the wall, as shown in **Figure 11.11**.

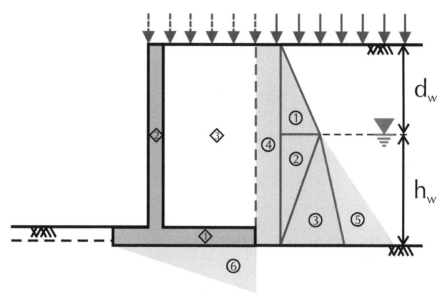

Figure 11.11. *T-shaped gravity wall retaining wet fill (for other dimensions, see* **Figure 11.10***)*

To counteract the additional horizontal pressure due to ground water, the wall's base has been widened.

Notes on Example 11.3

❶ The wall base has been widened in attempt to ensure that it is still satisfactory for DA1. Similar comments are applicable to aspects of the calculation to those given for Examples 11.1 and 11.2.

❷ Since the wall is now wider, both favourable and unfavourable vertical actions due to the wall's self -weight and backfill are increased.

❸ The presence of water in the backfill results in a more complicated pressure distribution on the virtual back and uplift on the base of the wall.

Example 11.3
T-shaped gravity wall retaining wet fill
Verification of drained strength (limit state GEO)

Design situation
Re-consider the design of the T-shaped gravity retaining wall from the previous worked example. Constraints during construction prevent a drain being placed at the heel of the wall. Therefore, the drain will be positioned behind the wall stem to reduce water level in the fill to a depth d_w = 1.5m below the retained surface. The base of the wall is increased to B = 4.3m wide but all other dimensions remain unchanged. Material properties are also unchanged. ❶

Design Approach 1

Geometrical parameters
Unplanned excavation $\Delta H = \min(10\% \ H, 0.5m) = 0.3 \ m$

Design retained height $H_d = H + \Delta H = 3.3 \ m$

Width of heel $b = B - t_s - x = 3.55 \ m$

Actions
Characteristic vertical actions and moments due to self-weight ❷

$\qquad$ Wall base: $W_{Gk_1} = \gamma_{ck} \times B \times t_b = 32.3 \ kN/m$

$\qquad$ Moment from base: $M_{k_1} = W_{Gk_1} \times \dfrac{B}{2} = 69.3 \ kNm/m$

$\qquad$ Wall stem: $W_{Gk_2} = \gamma_{ck} \times \left(H + d - t_b\right) \times t_s = 20 \ kN/m$

$\qquad$ Moment from stem: $M_{k_2} = W_{Gk_2} \times \left(\dfrac{t_s}{2} + x\right) = 12.5 \ kNm/m$

$\qquad$ Backfill: $W_{Gk_3} = \gamma_k \times b \times \left(H + d - t_b\right) = 204.5 \ kN/m$

$\qquad$ Moment from backfill: $M_{k_3} = W_{Gk_3} \times \left(\dfrac{b}{2} + t_s + x\right) = 516.3 \ kNm/m$

Total characteristic self-weight $W_{Gk} = \sum W_{Gk} = 256.7 \ kN/m$

Total characteristic stabilizing moment $M_{Ek,stb} = \sum M_k = 598.1 \ kNm/m$

Characteristic surcharge (variable) $Q_{Qk} = q_{Qk} \times (B - x) = 38 \ kN/m$

Soil stresses at depth of water table along virtual back of wall

vertical total stress $\sigma_{vk,w} = \gamma_k \times d_w = 27$ kPa

pore pressure $u_w = 0$ kPa

vertical effective stress $\sigma'_{vk,w} = \sigma_{vk,w} - u_w = 27$ kPa ❸

Soil stresses at wall heel

vertical total stress $\sigma_{vk,h} = \gamma_k \times (H + d) = 63$ kPa

height of water table $h_w = H + d - d_w = 2\,m$

pore pressure $u_h = \gamma_w \times h_w = 19.6$ kPa

vertical effective stress $\sigma'_{vk,h} = \sigma_{vk,h} - u_h = 43.4$ kPa ❸

Effects of actions

Partial factors, Sets $\begin{pmatrix} A1 \\ A2 \end{pmatrix}$: $\gamma_G = \begin{pmatrix} 1.35 \\ 1 \end{pmatrix}$, $\gamma_{G,fav} = \begin{pmatrix} 1 \\ 1 \end{pmatrix}$, and $\gamma_Q = \begin{pmatrix} 1.5 \\ 1.3 \end{pmatrix}$

Design vertical actions (unfavourable)

total $V_d = \gamma_G \times W_{Gk} + \gamma_Q \times Q_{Qk} = \begin{pmatrix} 403.6 \\ 306.1 \end{pmatrix} \dfrac{kN}{m}$

water upthrust $U_d = \gamma_G \times \dfrac{u_h}{2} \times B = \begin{pmatrix} 56.9 \\ 42.2 \end{pmatrix} \dfrac{kN}{m}$ ❹

effective $V'_d = V_d - U_d = \begin{pmatrix} 346.7 \\ 264 \end{pmatrix} \dfrac{kN}{m}$

Design vertical actions (favourable)

total $V_{d,fav} = \gamma_{G,fav} \times W_{Gk} = \begin{pmatrix} 256.7 \\ 256.7 \end{pmatrix} \dfrac{kN}{m}$

water upthrust $U_{d,fav} = \gamma_{G,fav} \times \dfrac{u_h}{2} \times B = \begin{pmatrix} 42.2 \\ 42.2 \end{pmatrix} \dfrac{kN}{m}$

effective $V'_{d,fav} = V_{d,fav} - U_{d,fav} = \begin{pmatrix} 214.6 \\ 214.6 \end{pmatrix} \dfrac{kN}{m}$

Active earth pressure coefficient $K_a = \dfrac{1 - \sin(\varphi_d)}{1 + \sin(\varphi_d)} = \begin{pmatrix} 0.26 \\ 0.331 \end{pmatrix}$

Design thrust on virtual back and destabilizing moments (about toe)

Dry backfill $\overrightarrow{P_{ad_1}} = \left(\dfrac{\gamma_G \times K_a \times \sigma'_{vk,w} \times d_w}{2} \right) = \begin{pmatrix} 7.1 \\ 6.7 \end{pmatrix} \dfrac{kN}{m}$

Moment from dry backfill $M_{d_1} = P_{ad_1} \times \left(h_w + \dfrac{d_w}{3} \right) = \begin{pmatrix} 17.7 \\ 16.8 \end{pmatrix} \dfrac{kNm}{m}$

Wet backfill (part) $P_{ad_2} = \overrightarrow{\left(\dfrac{\gamma_G \times K_a \times \sigma'_{vk,w} \times h_w}{2} \right)} = \begin{pmatrix} 9.5 \\ 8.9 \end{pmatrix} \dfrac{kN}{m}$

Moment from wet backfill (part) $M_{d_2} = P_{ad_2} \times \left(\dfrac{2\,h_w}{3} \right) = \begin{pmatrix} 12.6 \\ 11.9 \end{pmatrix} \dfrac{kNm}{m}$

Wet backfill (part) $P_{ad_3} = \overrightarrow{\left(\dfrac{\gamma_G \times K_a \times \sigma'_{vk,h} \times h_w}{2} \right)} = \begin{pmatrix} 15.2 \\ 14.4 \end{pmatrix} \dfrac{kN}{m}$

Moment from wet backfill (part) $M_{d_3} = P_{ad_3} \times \left(\dfrac{h_w}{3} \right) = \begin{pmatrix} 10.1 \\ 9.6 \end{pmatrix} \dfrac{kNm}{m}$

Surcharge $P_{ad_4} = \overrightarrow{\left[\gamma_Q \times K_a \times q_{Qk} \times (H+d) \right]} = \begin{pmatrix} 13.6 \\ 15.1 \end{pmatrix} \dfrac{kN}{m}$

Moment from surcharge $M_{d_4} = P_{ad_4} \times \left(\dfrac{H+d}{2} \right) = \begin{pmatrix} 23.9 \\ 26.4 \end{pmatrix} \dfrac{kNm}{m}$

Water $U_{ad} = \overrightarrow{\left(\dfrac{\gamma_G \times u_h \times h_w}{2} \right)} = \begin{pmatrix} 26.5 \\ 19.6 \end{pmatrix} \dfrac{kN}{m}$

Moment from water $M_{d_5} = U_{ad} \times \left(\dfrac{h_w}{3} \right) = \begin{pmatrix} 17.7 \\ 13.1 \end{pmatrix} \dfrac{kNm}{m}$

Moment from water uplift $M_{d_6} = U_d \times \left(\dfrac{2\,B}{3} \right) = \begin{pmatrix} 163.2 \\ 120.9 \end{pmatrix} \dfrac{kNm}{m}$

Total design horizontal thrust $H_{Ed} = \left(\sum_{i=1}^{4} \overrightarrow{P_{ad_i}} \right) + U_{ad} = \begin{pmatrix} 71.9 \\ 64.7 \end{pmatrix} \dfrac{kN}{m}$

Total design destabilizing moment $M_{Ed,dst} = \left(\sum_{i=1}^{6} \overrightarrow{M_{d_i}} \right) = \begin{pmatrix} 245.2 \\ 198.6 \end{pmatrix} \dfrac{kNm}{m}$

Sliding resistance

Partial factors from Sets $\begin{pmatrix} R1 \\ R1 \end{pmatrix}$: $\gamma_{Rh} = \begin{pmatrix} 1 \\ 1 \end{pmatrix}$ and $\gamma_{Rv} = \begin{pmatrix} 1 \\ 1 \end{pmatrix}$

Design drained sliding resistance (ignoring adhesion, as required by EN 1997-1

$$\text{exp. 6.3a) } H_{Rd} = \overline{\left[\frac{\left(V_{d,fav} - U_d\right) \times \tan\left(\delta_{d,fdn}\right)}{\gamma_{Rh}}\right]} = \begin{pmatrix} 72.7 \\ 78.1 \end{pmatrix} \frac{kN}{m} \; ⑤$$

Eccentricity of loads

Design stabilizing moment

$$M_{Ed,stb} = \gamma_G \times M_{Ek,stb} + \gamma_Q \times Q_{Qk} \times \frac{(B+x)}{2} = \begin{pmatrix} 944.3 \\ 716.7 \end{pmatrix} \frac{kNm}{m}$$

$$\text{Eccentricity of load } e_B = \overline{\left|\left(\frac{B}{2} - \frac{M_{Ed,stb} - M_{Ed,dst}}{V_d - U_d}\right)\right|} = \begin{pmatrix} 0.13 \\ 0.19 \end{pmatrix} m$$

Load is within middle-third of base if $e_B \leq \dfrac{B}{6} = 0.72\,m$

$$\text{Effective breadth is } B' = B - 2\,e_B = \overline{\begin{pmatrix} 4.03 \\ 3.93 \end{pmatrix}} m \quad \text{and area } A' = B'$$

Drained bearing capacity factors

$$N_q = \overline{\left[e^{\left(\pi \tan\left(\varphi_{d,fdn}\right)\right)}\left(\tan\left(45° + \frac{\varphi_{d,fdn}}{2}\right)\right)^2\right]} = \begin{pmatrix} 11.9 \\ 7.3 \end{pmatrix}$$

$$N_c = \overline{\left[\left(N_q - 1\right) \times \cot\left(\varphi_{d,fdn}\right)\right]} = \begin{pmatrix} 22.3 \\ 16.1 \end{pmatrix}$$

$$N_\gamma = \overline{\left[2\left(N_q - 1\right) \times \tan\left(\varphi_{d,fdn}\right)\right]} = \begin{pmatrix} 10.6 \\ 4.9 \end{pmatrix}$$

Drained inclination factors

$$\text{Effective length } L' = \infty\,m, \text{ hence exponent } m_B = \frac{\left(2 + \dfrac{B'}{L'}\right)}{\left(1 + \dfrac{B'}{L'}\right)} = \begin{pmatrix} 2 \\ 2 \end{pmatrix}$$

$$i_q = \overline{\left[1 - \left(\frac{H_{Ed}}{V'_d + A' \times c'_{d,fdn} \times \cot\left(\varphi_{d,fdn}\right)}\right)\right]}^{m_B} = \begin{pmatrix} 0.66 \\ 0.62 \end{pmatrix}$$

$$i_c = \left[i_q - \left[\frac{(1 - i_q)}{N_c \times \tan(\varphi_{d,fdn})}\right]\right] = \binom{0.63}{0.56}$$

$$i_\gamma = \left[1 - \left(\frac{H_{Ed}}{V'_d + A' \times c'_{d,fdn} \times \cot(\varphi_{d,fdn})}\right)\right]^{m_B+1} = \binom{0.54}{0.49}$$

Drained bearing resistance

Drained overburden at foundation base $\sigma'_{vk,b} = \gamma_{k,fdn} \times (d - \Delta H) = 4.4$ kPa

Ultimate resistance...

from overburden $q_{ult_1} = \overrightarrow{\left(N_q \times i_q \times \sigma'_{vk,b}\right)} = \binom{34.6}{19.9}$ kPa

from cohesion $q_{ult_2} = \overrightarrow{\left(N_c \times i_c \times c'_{d,fdn}\right)} = \binom{70.4}{36.1}$ kPa

from self-weight $q_{ult_3} = \overrightarrow{\left[N_\gamma \times i_\gamma \times \left(\gamma_{k,fdn} - \gamma_w\right) \times \frac{B'}{2}\right]} = \binom{140.8}{57.4}$ kPa

total $q_{ult} = \sum_{i=1}^{3} q_{ult_i} = \binom{245.9}{113.5}$ kPa

Design resistance $q'_{Rd} = \dfrac{q_{ult}}{\gamma_{Rv}} = \binom{245.9}{113.5}$ kPa

Verifications

For drained sliding $H_{Ed} = \binom{71.9}{64.7} \dfrac{kN}{m}$ and $H_{Rd} = \binom{72.7}{78.1} \dfrac{kN}{m}$

Degree of utilization $\boxed{\Lambda_{GEO,1} = \dfrac{H_{Ed}}{H_{Rd}} = \binom{99}{83} \%}$ ⑥

For drained bearing $q'_{Ed} = \dfrac{V'_d}{B'} = \binom{85.9}{67.2}$ kPa and $q'_{Rd} = \binom{245.9}{113.5}$ kPa

Degree of utilization $\boxed{\Lambda_{GEO,1} = \dfrac{q'_{Ed}}{q'_{Rd}} = \binom{35}{59} \%}$

Design is unacceptable if degree of utilization is > 100%

Design Approach 2 (summary)

Verification of resistance

For drained sliding $H_{Ed} = 71.9 \dfrac{kN}{m}$ and $H_{Rd} = \blacksquare \dfrac{kN}{m}$

Degree of utilization $\left| \Lambda_{GEO,2} = \dfrac{H_{Ed}}{H_{Rd}} = 109\% \right.$ ⑦

For drained bearing $q'_{Ed} = 85.9\ kPa$ and $q'_{Rd} = 175.6\ kPa$

Degree of utilization $\left| \Lambda_{GEO,2} = \dfrac{q'_{Ed}}{q'_{Rd}} = 49\% \right.$

For toppling $M_{Ed,dst} = 245.2 \dfrac{kNm}{m}$ and $M_{Ed,stb} = 944.3 \dfrac{kNm}{m}$

Degree of utilization $\left| \Lambda_{GEO,2} = \dfrac{M_{Ed,dst}}{M_{Ed,stb}} = 26\% \right.$

Design is unacceptable if degree of utilization is > 100%

Design Approach 3 (summary)

Verification of resistance

For drained sliding $H_{Ed} = 64.7 \dfrac{kN}{m}$ and $H_{Rd} = 78.1 \dfrac{kN}{m}$

Degree of utilization $\left| \Lambda_{GEO,3} = \dfrac{H_{Ed}}{H_{Rd}} = 83\% \right.$ ⑦

For drained bearing $q'_{Ed} = 79.3\ kPa$ and $q'_{Rd} = 113.7\ kPa$

Degree of utilization $\left| \Lambda_{GEO,3} = \dfrac{q'_{Ed}}{q'_{Rd}} = 70\% \right.$

For toppling $M_{Ed,dst} = 198.6 \dfrac{kNm}{m}$ and $M_{Ed,stb} = 716.7 \dfrac{kNm}{m}$

Degree of utilization $\left| \Lambda_{GEO,3} = \dfrac{M_{Ed,dst}}{M_{Ed,stb}} = 28\% \right.$

Design is unacceptable if degree of utilization is > 100%

❹ It is assumed that the uplift on the base of the wall can be represented by a simple triangular distribution, reducing from a maximum pressure at the heel to zero at the toe.

❺ For sliding resistance, vertical actions on the wall heel are favourable and so variable actions are ignored. However, the uplift due to the water pressure is unfavourable and results in a lower effective vertical force for DA1-1 compared with DA1-2. This results in a lower design resistance for DA1-1 compared with DA1-2, even though the design shear strength is lower for DA1-2.

❻ The introduction of ground water in the back fill results in a wider wall (B = 4.3m) being required to provide adequate sliding and bearing resistance compared with Examples 11.1 and 11.2 (for which B = 3.0m). For this example, sliding governs the design according to Combination 1. To improve the design account could be taken of passive resistance in front of the wall and/or a shear key provided.

❼ The results for Design Approaches 2 and 3 are presented in summary only. The full calculations are available from the book's website at www.decodingeurocode7.com.

Design Approach 2 applies factors greater than 1.0 to actions and resistance. Sliding is critical in DA2 and the degree of utilization (109%) is more than Eurocode 7 allows.

Design Approach 3 applies factors greater than 1.0 to structural actions (i.e. the self-weight of the concrete) and material properties. Sliding is again critical with a degree of utilization (83%), but is less onerous than for Design Approach 1.

11.11.4 Mass concrete wall retaining dry fill

Example 11.4 considers the design of a mass concrete wall retaining dry ground, as shown in **Figure 11.12**.

The wall, which has sloping front and back faces, sits on a strong rock and retains granular backfill. The ground surface behind the wall slopes upwards and carries a surcharge loading.

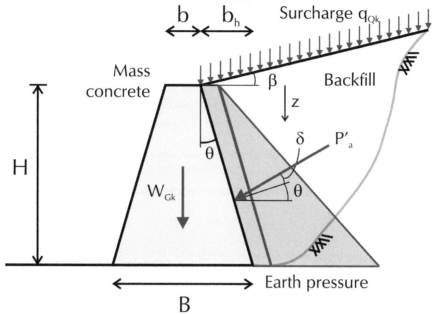

Figure 11.12. *Mass concrete wall retaining dry ground*

Notes on Example 11.4

❶ Similar comments are applicable to aspects of the calculation to those given for the undrained example, the reader is referred to this example for additional guidance.

❷ The selection of an appropriate angle of interface friction between the wall and the backfill is a matter of engineering judgement. The constant volume angle of friction, φ_{cv}, is the minimum likely value for a soil and thus it is considered that it would be unreasonable to adopt a design value of φ lower than φ_{cv}.

❸ The selection of an appropriate angle of interface friction between the wall base and the underlying ground is also a matter of engineering judgement. For a rock/concrete interface this may be taken as the design angle of friction for the rock.

❹ The values of earth pressure coefficients have been calculated using the equations given in Annex C of EN 1997-1 and discussed in Chapter 12.

Example 11.4
Mass concrete wall retaining granular fill
Verification of strength (limit state GEO)

Design situation

Consider a mass concrete gravity wall, B = 2.0m wide, which retains
H = 4.0m of granular fill and sits upon a strong rock (so bearing failure is not
a design issue). The top of the wall (which is symmetrical) is b = 1.0m
wide. The weight density of unreinforced concrete is γ_{ck} = 24 $\dfrac{kN}{m^3}$ (as per

EN 1991-1-1 Table A.1). The backfill has characteristic drained strength
parameters φ_k = 36°, c'_k = 0kPa, and weight density γ_k = 19 $\dfrac{kN}{m^3}$. The fill's

constant volume angle of shearing resistance is $\varphi_{cv,k}$ = 30°. The

characteristic angle of shearing resistance of the rock beneath the wall base
is $\varphi_{k,fdn}$ = 40°. The ground behind the wall slopes upwards at a slope of 1m

vertically to h = 4m horizontally, i.e. at an angle $\beta = \tan^{-1}\left(\dfrac{1m}{h}\right)$ = 14°. A

variable surcharge q_{Qk} = 10kPa acts on this ground surface during

persistent and transient situations. ❶

Design Approach 1

Geometrical parameters

There is no need to consider an unplanned excavation

Inclination of wall surface (virtual plane) $\theta = \dfrac{B-b}{2H}$ = 7.2°

Width of heel $b_h = \dfrac{B-b}{2}$ = 0.5 m

Actions

Characteristic self-weight of wall $W_{Gk} = \gamma_{ck} \times \left(\dfrac{B+b}{2}\right) \times H = 144\dfrac{kN}{m}$

Characteristic moment about toe (stabilizing)

$M_{Ek,stb} = W_{Gk} \times \dfrac{B}{2} = 144\dfrac{kNm}{m}$

Material properties

Partial factors from Sets $\begin{pmatrix} M1 \\ M2 \end{pmatrix}$ are $\gamma_\varphi = \begin{pmatrix} 1 \\ 1.25 \end{pmatrix}$ and $\gamma_c = \begin{pmatrix} 1 \\ 1.25 \end{pmatrix}$

Design shearing resistance of backfill $\varphi_d = \tan^{-1}\left(\dfrac{\tan\left(\varphi_k\right)}{\gamma_\varphi} \right) = \begin{pmatrix} 36 \\ 30.2 \end{pmatrix}°$

Design effective cohesion of backfill $c'_d = \dfrac{c'_k}{\gamma_c} = \begin{pmatrix} 0 \\ 0 \end{pmatrix}$ kPa

UK NA to BS EN 1997-1 allows $\varphi_{cv,d}$ to be selected directly. Here, take the

smaller of φ_d and $\varphi_{cv,k}$, i.e. $\varphi_{cv,d} = \overrightarrow{\min\left(\varphi_d, \varphi_{cv,k}\right)} = \begin{pmatrix} 30 \\ 30 \end{pmatrix}°$

For cast in place concrete $k = 1$

Interface friction between backfill and wall is $\delta_d = k \times \varphi_{cv,d} = \begin{pmatrix} 30 \\ 30 \end{pmatrix}°$ ❷

Design shearing resistance of rock $\varphi_{d,fdn} = \tan^{-1}\left(\dfrac{\tan\left(\varphi_{k,fdn}\right)}{\gamma_\varphi} \right) = \begin{pmatrix} 40 \\ 33.9 \end{pmatrix}°$

Interface friction between rock and wall is $\delta_{d,fdn} = k \times \varphi_{d,fdn} = \begin{pmatrix} 40 \\ 33.9 \end{pmatrix}°$ ❸

Effects of actions

Active earth pressure coefficients (giving normal components of stress)

$K_{a\gamma} = \begin{pmatrix} 0.304 \\ 0.385 \end{pmatrix}$, $K_{aq} = \begin{pmatrix} 0.297 \\ 0.377 \end{pmatrix}$, and $K_{ac} = \begin{pmatrix} 0.942 \\ 1.032 \end{pmatrix}$ ❹

Partial factors from Sets $\begin{pmatrix} A1 \\ A2 \end{pmatrix}$: $\gamma_G = \begin{pmatrix} 1.35 \\ 1 \end{pmatrix}$ $\gamma_{G,fav} = \begin{pmatrix} 1 \\ 1 \end{pmatrix}$ $\gamma_Q = \begin{pmatrix} 1.5 \\ 1.3 \end{pmatrix}$

From backfill:

design thrust: $P_{ahd_1} = \overrightarrow{\left(\gamma_G \times K_{a\gamma} \dfrac{\cos\left(\theta + \delta_d\right)}{\cos\left(\delta_d\right)} \times \dfrac{\gamma_k H^2}{2} \right)} = \begin{pmatrix} 57.4 \\ 53.9 \end{pmatrix} \dfrac{kN}{m}$

vertical thrust: $P_{avd_1} = \overrightarrow{\left(P_{ahd_1} \times \tan\left(\theta + \delta_d\right) \right)} = \begin{pmatrix} 43.5 \\ 40.9 \end{pmatrix} \dfrac{kN}{m}$ ❺

moment about toe: $M_{d_1} = P_{ahd_1} \times \dfrac{H}{3} = \begin{pmatrix} 76.5 \\ 71.9 \end{pmatrix} \dfrac{kNm}{m}$

From surcharge:

design thrust $P_{ahd_2} = \left(\gamma_Q \times K_{aq} \dfrac{\cos(\theta + \delta_d)}{\cos(\delta_d)} \times q_{Qk} H \right) = \begin{pmatrix} 16.4 \\ 18 \end{pmatrix} \dfrac{kN}{m}$

vertical thrust: $P_{avd_2} = \left(P_{ahd_2} \times \tan(\theta + \delta_d) \right) = \begin{pmatrix} 12.4 \\ 13.7 \end{pmatrix} \dfrac{kN}{m}$ ⑤

from surcharge $M_{d_2} = P_{ahd_2} \times \dfrac{H}{2} = \begin{pmatrix} 32.8 \\ 36.1 \end{pmatrix} \dfrac{kNm}{m}$

Total design horizontal thrust $H_{Ed} = \displaystyle\sum_{i=1}^{2} P_{ahd_i} = \begin{pmatrix} 73.8 \\ 71.9 \end{pmatrix} \dfrac{kN}{m}$

Total design vertical thrust $P_{avd} = \displaystyle\sum_{i=1}^{2} P_{avd_i} = \begin{pmatrix} 55.9 \\ 54.5 \end{pmatrix} \dfrac{kN}{m}$

Total design destabilizing moment $M_{Ed,dst} = \displaystyle\sum_{i=1}^{2} M_{d_i} = \begin{pmatrix} 109.3 \\ 107.9 \end{pmatrix} \dfrac{kNm}{m}$

Vertical action (unfavourable) $V_d = \gamma_G \times W_{Gk} + P_{avd} = \begin{pmatrix} 250.3 \\ 198.5 \end{pmatrix} \dfrac{kN}{m}$ ⑥

Vertical action (favourable) $V_{d,fav} = \gamma_{G,fav} \times W_{Gk} + P_{avd} = \begin{pmatrix} 199.9 \\ 198.5 \end{pmatrix} \dfrac{kN}{m}$ ⑥

Sliding resistance

Partial factors from Sets $\begin{pmatrix} R1 \\ R1 \end{pmatrix}$: $\gamma_{Rh} = \begin{pmatrix} 1 \\ 1 \end{pmatrix}$ and $\gamma_{Rv} = \begin{pmatrix} 1 \\ 1 \end{pmatrix}$

Design drained sliding resistance (ignoring adhesion, as required by EN 1997-1

exp. 6.3a) $H'_{Rd} = \left(\dfrac{V_{d,fav} \times \tan(\delta_{d,fdn})}{\gamma_{Rh}} \right) = \begin{pmatrix} 167.7 \\ 133.3 \end{pmatrix} \dfrac{kN}{m}$

Toppling resistance
Design stabilizing moments (about toe):

From backfill: $M_{d_1} = \left[P_{ahd_1} \times \tan(\theta + \delta_d) \times \left(B - \dfrac{b_h}{3} \right) \right] = \begin{pmatrix} 79.7 \\ 74.9 \end{pmatrix} \dfrac{kNm}{m}$ ⑦

From surcharge: $M_{d_2} = \left[\overrightarrow{P_{ahd_2}} \times \left[\tan(\theta + \delta_d) \times \left(B - \dfrac{b_h}{2} \right) \right] \right] = \begin{pmatrix} 21.7 \\ 23.9 \end{pmatrix} \dfrac{kNm}{m}$ **❼**

From wall $M_{d_3} = \overrightarrow{\left(\gamma_{G,fav} \times M_{Ek,stb} \right)} = \begin{pmatrix} 144 \\ 144 \end{pmatrix} \dfrac{kNm}{m}$

Total design stabilizing moment $M_{Ed,stb} = \displaystyle\sum_{i=1}^{3} \overrightarrow{M_{d_i}} = \begin{pmatrix} 245.5 \\ 242.8 \end{pmatrix} \dfrac{kNm}{m}$

Eccentricity of load $e_B = \left| \overrightarrow{\left[\dfrac{B}{2} - \dfrac{M_{Ed,stb} - M_{Ed,dst}}{V_d} \right]} \right| = \begin{pmatrix} 0.46 \\ 0.32 \end{pmatrix} m$

To be within middle third of base, e_B must be not be $> \dfrac{B}{6} = 0.33 \, m$ **❽**

Verifications

For drained sliding and $H_{Ed} = \begin{pmatrix} 73.8 \\ 71.9 \end{pmatrix} \dfrac{kN}{m}$ and $H'_{Rd} = \begin{pmatrix} 167.7 \\ 133.3 \end{pmatrix} \dfrac{kN}{m}$

Degree of utilization $\Lambda_{GEO,1} = \dfrac{H_{Ed}}{H'_{Rd}} = \begin{pmatrix} 44 \\ 54 \end{pmatrix} \%$ **❾**

For toppling $M_{Ed,dst} = \begin{pmatrix} 109.3 \\ 107.9 \end{pmatrix} \dfrac{kNm}{m}$ and $M_{Ed,stb} = \begin{pmatrix} 245.5 \\ 242.8 \end{pmatrix} \dfrac{kNm}{m}$

Degree of utilization $\Lambda_{GEO,1} = \dfrac{M_{Ed,dst}}{M_{Ed,stb}} = \begin{pmatrix} 45 \\ 44 \end{pmatrix} \%$ **❾**

Design is unacceptable if the degree of utilization is > 100%

Design Approach 2 (summary)

Verification of resistance

For drained sliding $H_{Ed} = 73.8 \dfrac{kN}{m}$ and $H'_{Rd} = 152.5 \dfrac{kN}{m}$

Degree of utilization $\boxed{\Lambda_{GEO,2} = \dfrac{H_{Ed}}{H'_{Rd}} = 48\,\%}$ ⑩

For toppling $M_{Ed,dst} = 109.3 \dfrac{kNm}{m}$ and $M_{Ed,stb} = 245.5 \dfrac{kNm}{m}$

Degree of utilization $\boxed{\Lambda_{GEO,2} = \dfrac{M_{Ed,dst}}{M_{Ed,stb}} = 45\,\%}$

Design is unacceptable if degree of utilization is > 100%

Design Approach 3 (summary)

Verification of resistance

For drained sliding $H_{Ed} = 71.9 \dfrac{kN}{m}$ and $H'_{Rd} = 133.3 \dfrac{kN}{m}$

Degree of utilization $\boxed{\Lambda_{GEO,3} = \dfrac{H_{Ed}}{H'_{Rd}} = 54\,\%}$ ⑩

For toppling $M_{Ed,dst} = 107.9 \dfrac{kNm}{m}$ and $M_{Ed,stb} = 242.8 \dfrac{kNm}{m}$

Degree of utilization $\boxed{\Lambda_{GEO,3} = \dfrac{M_{Ed,dst}}{M_{Ed,stb}} = 44\,\%}$

Design is unacceptable if degree of utilization is > 100%

❺ Since the back of the wall is inclined and friction is assumed to develop along its face, there are vertical as well as horizontal components of thrust from the backfill and surcharge.

❻ The design vertical action due to surcharge and backfill is treated as unfavourable for both sliding and toppling, since it comes from the same sources (backfill and surcharge) – both of which provide unfavourable horizontal actions.

❼ Additional stabilizing moments are derived from the vertical thrusts from the backfill and surcharge.

❽ The eccentricity is outside of the middle third and would normally not be acceptable for a mass gravity wall, as it implies tension within the structure. Note that Eurocode 7 only requires the eccentricity to lie withing the middle two-thirds of the base and hence this design satisfies its requirements.

❾ Sliding governs the calculation for Design Approach 1, with Combination 2 being critical. The degree of utilization (54%) is relatively low, suggesting that the wall may be over-designed. However, water pressure may develop behind the wall which will reduce the degree of utilization.

❿ The results for Design Approaches 2 and 3 are presented in summary only. The full calculations are available from the book's website at www.decodingeurocode7.com.

Design Approach 2 applies factors greater than 1.0 to actions and resistance. Sliding governs the design for DA2 with a degree of utilization (48%) that is slightly lower than for DA1 and well within the limits of Eurocode 7.

Design Approach 3 applies factors greater than 1.0 to structural actions (i.e. the self-weight of the concrete) and material properties. Once again, sliding governs with a degree of utilization (54%) identical to DA1.

11.12 Notes and references

1. See, for example, Clayton, C. R. I., Milititsky, J., and Woods, R. I. (1993) *Earth pressure and earth-retaining structures* (2nd edition), Glasgow, Blackie Academic & Professional, 398pp.

2. See §3.2.2.2 of BS 8002: 1994, Code of practice for earth retaining structures, British Standards Institution, with Amendment 2 (dated May 2001).

3. See BS 8102: 1990, Code of practice for protection of structures against water from the ground, British Standards Institution.

4. See Clayton et al., ibid., for further discussion of this point.

5. The procedure is not attributed in Annex C, but appears to have been developed by Brinch-Hansen. See Christensen, N. H., (1961) *Model tests on plane active earth pressures in sand*, Bulletin No. 10, Copenhagen: Geoteknisk Institut, 19pp. A new derivation is provided by Hansen, B. (2001) *Advanced theoretical soil mechanics*, Bulletin No. 20, Lyngby: Dansk Geoteknisk Forening, 541pp.

6. EN 14475: 2006, Execution of special geotechnical works – Reinforced fill, European Committee for Standardization, Brussels.

7. BS 8006: 1995, Code of practice for strengthened/reinforced soils and other fills, British Standards Institution.

8. Forward to EN 14475, ibid.

9. Chapman, T., Taylor, H., and Nicholson, D. (2000) *Modular gravity retaining walls – design guidance*, CIRIA C516, London: CIRIA, 202pp.

10. EN 1992, Eurocode 2 – Design of concrete structures, European Committee for Standardization, Brussels.

11. EN 1996, Eurocode 6 – Design of masonry structures, European Committee for Standardization, Brussels.

12. Bond A. J., Brooker, O., Harris, A. J., Harrison, T., Moss, R. M., Narayanan, R. S., and Webster, R. (2006) *How to design concrete structures using Eurocode 2*, The Concrete Centre, Camberley, Surrey, 98pp.

13. Institution of Structural Engineers (2006), *Manual for the design of concrete building structures to Eurocode 2*, London; and (2008) *Manual for the design of plain masonry in building structures to Eurocode*, London.

14. For example BS 8004: 1986, Code of practice for foundations, British Standards Institution.

Chapter 12

Design of embedded walls

The design of embedded retaining walls is covered by Section 9 of Eurocode 7 Part 1, 'Retaining structures', whose contents are as follows:

§9.1   General (6 paragraphs)
§9.2   Limit states (4)
§9.3   Actions, geometrical data and design situations (26)
§9.4   Design and construction considerations (10)
§9.5   Determination of earth pressures (23)
§9.6   Water pressures (5)
§9.7   Ultimate limit state design (26)
§9.8   Serviceability limit state design (14)

An embedded wall is a relatively thin structure whose bending capacity plays a significant role in the support of the retained material. Such walls are usually made of steel, reinforced concrete, or timber and supported by anchorages, struts, and passive earth pressure.   *[EN 1997-1 §9.1.2.2]*

Structures composed of elements from both gravity and embedded walls – for example, double sheet pile wall cofferdams – are called 'composite walls' in Eurocode 7. Composite walls should be designed according to the rules discussed in Chapter 11 and in this chapter.   *[EN 1997-1 §9.1.2.3]*

Section 9 of EN 1997-1 applies to structures which retain ground (soil, rock, or backfill) and water, where 'retained' means 'kept at a slope steeper than it would eventually adopt if no structure were present'.   *[EN 1997-1 §9.1.1(1)P]*

The design of silos is covered by ENs 1991-4[1] and 1993-4-1[2], not EN 1997-1.

12.1   Ground investigation for embedded walls

Annex B.3 of Eurocode 7 Part 2 provides outline guidance on the depth of investigation points for retaining structures, as illustrated in **Figure 12.1**. (See Chapter 4 for guidance on the spacing of investigation points.)

The recommended minimum depth of investigation, z_a, for excavations where the groundwater table is below formation level is the greater of:

$z_a \geq 0.4h$ and $z_a \geq (t+2m)$

and, where the groundwater is above formation level, the greater of:

$z_a \geq (H+2m)$ and $z_a \geq (t+2m)$

If H is small and t under-estimated, there is a danger that the latter rule will result in an investigation that does not go deep enough. It therefore seems advisable to ensure also that $z_a \geq 0.4h$.

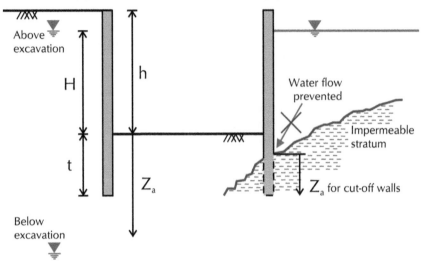

Figure 12.1. *Recommended depth of investigation points for excavations*

If groundwater is above formation and no impermeable stratum is encountered within the depth of investigation suggested above, then the depth should be increased to:

$z_a \geq (t+5m)$

If the wall is designed to cut off water flow into the excavation (see right-hand side of **Figure 12.1**), then the investigation should extend at least 2m into the impermeable stratum.

The depth z_a may be reduced to 2m if the wall is built on competent strata[†] with 'distinct' (i.e. known) geology. With 'indistinct' geology, at least one borehole should go to at least 5m. If bedrock is encountered, it becomes the reference level for z_a.   *[EN 1997-2 §B.3(4)]*

[†] i.e. weaker strata are unlikely to occur at depth, structural weaknesses such as faults are absent, and solution features and other voids are not expected.

Greater depths of investigation may be needed for very large or highly complex projects or where unfavourable geological conditions are encountered. *[EN 1997-2 §B.3(2)NOTE and B.3(3)]*

12.2   Design situations and limit states

Examples of limit states that can affect embedded retaining walls are shown in **Figure 12.2**. From left to right, these include: (top) wall head kicking into excavation, owing to rotation about a fixed point in the ground, and wall toe kicking into excavation, owing to rotation about a single prop; (middle) failure of the wall stem in bending and pull-out of a supporting anchor; and (bottom) basal heave and prop failure.

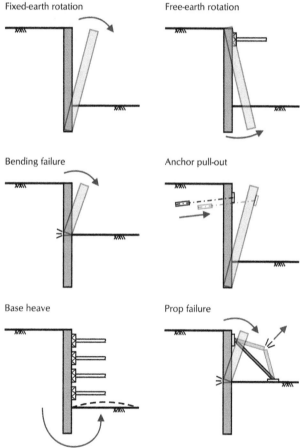

Figure 12.2. Examples of limit states for embedded retaining walls

§9 of Eurocode 7 Part 1 includes a series of illustrations showing limit modes for embedded walls, covering overall stability; rotational, vertical, and structural failure; and failure by pullout of anchors.

12.3   Basis of design

Eurocode 7 requires embedded walls to be designed with sufficient embedment to prevent rotational and vertical failure. The wall's cross-section and any supports it relies upon must be verified against structural failure (see Section 12.7). Furthermore, embedded walls must not fail due to overall instability of the ground in which they are installed.

[EN 1997-1 §9.7.2(1)P, 9.7.4(1)P, 9.7.5(1)P, and 9.7.6(1)P]

Regrettably, Eurocode 7 gives little detailed guidance on the design of embedded retaining walls. Traditional national practice will therefore still have a large role to play in their design for the foreseeable future.

This book does not attempt to provide complete guidance on the design of retaining walls, for which the reader should refer to any well-established text on the subject.[3]

12.3.1   Role of CIRIA C580 in UK practice

The UK National Annex to BS EN 1997-1[4] lists CIRIA report C580[5] as a source of 'non-conflicting, complementary information' (NCCI) which can be used to provide detailed guidance on the design of embedded retaining walls. However, CIRIA C580 was written before publication of the final version of EN 1997-1 and several of its recommendations contradict Eurocode 7.[†]

For example, C580 presents three design approaches (A, B, and C) which are unrelated to the three Design Approaches (1, 2, and 3) in Eurocode 7 Part 1. The partial factors recommended in C580 differ from those given in EN 1997-1 and the UK National Annex. CIRIA's guidance on the structural design of embedded walls is based on existing British Standards, such as BS 5950[6] and BS 8110,[7] and not on their Eurocode replacements.

C580's Design Approach A employs 'moderately conservative' soil parameters, groundwater pressures, loads, and geometries. As discussed in Chapter 5, the moderately conservative approach – which was originally documented in CIRIA Report 104 (the predecessor to C580) – adopts parameters that have similar reliability to Eurocode 7's characteristic values

---

[†]Proposals have been made to update CIRIA C580 to remove these conflicts.

(based on a 'cautious estimate'). Approach B employs 'worst credible' parameters, which have significantly higher reliability than characteristic values. Approach C, which should only be used as part of the observational method of design,[8] employs 'most probable' parameters with slightly lower reliability than characteristic values.

The differences between the partial factors recommended by CIRIA C580 and those specified in the UK National Annex to Eurocode 7 – for ultimate limit states (ULSs) GEO and STR and serviceability limit states (SLSs) – are summarized in the table below. The requirements of Eurocode 7 take precedence over those of CIRIA C580.

Limit state*		Design Approach		Partial factors				
				γ_G	γ_Q	γ_φ	γ_c	γ_{cu}
CIRIA C580	ULS	A	Moderately conservative	1.0	1.0‡	1.2	1.2	1.5
		B	Worst credible	1.0	1.0‡	1.0	1.0	1.0
		C	Most probable	1.0	1.0‡	1.2	1.2	1.5
	SLS	A and C†		All factors = 1.0				
UK NA	ULS	1	Combination 1	1.35	1.5	1.0	1.0	1.0
			Combination 2	1.0	1.3	1.25	1.25	1.4
	SLS			All factors = 1.0				

*ULS = ultimate; SLS = serviceability limit state
†Design Approach B is not suitable for SLS calculations
‡A minimum surcharge of 10kPa should be applied to walls with retained height of 3m or more

If worst credible parameters are chosen for retaining wall design (C580's Design Approach B), then these could be regarded as design values for the purposes of Eurocode 7.

12.3.2 Unplanned excavation

EN 1997-1 requires an allowance to be made for the possibility of an unplanned excavation reducing formation level on the restraining side of a retaining wall (see **Figure 12.3**).

When normal levels of site control are employed, verification of ultimate limit states should assume an increase in retained height ΔH given by:

$$\Delta H = \frac{H}{10} \leq 0.5m$$

where H is the retained height of a cantilever wall (left hand side of **Figure 12.3**) or the height below the lowest support for a propped wall (right hand side of **Figure 12.3**).   *[EN 1997-1 §9.3.2.2(2)]*

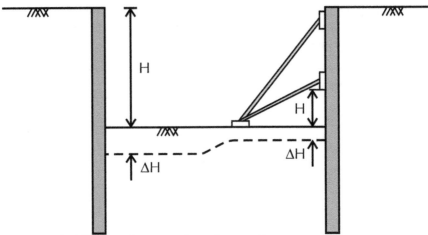

***Figure 12.3**. Allowance for an unplanned excavation*

Eurocode 7 also warns that, where the surface level is particularly uncertain, larger values of ΔH should be used. However, it also allows smaller values (including ΔH = 0) to be assumed when measures are put in place throughout the execution period to control the formation reliably.
                                              [EN 1997-1 §9.3.2.2(3) and (4)]

The rules for unplanned excavations apply to ultimate limit states only and not to serviceability limit states. Anticipated excavations in front of the wall should be considered specifically – they are not, by definition, *unplanned*. Planned excavations include French drains, pipe trenches, buried close-circuit television cables, etc.

These rules provide the designer with considerable flexibility in dealing with the risk of over-digging. A more economical design may be obtained by adopting ΔH = 0, but the risk involved must be controlled during construction – leading to a supervision requirement that must be specified in the Geotechnical Design Report (see Chapter 16). A designer who wants to minimize the need for supervision must guard against the effects of over-digging by adopting ΔH = 10%H.

The rules for unplanned excavations in BS 8002[9] were changed in 2001 to match those of (the then draft) Eurocode 7 Part 1.

12.3.3   Selection of water levels

Selection of appropriate water levels is particularly important in retaining wall design and Eurocode 7 makes specific demands in this regard.

When the wall retains medium or low permeability soils (mainly fine soils), then the wall should be designed for a water level above formation level. If a reliable drainage system is provided, then the water level may be assumed to occur below the top of the wall. Without reliable drainage, the water should be taken at the surface of the retained material. *[EN 1997-1/AC §9.6(3)P]*

It would be unusual to install a drainage system behind an embedded retaining wall (weep holes provided through a sheet pile wall should not be considered a 'reliable drainage system'). Eurocode 7's requirements appear far more onerous than would normally be adopted in current practice. It is rare for the natural water table to rise to ground surface and so, when it can be demonstrated to come below formation level, it seems excessive to design for such large sustained water pressures behind the wall.

As discussed in Chapter 11, current British practice[10] recommends taking groundwater level, when drainage is absent, at a depth d_w given by:

$$d_w = \frac{H}{4} \text{ for } H \le 4m, d_w = 1m \text{ for } H > 4m$$

where H is the wall's retained height.

12.3.4   Wall friction and adhesion

Eurocode 7 Part 1 states that the amount of wall friction and adhesion mobilized depends on the ground's strength, friction properties of the wall/ground interface, direction of wall movement relative to the ground, and the wall's ability to support vertical actions.   *[EN 1997-1 §9.5.1(4)]*

The design angle of wall friction δ_d may be taken as:
$$\delta_d = k \times \varphi_{cv,d}$$

where $\varphi_{cv,d}$ is the design value of the soil's constant volume angle of shearing resistance. For steel sheet piling supporting sand or gravel, $k = 2/3$ and, for cast-in-place concrete walls, $k = 1$.   *[EN 1997-1 §9.5.1(6) and 9.5.1(7)]*

Eurocode 7 does not give any advice regarding suitable values of wall adhesion to use in design. The absence of wall adhesion from the expressions for earth pressures given in Annex C of EN 1997-1 suggests wall adhesion should be ignored – although the UK National Annex has 're-introduced' wall adhesion into these equations (see Section 12.4.4 below).

Values of wall friction may be less than the equation above suggests, particularly when the wall carries large vertical loads or coatings are applied to the wall surface. It is common practice to ignore wall friction and adhesion in such circumstances (or even apply negative values of wall friction, for example when extremely high vertical loads are applied to the wall). The wall should be checked for vertical equilibrium to confirm an appropriate value of wall friction has been used (especially for sheet pile walls which have limited base resistance).   [EN 1997-1 §9.7.5(4)P]

The table below compares the recommendations for wall friction given in Eurocode 7 with those from traditional practice.

Publication	Surface	Wall friction tan δ	
		Active	Passive
Terzaghi[11]/Clayton et al.[12]	steel	$\tan(\frac{1}{2}\varphi)$	$\tan(\frac{2}{3}\varphi)$
CIRIA 104[13]	any	$\tan(\frac{2}{3}\varphi)$	$\tan(\frac{1}{2}\varphi)$
EAU[14]	steel	$\tan(\frac{2}{3}\varphi)$	$\tan(\frac{2}{3}\varphi)$
Piling Handbook[15]	steel	usually ignored	$\frac{2}{3}\tan\varphi$
BS 8002[16]	any	$\frac{3}{4}\tan\varphi$	$\frac{3}{4}\tan\varphi$
Canadian Foundation Engineering Manual[17]	steel	$\tan(11\text{--}22°)$	
	cast concrete	$\tan(17\text{--}35°)$	
	pre-cast concrete	$\tan(14\text{--}26°)$	
CIRIA C580[18] Eurocode 7	steel	$\tan(\frac{2}{3}\varphi_{cv})$	$\tan(\frac{2}{3}\varphi_{cv})$
	cast concrete	$\tan(\varphi_{cv})$	$\tan(\varphi_{cv})$
	pre-cast concrete	$\tan(\frac{2}{3}\varphi_{cv})$	$\tan(\frac{2}{3}\varphi_{cv})$

As discussed in Chapter 10, it may be preferable to select the design value of φ_{cv} directly rather than to calculate it from its characteristic value ($\varphi_{cv,k}$) by applying a partial factor on shearing resistance (γ_φ).

12.4   Limiting equilibrium methods

Limiting equilibrium methods are commonly used to assess the required penetration of embedded retaining walls, associated shear forces and bending moments in their cross-sections, and the forces in any props or anchors used to support them. Limiting equilibrium methods assume that the full strength of the ground is mobilized uniformly around the wall, so that the wall is at the point of collapse (or 'limiting equilibrium').

Cantilever walls and walls propped near their top are statically determinate structures that can be analysed using limiting equilibrium methods, as described in Sections 12.4.1 and 12.4.2 below. Multi-propped walls are statically indeterminate structures that can only be analysed using limiting equilibrium methods by introducing significant simplifications to make the structure statically determinate.

12.4.1   Fixed-earth conditions

The stability of an embedded cantilever wall can be verified by assuming 'fixed-earth' conditions, as illustrated in **Figure 12.4**. The wall, which is assumed to rotate about the fixed point 'O', relies on the support of the ground to maintain horizontal and moment equilibrium.

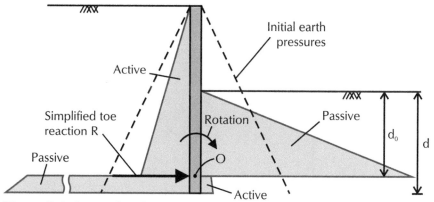

Figure 12.4. Assumed earth pressures acting on an embedded cantilever retaining wall at limiting equilibrium (fixed earth conditions)

Above the point of fixity, ground on the retained (left-hand) side of the wall goes into an active state and that on the restraining (right-hand) side into a passive state. The earth pressures bearing on the wall decrease from their initial, at-rest (K_0) values to active (K_a) values on the left and increase towards fully passive (K_p) on the right.

Below the point of fixity, ground on the retained side goes into a passive state and that on the restraining side into an active state. The earth pressures below O therefore increase towards fully passive values on the left and decrease to active values on the right.

The situation shown in **Figure 12.4** is often simplified by replacing the earth pressures below O with an equivalent reaction R. The depth of embedment (d_O) required to ensure moment equilibrium about the point of fixity is then increased by 20% to compensate for this assumption,[19] i.e. $d = 1.2 \, d_O$.

12.4.2   Free-earth conditions

The stability of an embedded wall propped near its top can be verified by assuming 'free-earth' conditions, illustrated in **Figure 12.5**. The wall, which is assumed to rotate about the fixed point 'O', relies on the support of the ground and prop to maintain moment and horizontal equilibrium.

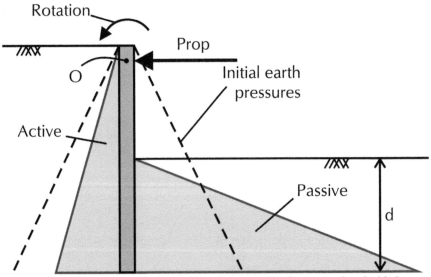

Figure 12.5. Assumed earth pressures acting on a singly-propped embedded retaining wall at limiting equilibrium conditions (free earth conditions)

Below the point of fixity, ground on the retained (left-hand) side of the wall goes into an active state and that on the restraining (right-hand) side into a passive state. The earth pressures bearing on the wall decrease from their initial, at-rest (K_0) values to active (K_a) values on the left and increase towards fully passive (K_p) on the right.

Above the point of fixity, the retained ground goes into a passive state and the restraining ground into an active state. However, to avoid over-complicating the calculations, earth pressures are treated in the same way above the point of fixity as they are below. This simplification is less easily justified when the prop is located in the lower half of the retained height.

Propped embedded walls may also be verified assuming fixed-earth conditions, but the calculations are more complicated. Fixed-earth conditions suggest a longer pile is needed than free-earth conditions, but the assumed fixity reduces the theoretical bending moment in the pile. Hence if driving is difficult and the full design penetration is not achieved, the stability requirements for free-earth conditions may be satisfied but the moment capacity of the pile may be compromised.

12.4.3   At-rest values of earth pressure

When the ground behind a retaining wall is in an at-rest state, the horizontal total stress σ_h acting on the wall at depth z below ground surface is given by:

$$\sigma_h = K_0 \left(\int_0^z \gamma dz + q - u \right) + u$$

where K_0 is the at-rest earth pressure coefficient; γ the soil's weight density; q any vertical surface loading; and u the pore water pressure in the ground.

Eurocode 7 suggests that at-rest conditions should be assumed behind retaining walls in normally consolidated soils when the horizontal movement of the wall is less than 0.05% of its retained height (see **Figure 12.14**).

[EN 1997-1 §9.5.2(2)]

The earth pressure coefficient K_0 is given in Eurocode 7 as:
$$K_0 = (1 - \sin\varphi) \times \sqrt{OCR} \times (1 + \sin\beta)$$
(which is Meyerhof's[20] formulation combined with Kezdi's[21] modification for sloping ground) where φ is the soil's angle of shearing resistance; OCR its over-consolidation ratio; and β the slope angle of the ground surface (see **Figure 12.6** for sign convention).

12.4.4   Limiting values of earth pressure

Annex C of Eurocode 7 Part 1 gives guidance on determining limiting values of earth pressure for active and passive states. Unfortunately, expressions C.1 and C.2 in EN 1997-1 are solely for dry ground (and contain errors, as noted in the UK National Annex to EN 1997-1[22]). The correct expressions are:

$$\sigma_a = K_a \left(\int_0^z \gamma dz + q - u \right) - 2c\sqrt{K_a(1+a/c)} + u$$

and

$$\sigma_p = K_p \left(\int_0^z \gamma dz + q - u \right) + 2c\sqrt{K_p(1+a/c)} + u$$

where σ_a and σ_p are the active and passive total stresses normal to the wall; K_a and K_p are active and passive earth pressure coefficients; γ, q, and u are as defined in Section 12.4.3; c is the soil's effective cohesion; a the adhesion between ground and wall (limited to 0.5c); and z the depth below ground.

Annex C provides charts for determining the active and passive earth pressure coefficients for use in these expressions. They are identical to those that appeared in BS 8002: 1994[23] and are based on the work of Kerisel and Absi[24].

Annex C also provides a numerical procedure[25] for determining active and passive *effective* earth pressures, σ'_a and σ'_p respectively, from the following equations:

$$\sigma'_a = K_{a\gamma} \left(\int_0^z \gamma dz - u \right) + K_{aq}q - K_{ac}c \text{ and } \sigma'_p = K_{p\gamma} \left(\int_0^z \gamma dz - u \right) + K_{pq}q + K_{pc}c$$

where $K_{a\gamma}$, K_{aq}, and K_{ac} are active earth pressure coefficients for soil weight, surcharge loading, and effective cohesion; $K_{p\gamma}$, K_{pq}, and K_{pc} are their passive counterparts; and the other symbols are as defined above.

The coefficients themselves are given by:

$$\left. \begin{array}{c} K_{a\gamma} \\ K_{p\gamma} \end{array} \right\} = K_n \times \cos\beta \times \cos(\beta - \theta)$$

$$\left. \begin{array}{c} K_{aq} \\ K_{pq} \end{array} \right\} = K_n \times \cos^2\beta = \left\{ \begin{array}{c} K_{a\gamma} \\ K_{p\gamma} \end{array} \right\} \times \frac{\cos\beta}{\cos(\beta - \theta)}$$

$$\left. \begin{array}{c} K_{ac} \\ K_{pc} \end{array} \right\} = \pm(K_n - 1) \times \cot\varphi = \left(\frac{1}{\cos\beta \times \cos(\beta - \theta)} \times \left\{ \begin{array}{c} K_{a\gamma} \\ K_{p\gamma} \end{array} \right\} - 1 \right) \times \cot\varphi$$

where the derived expressions allow K_{aq}, K_{pq}, K_{ac}, and K_{pc} to be derived from values of $K_{a\gamma}$ and $K_{p\gamma}$, which are readily available from charts.

The auxiliary coefficient K_n in these equations is given by:

$$K_n = \frac{1 \pm \sin\varphi \times \sin(2m_w \pm \varphi)}{1 \mp \sin\varphi \times \sin(2m_t \pm \varphi)} e^{\pm 2(m_t + \beta - m_w - \theta)\tan\varphi}$$

and the terms m_t and m_w are:

$$2m_t = \cos^{-1}\left(\frac{-\sin\beta}{\pm\sin\varphi}\right) \mp \varphi - \beta \text{ and } 2m_w = \cos^{-1}\left(\frac{\sin\delta}{\sin\varphi}\right) \mp \varphi \mp \delta$$

Active coefficients are obtained by using the lower sign (+ or –) where a choice is shown (± or ∓) in the expressions for K_{ac}, K_{pc}, K_n, m_t, and m_w; and passive coefficients by using the upper signs. Note that, for vertical walls, $K_{aq} = K_{ay}$ and $K_{pq} = K_{py}$, since $\theta = 0°$.

In all these equations, φ is the soil's angle of shearing resistance; δ the angle of friction between the wall and the soil; β the slope angle of the ground surface; and θ the inclination of the wall to the vertical (see **Figure 12.6**).

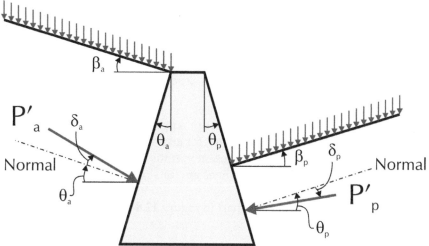

Figure 12.6. *Sign convention for determining active and passive earth pressures from numerical procedure in Eurocode 7*

We have provided in Appendix 2 a series of charts for vertical walls (i.e. with $\theta = 0°$) based on these expressions, showing the variation in K_{ay} and K_{py} with angle of shearing resistance φ for various values of interface friction δ and slope angle β. Most charts of this kind (e.g. the ones published in CIRIA C580[26]) plot values of K_{ay} and K_{py} against φ for various ratios of δ/φ and β/φ. The format of our charts is therefore unusual but – in our opinion – preferable, since it emphasizes the valid range of values for δ and β. The following example illustrates the use of these charts.

Consider the cantilever retaining wall shown in **Figure 12.7**, which is embedded in sand with characteristic angles of shearing resistance $\varphi_k = 32°$ (peak) and $\varphi_{cv,k} = 30°$ (constant volume). The ground behind the wall slopes upwards at a gradient of 1:3 (i.e. 18.4°) and the formation is horizontal. The characteristic angle of wall friction δ_k is calculated as: $\delta_k = \frac{2}{3}\varphi_{cv,k} = 20°$.

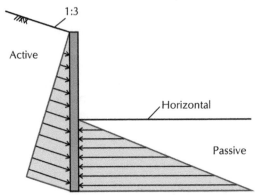

Figure 12.7. Earth pressures acting on embedded wall supporting sloping ground

Figure 12.8 shows one of the charts from Appendix 2, giving values of K_a versus angle of shearing resistance (φ), for an angle of interface friction $\delta = 20°$. The numbers on the lines represent various slope gradients, $\tan \beta = 0$ (i.e. flat), 1:10, 1:5, etc. The dashed lines are for slope angles < 0°.

For $\varphi = 32°$ and $\tan \beta = 1:3$, the chart in **Figure 12.8** gives $K_{ay} = 0.34$.

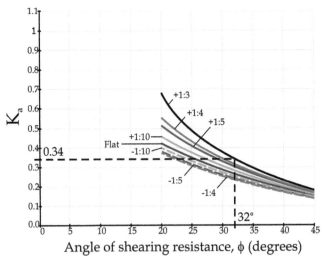

Figure 12.8. Active earth pressure coefficients for $\delta = 20°$

Figure 12.9 shows another chart from Appendix 2, this one giving values of K_p against φ for δ = 20°. Note that the scale of the y-axis on this chart is different to that on **Figure 12.8**. For φ = 32° and β = 0°, $K_{pγ}$ = 5.18.

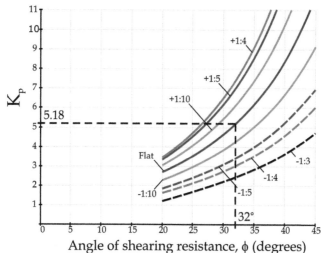

Angle of shearing resistance, φ (degrees)

Figure 12.9. Passive earth pressure coefficients for δ = 20 °

12.4.5   Passive earth pressure: resistance or action?

A question of particular importance in the design of retaining walls is: should passive earth pressures be regarded as a resistance or as an action?

Reliability is introduced into the design of embedded cantilever walls by applying partial factors to actions, material properties, or resistance (or to a combination of these variables). The table below summarizes the values of these factors for each of Eurocode 7's Design Approaches (see Chapter 6).

Partial factor	Design Approach			
	1		2	3
	Comb. 1	Comb. 2		
Unfavourable action (γ_G)	1.35	1.0	1.35	1.35†/1.0‡
Favourable action $(\gamma_{G,fav})$	1.0	1.0	1.0	1.0
Resistance (γ_{Re})	1.0	1.0	1.4	1.0

Factors from †Set A1 on structural actions; ‡Set A2 on geotechnical

Few engineers would dispute that active earth pressures on the retained side of a wall are an unfavourable action and hence should be multiplied by γ_G. Thus in Design Approach 1, Combination 1 and Design Approach 2, active earth pressures are *increased* from their characteristic values by 35% (since γ_G = 1.35), but are left unchanged in Design Approach 1, Combination 2 (since γ_G = 1.0). In Design Approach 3, earth pressures from the ground's self-weight are treated as a geotechnical action, for which γ_G = 1.0.

The question arises, however: are passive earth pressures acting on the restraining side of the wall a favourable action or a resistance? The answer to this question has a significant bearing on the verification of limit states GEO and STR. **Figure 12.10** shows some of the possible outcomes depending on the answer to this question.

The top half of the diagram shows the bending moment (M) and shear force (V) that occur along the length of a cantilever embedded wall retaining 5m

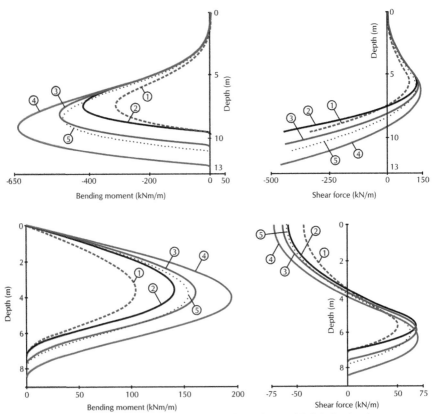

Figure 12.10. Bending moments (left) and shear forces (right) for (top) a cantilever and (bottom) a propped embedded wall, both retaining 5m of soil

of coarse soil with characteristic angle of shearing resistance $\varphi_k = 30°$ and characteristic weight density $\gamma_k = 20\,kN/m^3$. The bottom half of the diagram shows M and V for a propped wall also retaining 5m of coarse soil (with identical properties). The numbers on the curves refer to five different assumptions (or 'cases') summarized in the table below.

Assumption made about passive earth pressure (σ_p)	Partial factors applied to...		
	Shearing resistance $(\tan \varphi_k)$	Earth pressure	
		Active (σ_{ak})	Passive (σ_{pk})
1 Serviceability	$\div 1.0$	$\times 1.0$	$\times 1.0$
2 Unfavourable action	$\div \gamma_\varphi = 1.0$	$\times \gamma_G = 1.35$	$\times \gamma_G = 1.35$
3 Favourable action	$\div \gamma_\varphi = 1.0$	$\times \gamma_G = 1.35$	$\times \gamma_{G,fav} = 1.0$
4 Resistance	$\div \gamma_\varphi = 1.0$	$\times \gamma_G = 1.35$	$\div \gamma_{Re} = 1.4$
(Variation on 4)	$\div \gamma_\varphi = 1.0$	$\times \gamma_G = 1.35$	$\times \gamma_G \div \gamma_{Re} = \div 1.04$
5 Any of above	$\div \gamma_\varphi = 1.25$	$\times \gamma_G = 1.0$	$\times 1.0$

Curve 1 on each graph shows the results obtained for a serviceability limit state calculation, with all partial factors set to 1.0 – i.e. with all parameters at their characteristic values. The depths of embedment needed to ensure stability for this situation are 9.63m and 7.00m respectively for the two walls.

Curve 2 shows the results obtained when passive earth pressures are treated as an *unfavourable action*, as allowed by the Single-Source Principle discussed in Chapter 3. A single partial factor $\gamma_G = 1.35$ is applied to active and passive earth pressures. The depths of embedment needed to ensure stability for this situation are identical to those obtained for the serviceability limit state (SLS) calculation and the structural effects (bending moments, shear forces, and prop force) are 35% higher. Hence this result could have been obtained directly from the SLS calculation by applying the factor $\gamma_G = 1.35$ to the *effects* of actions, rather than to the actions themselves (as suggested previously[27]).

Curve 3 shows the results obtained when passive earth pressures are treated as a *favourable action*. The partial factor $\gamma_G = 1.35$ is applied solely to active earth pressures and $\gamma_{G,fav} = 1.0$ to passive. With this assumption, both forms of earth pressure are treated consistently (as actions), but their effects are taken into account when deciding which load factor to apply. The embedments needed to ensure stability (10.67m and 7.52m) are larger than before and, as a consequence, so too are the structural effects.

Curve 4 shows the results obtained when passive earth pressures are treated as a *resistance*, using partial factors from Design Approach 2. Active earth pressures are multiplied by $\gamma_G = 1.35$, but passive pressures are *divided* by $\gamma_{Re} = 1.4$. This is the most 'natural' way in which to apply partial factors in embedded retaining wall design[28] – intuitively, most engineers would regard passive earth pressures as a resistance. However, the depths of embedment needed to ensure stability are greatest with this assumption (12.32m and 8.36m) and so too are structural effects. (Note that, if the partial factors from Design Approaches 1 or 3 had been used, then Curve 4 would have coincided with curve 3, since $\gamma_{Re} = 1.0$.)

A variation on Case 4 is to treat the passive earth pressures as both an unfavourable action and a resistance *at the same time*. Hence, they would be multiplied by $\gamma_G = 1.35$ *and* divided by $\gamma_{Re} = 1.4$, i.e. multiplied by a net value $\gamma_G/\gamma_{Re} = 0.96$. The resulting action effects would be very similar to Curve 3.

Finally, Curve 5 shows the results obtained when partial factors from Design Approach 1, Combination 2 are used, i.e. earth pressures are unfactored but a material factor $\gamma_\varphi = 1.25$ is applied to the soil's shearing resistance on both sides of the wall. The effect is to increase active earth pressures and simultaneously to decrease passive. The embedments needed to ensure stability with this assumption (11.14m and 7.76m) are similar to those for Curve 3, as are the structural effects.

Our conclusion from the above analysis is that, for consistent results to be obtained regardless of the calculation model assumed, passive earth pressures should be treated as both an unfavourable action *and* a resistance *simultaneously*. Adopting this philosophy ensures that all three Design Approaches yield similar results.

12.4.6   Net pressures

Figure 12.11 shows another inconsistency that arises when passive earth pressures against embedded walls are treated as a resistance.

The top diagram shows *gross* values of active and passive pressures, to which the partial factors γ_G and γ_{Re} are applied. (The discussion in Section 12.4.5 implicitly assumes factors are applied to gross pressures.)

The middle diagram shows *net* pressures for the same situation, i.e. the difference between the active and passive pressures at each point along the wall. The active earth pressures that are ignored are equal in magnitude to the ignored passive pressures, *before partial factors are applied.* The only way for bending moments and shear forces for these two situations to be identical is if the *same* factor is applied to pressures on both sides of the wall. This is only true if passive earth pressures are treated as unfavourable actions (or all the factors are 1.0, as they are in Design Approach 1, Combination 2 and Design Approach 3).

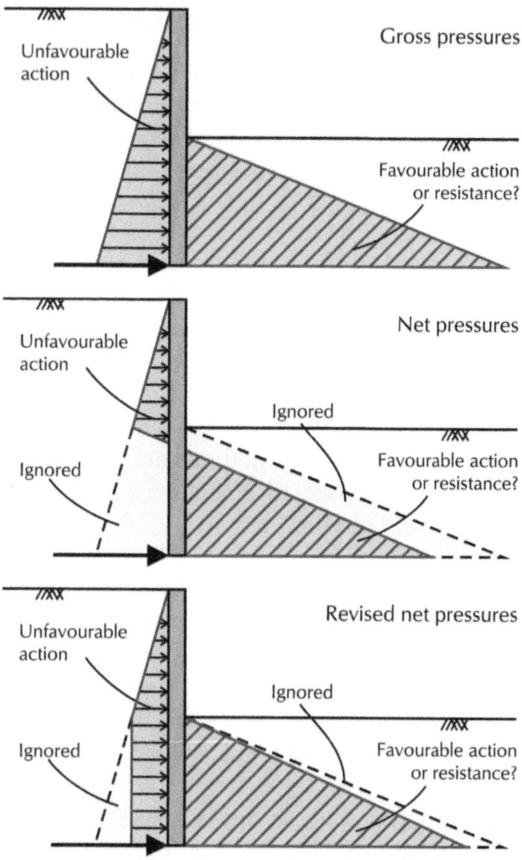

Figure 12.11. Treatment of active and passive earth pressures by common limiting equilibrium design methods: (top) gross, (middle) net, and (bottom) revised net pressure methods

The same argument applies — but to a lesser extent — with the use of *revised net* earth pressures, shown at the bottom of **Figure 12.11**.

Revised net pressures (which are promoted in CIRIA 104[29]) are obtained by deducting, from the active and passive earth pressures below formation level, the *increase* in active earth pressure that occurs below that level.

This discussion highlights the complications involved in applying partial factors to established design methods that were not originally developed with partial factors in mind. It is uncertain whether net pressure methods, in particular, can be used reliably with Eurocode 7 and hence the considerable experience that has been acquired from their use may be lost.

12.5   Soil-structure interaction analysis

Eurocode 7 notes that, for anchored or strutted flexible walls, the magnitude and distribution of earth pressures, internal structural forces, and bending moments depend to a great extent on the stiffness of the structure, the stiffness and strength of the ground, and the state of stress in the ground.

[EN 1997-1 §2.4.1(14)]

If structural stiffness is significant, soil-structure interaction analysis should be performed to determine the distribution of actions. The stress-strain relationships used in such analyses should be sufficiently representative to give a safe result.   *[EN 1997-1 §6.3(3) and 2.4.1(15)]*

The main advantages of soil-structure interaction analysis over the simpler limiting equilibrium methods are that wall movements are calculated, the influence of construction sequence can be assessed, and the beneficial effects of force and moment redistribution taken into account. However:

> 'There is no point in using [soil-structure interaction] analysis unless the level of detail to which results are obtained is really needed and appropriate input data are available.'[30]

This statement emphasizes the importance given by Eurocode 7 Part 2 to performing sufficient and reliable ground investigations (see Chapter 4).

Soil-structure interaction analysis is normally performed using a sub-grade reaction model (discussed in Section 12.5.1) or a more advanced numerical method (see Section 12.5.2). In both cases, the analysis depends not only on strength parameters for the soil and structure, but also on suitable evaluation of the stiffness of the structure and the ground in which it is constructed.

Reliable measurements of ground stiffness are often very difficult to obtain from field or laboratory tests. The latter often underestimate the soil's in-situ stiffness. Eurocode 7 recommends that observations of previous construction behaviour should be analysed wherever available.   *[EN 1997-1 §3.3.7(2)]*

12.5.1   Sub-grade reaction models

Sub-grade reaction theory idealizes the soil as a series of linear-elastic/perfectly-plastic springs, as shown in **Figure 12.12**. The forces on the wall and in any props or anchors supporting it are calculated from deformations along the wall. Iteration brings forces into equilibrium while keeping movements compatible with the elastic properties of the wall.

The springs' sub-grade reaction coefficients k are estimated from field and laboratory measurements of soil stiffness (when available), otherwise from crude rules-of-thumb. The springs' load capacities are normally defined using limiting earth pressure coefficients (K_a for tension, K_p for compression – see Section 12.4.4).

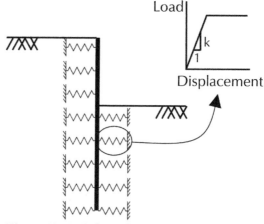

Figure 12.12. *Sub-grade reaction model of embedded retaining wall*

To verify that an ultimate limit state is not exceeded, Eurocode 7 requires partial factors to be applied to actions, material properties, and resistances. Their values depend on which Design Approach is adopted (as discussed in Chapter 6). No partial factors are given in Eurocode 7 for stiffness and hence the design value of the springs' sub-grade reaction coefficients should be identical to their characteristic values. However, CIRIA C580[31] recommends that spring stiffnesses for ultimate limit state calculations should be taken as 50% of their serviceability values (to account for the soil's greater compressibility at large strain). This can be achieved by dividing sub-grade reaction coefficients k by a model factor $\gamma_{Rd} = 2.0$.

The application of partial factors to soil strength changes the values of the active and passive earth pressure coefficients used to define the ultimate resistance of the soil springs. As a consequence, the interaction between the ground and the structure will differ from that under serviceability loads, particularly if some of the springs reach their load capacity prematurely. The displacements obtained from ultimate limit state calculations using sub-grade reaction models should be ignored, since they do not represent the true behaviour of the structure.

It is not clear how partial factors can be applied to actions or resistance when sub-grade models are employed. The logic necessary to determine whether a particular component of earth pressure should be treated as a favourable action, an unfavourable action, or a resistance is extremely complicated — even if there were a universally agreed interpretation of the Eurocode. If part of the ground starts to unload, would that signal a switch from one interpretation of earth pressure to another? Unless the computer program has been specially written to include the relevant factors at the appropriate points in the calculation, then the only way to achieve their intended effect is to adjust input parameters instead. This could be attempted by increasing weight densities by 1.35 to simulate the application of γ_G. However, this is generally not a wise thing to do, since it may lead to unintended side-effects in other parts of the calculation.

The table below summarizes one possible way[†] of using a sub-grade reaction model to verify embedded retaining walls for ultimate limit states, according to Eurocode 7.[32]

Step	Factor	Design Approach			
		1		2	3
		C1	C2		
1. Multiply variable actions by ratio γ_Q/γ_G	γ_Q/γ_G	1.11	1.3	1.11	1.3†
2. Apply partial factors to soil strengths	$\gamma_\varphi = \gamma_c$	1.0	1.25	1.0	1.25
	γ_{cu}	1.0	1.4	1.0	1.4
3. Perform soil structure interaction analysis					
4. Check ratio of restoring to overturning moment $M_R/M_O \geq \gamma_G \times \gamma_{Re}$	$\gamma_G \times \gamma_{Re}$	1.35	1.0	1.89	1.0
5. Apply partial factor to action effects	γ_G	1.35	1.0	1.35	1.0†

†Partial factors from Set A2 for geotechnical actions

First, variable actions are 'pre-factored' by the ratio $\gamma_Q/\gamma_G > 1$ so that subsequent parts of the calculation can treat them as permanent actions. Second, soil strengths are factored down by $\gamma_M \geq 1$. The resulting design values of surcharge and material properties are entered into the computer program and the soil structure interaction analysis is performed (Step 3).

---

†This method implicitly treats passive earth pressure as both an unfavourable action and a resistance.

For cantilever and single-propped walls, toe embedment is then verified (in Step 4), by checking that the ratio of the restoring moment about the point of fixity M_R to the overturning moment M_O about the same point is at least equal to the product of γ_G (the partial factor on unfavourable actions) and γ_{Re} (the partial factor on passive resistance). If the wall passes this check, then design bending moments and shear forces in the wall (and design forces in any props or anchors) may be obtained from the calculated action effects by multiplying by γ_G.

12.5.2   Numerical methods

Embedded walls may also be designed to Eurocode 7 using numerical methods based on finite elements (for example, see **Figure 12.13**, left), boundary elements, or finite difference techniques. Some of the issues that arise when using numerical methods for Eurocode 7 designs are similar to those discussed in Section 12.5.1 for sub-grade reaction models.

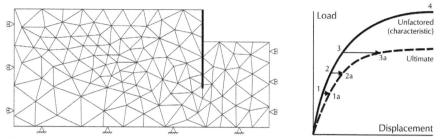

Figure 12.13. *(Left) Finite element model of embedded retaining wall; (right) c-φ reduction*

Verification of ultimate limit states using numerical methods can be achieved most easily by using the material factor approach, which is embodied in Design Approach 1, Combination 2 and Design Approach 3 (as discussed in Chapter 6).[33]

There are two main ways of introducing material factors into numerical models. In the first, partial factors $\gamma_M \geq 1$ are applied to material properties before they are entered into the computer program. The analysis is then performed and the resulting action effects (i.e. bending moments and shear forces in the wall and forces in any props or anchors) are regarded as design values for verification of structural strength. A potential problem with this method is that premature yielding of soil in highly stressed regions may lead to the wrong failure mechanism being predicted.

In the second (so-called 'c-φ reduction') method, the analysis is performed using unfactored (i.e. characteristic) values of material properties and is saved at various points up to failure (points 1, 2, and 3 on **Figure 12.13**, right). At each of these points, a separate analysis is undertaken, starting from the saved conditions but with material strengths reduced by the appropriate partial factor γ_φ, γ_c, or γ_{cu}. The additional movement caused by the reduction in soil strength (with the same external loads) defines the load vs displacement curve for ultimate conditions (points 1a, 2a, and 3a on **Figure 12.13**, right). The ultimate load is given by the peak of this curve. An advantage of this method is that the introduction of partial factors, per se, is unlikely to trigger a wrong failure mechanism. On the downside, analyses based on c-φ reduction do take considerably longer to complete.

12.6   Design for serviceability

Design values of earth pressures for the verification of serviceability limit states must be derived using characteristic soil parameters, taking account of the initial stress, stiffness, and strength of the ground; the stiffness of structural elements; and the allowable deformation of the structure. These earth pressures may not reach limiting (i.e. fully active or passive) values.
[EN 1997-1 §9.8.1(2)P, (4), and (5)]

Limiting values for the allowable wall and ground displacements must take into account the tolerance to those displacements of supported structures and services.   *[EN 1997-1 §9.8.2(1)P]*

Figure 12.14 shows the limiting movements necessary to develop active, half passive, and full passive earth pressures in loose and dense soils, according to Annex C3 of EN 1997-1. Values of horizontal movement v_a and v_p (for active and passive conditions, respectively) – normalized by the wall's retained height, h – are given for (up to) four modes of wall movement, as illustrated on the figure. Also shown is the limit of movement for which at-rest pressures should be considered.   *[EN 1997-1 §9.5.2(2)]*

For example, the horizontal movement needed for earth pressures to drop from at-rest to active values behind a wall retaining 5m of dense soil, is:
$$v_a \approx 0.05 - 0.5\% \times h = 2.5 - 25mm$$
For the common modes of movement under active conditions, movements less than 0.2% (or 10mm) are required for a 5m retained height. As far as the wall is concerned, this level of movement is not that significant. However, if limiting deformations for adjacent structures or structures constructed on top of the wall are less than these values, then it may be necessary to design for K_0 rather than K_a pressures. Designing for K_0 pressures will significantly

increase the resulting bending moments and shear stresses in the wall and should be avoided if at all possible.

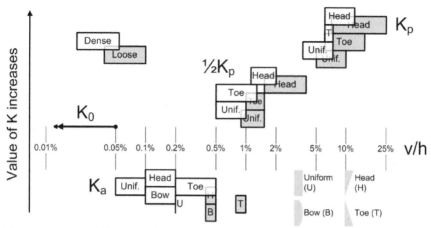

Figure 12.14. *Mobilization of active and passive earth pressures against embedded retaining walls*

It should also be noted that the required movements to develop half passive pressures are between two and ten (typically five) times those required to develop active pressures. The lateral movement of a wall under working conditions may therefore be expected to be significantly greater than that required to reduce earth pressures to the active state.

Calculations of wall movement are not necessarily required to verify the avoidance of serviceability limit states. Eurocode 7 Part 1 states that:

> *a cautious estimate of the distortion and displacement of retaining walls ... shall always be made on the basis of comparable experience ... [including] the effects of construction of the wall.*   *[EN 1997-1 §9.8.2(2)P]*

Thus, a semi-empirical approach to estimating wall movement, such as that described in CIRIA C580,[34] may be sufficient to satisfy the requirements of Eurocode 7.

However, if this 'initial' cautious estimate of displacement exceeds the limiting values, more detailed displacement calculations must be performed. Such calculations are also necessary when nearby structures and services are unusually sensitive to displacement or comparable experience is not well established. The calculations must take account of ground and structural stiffness and the construction sequence.   *[EN 1997-1 §9.8.2(3)P, (5)P, and (7)P]*

Displacement calculations should also be considered when the wall retains more than 6m of low plasticity or more than 3m of high plasticity cohesive soil – or when soft clay comes within the height or beneath the base of the wall.   [EN 1997-1 §9.8.2(6)]

12.7   Structural design

Embedded walls, and any supports they rely upon, must be verified against structural failure according to the provisions of Eurocodes 2 (for concrete),[35] 3 (for steel),[36] 5 (for timber),[37] and 6 (for masonry).[38]   [EN 1997-1 §9.7.6(1)P]

Ultimate limit states that should be verified include failure in bending, shear, compression, tension, and buckling. The strength of the structure assumed in these verifications should be demonstrated to be compatible with the expected ground deformations. Reductions in strength owing to cracking of unreinforced concrete, rotation of plastic hinges in steel, or buckling of thin steel sections should also be considered in accordance with the appropriate structural Eurocode.   [EN 1997-1 §9.7.6(2), (3), and (4)]

12.8   Supervision, monitoring, and maintenance

There are no clauses in EN 1997-1 that specifically cover supervision, monitoring, and maintenance of retaining structures. However, relevant information is available in the following execution standards (which are discussed at length in Chapter 15):

EN 1536	bored piles
EN 1537	ground anchors
EN 1538	diaphragm walls
EN 12063	sheet pile walls
EN 12699	displacement piles

12.9   Summary of key points

The design of embedded walls to Eurocode 7 involves checking that the wall has sufficient embedment to prevent the wall rotating about a fixed point (for example, a point of fixity below formation or a single row of anchors), sufficient strength to mobilize resistance over the full length of the wall, and sufficient stiffness to keep wall displacements and settlement behind the wall within acceptable limits. The design must also demonstrate that the wall has sufficient bearing resistance to withstand any significant vertical load.

Verification of ultimate limit states is demonstrated by satisfying:

$$V_d \leq R_d \text{, and } H_d \leq R_d + R_{pd} \text{, and } M_{Ed,dst} \leq M_{Ed,stb}$$

(where the symbols are defined in Section 10.3). These equations are merely specific forms of:

$$E_d \leq R_d$$

which is discussed at length in Chapter 6.

12.10 Worked examples

The worked examples in this chapter consider the design of an embedded cantilever wall (Example 12.1); and the same wall but singly propped (Example 12.2).

Specific parts of the calculations are marked ❶, ❷, ❸, etc., where the numbers refer to the notes that accompany each example.

12.10.1 Cantilever embedded wall

Example 12.1 considers the design of a cantilever embedded wall retaining 4m of medium dense sand overlying low to medium strength clay as shown in **Figure 12.15**. The sand is assumed to be drained for the purposes of this example, with the groundwater table at least below excavation level. The analysis considers a short-term situation and therefore treats the clay as undrained. This allows the principles of Eurocode 7 to be demonstrated for undrained conditions.

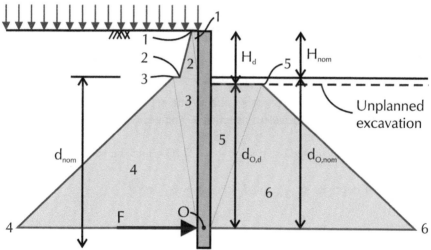

Figure 12.15. *Cantilever embedded wall in sand and clay*

Notes on Example 12.1

❶ Retaining walls are the one geotechnical structure where a dimensional tolerance is specified in Eurocode 7.

Example 12.1
Embedded cantilever wall
Verification of strength (limit state GEO)

Design situation

Consider an embedded sheet pile retaining wall which is retaining H_{nom} = 4m

of medium dense sand overlying low/medium strength clay. A characteristic variable surcharge q_{Qk} = 10kPa acts at the head of the wall. The sand has

characteristic weight density $\gamma_{k_1} = 18\,\dfrac{kN}{m^3}$, angle of shearing resistance

φ_k = 36°, and effective cohesion c'_k = 0kPa. Its angle of shearing

resistance under constant volume conditions is estimated to be $\varphi_{cv,k}$ = 32°.

The clay has characteristic weight density $\gamma_{k_2} = 20\,\dfrac{kN}{m^3}$ and undrained

strength $c_{u,k}$ = 40kPa. The wall toe is at a nominal depth d_{nom} = 9.8m below formation level. The ground is dry throughout.

Design Approach 1

Geometry

Unplanned 'overdig' $\Delta H = min\left(10\% \times H_{nom}, 0.5m\right) = 0.4\,m$

Unplanned height of excavation $H_d = H_{nom} + \Delta H = 4.4\,m$ ❶
Reduced depth of embedment $d_d = d_{nom} - \Delta H = 9.4\,m$

For cantilever walls, earth pressures below the effective wall toe (point O) can be replaced by an equivalent reaction R. The design embedment depth is

conservatively calculated as: $d_{O,d} = \dfrac{d_d}{1.2} = 7.83\,m$ and the nominal

embedment depth as $d_{O,nom} = d_{O,d} + \Delta H = 8.23\,m$

Actions

Vertical total stresses on retained side of wall (from soil self-weight only):

at top of sand σ_{v,k_1} = 0kPa

at bottom of sand $\sigma_{v,k_2} = \gamma_{k_1} \times H_{nom} = 72\,kPa$

at top of clay $\sigma_{v,k_3} = \sigma_{v,k_2} = 72\,kPa$

at point 'O' $\sigma_{v,k_4} = \sigma_{v,k_3} + \left(\gamma_{k_2} \times d_{O,nom}\right) = 236.7\,kPa$

Vertical total stresses on restraining side of wall:

at formation level $\sigma_{v,k_5} = 0\,kPa$

at point 'O' $\sigma_{v,k_6} = \sigma_{v,k_5} + \left(\gamma_{k_2} \times d_{O,d}\right) = 156.7\,kPa$

Material properties

Partial factors from Sets $\begin{pmatrix} M1 \\ M2 \end{pmatrix}$: $\gamma_\varphi = \begin{pmatrix} 1 \\ 1.25 \end{pmatrix}$ and $\gamma_{cu} = \begin{pmatrix} 1 \\ 1.4 \end{pmatrix}$ ❸

Design angle of shearing resistance of sand $\varphi_d = \tan^{-1}\left(\dfrac{\tan\left(\varphi_k\right)}{\gamma_\varphi}\right) = \begin{pmatrix} 36 \\ 30.2 \end{pmatrix}°$

Design constant volume angle of shearing resistance of sand

$\varphi_{cv,d} = \tan^{-1}\left(\dfrac{\tan\left(\varphi_{cv,k}\right)}{\gamma_\varphi}\right) = \begin{pmatrix} 32 \\ 26.6 \end{pmatrix}°$ ❹

For soil/steel interface, $k = \dfrac{2}{3}$

Design angle of wall friction $\delta_d = k \times \varphi_{cv,d} = \begin{pmatrix} 21.3 \\ 17.7 \end{pmatrix} deg$ ❺

Design undrained strength of clay $c_{u,d} = \dfrac{c_{u,k}}{\gamma_{cu}} = \begin{pmatrix} 40 \\ 28.6 \end{pmatrix} kPa$

Effects of actions

Partial factors from Sets $\begin{pmatrix} A1 \\ A2 \end{pmatrix}$: $\gamma_G = \begin{pmatrix} 1.35 \\ 1 \end{pmatrix}$ and $\gamma_Q = \begin{pmatrix} 1.5 \\ 1.3 \end{pmatrix}$

Active earth pressure coeffs for sand $K_{a\gamma} = \begin{pmatrix} 0.222 \\ 0.287 \end{pmatrix}$ $K_{aq} = \begin{pmatrix} 0.222 \\ 0.287 \end{pmatrix}$ ❻

Horizontal stresses on retained side of wall...

At top of sand $\overrightarrow{\sigma_{a,d_1}} = \left(\gamma_G \times K_{a\gamma} \times \sigma_{v,k_1} + \gamma_Q \times K_{aq} \times q_{Qk}\right) = \begin{pmatrix} 3.3 \\ 3.7 \end{pmatrix} kPa$

At bottom of sand $\overrightarrow{\sigma_{a,d_2}} = \left(\gamma_G \times K_{a\gamma} \times \sigma_{v,k_2} + \gamma_Q \times K_{aq} \times q_{Qk}\right) = \begin{pmatrix} 25 \\ 24.4 \end{pmatrix} kPa$

At top of clay $\overrightarrow{\sigma_{a,d_3}} = \left[\gamma_G \times \left(\sigma_{v,k_3} - 2 \times c_{u,d}\right) + \gamma_Q \times q_{Qk}\right] = \begin{pmatrix} 4.2 \\ 27.9 \end{pmatrix} kPa$

At point 'O' $\sigma_{a,d_4} = \overline{\left[\gamma_G \times \left(\sigma_{v,k_4} - 2 \times c_{u,d} \right) + \gamma_Q \times q_{Qk} \right]} = \begin{pmatrix} 226.5 \\ 192.5 \end{pmatrix}$ kPa

Horizontal stresses on restraining side of wall

At formation level $\sigma_{p,d_5} = \overline{\left[\gamma_G \times \left(\sigma_{v,k_5} + 2 \times c_{u,d} \right) \right]} = \begin{pmatrix} 108 \\ 57.1 \end{pmatrix}$ kPa **⑦**

At point 'O' $\sigma_{p,d_6} = \overline{\left[\gamma_G \times \left(\sigma_{v,k_6} + 2 \times c_{u,d} \right) \right]} = \begin{pmatrix} 319.5 \\ 213.8 \end{pmatrix}$ kPa **⑦**

Horizontal thrust

from sand $H_{Ed_1} = \left(\dfrac{\sigma_{a,d_1} + \sigma_{a,d_2}}{2} \right) \times H_{nom} = \begin{pmatrix} 56.6 \\ 56.3 \end{pmatrix} \dfrac{kN}{m}$

from clay $H_{Ed_2} = \overline{\left[\left(\dfrac{\sigma_{a,d_3} + \sigma_{a,d_4}}{2} \right) \times d_{O,nom} \right]} = \begin{pmatrix} 949.7 \\ 907.2 \end{pmatrix} \dfrac{kN}{m}$

total $H_{Ed} = \displaystyle\sum_{i=1}^{2} H_{Ed_i} = \begin{pmatrix} 1006.3 \\ 963.6 \end{pmatrix} \dfrac{kN}{m}$

Overturning moments about point 'O' (subscripts refer to numbers on diagram)

$M_{Ed_1} = \overline{\left[\dfrac{\sigma_{a,d_1}}{2} \times H_{nom} \times \left(\dfrac{2}{3} H_{nom} + d_{O,nom} \right) \right]} = \begin{pmatrix} 72.7 \\ 81.4 \end{pmatrix} \dfrac{kNm}{m}$

$M_{Ed_2} = \overline{\left[\dfrac{\sigma_{a,d_2}}{2} \times H_{nom} \times \left(\dfrac{1}{3} H_{nom} + d_{O,nom} \right) \right]} = \begin{pmatrix} 477.5 \\ 467.3 \end{pmatrix} \dfrac{kNm}{m}$

$M_{Ed_3} = \overline{\left(\dfrac{\sigma_{a,d_3}}{2} \times d_{O,nom} \times \dfrac{2}{3} d_{O,nom} \right)} = \begin{pmatrix} 94.9 \\ 629.5 \end{pmatrix} \dfrac{kNm}{m}$

$M_{Ed_4} = \overline{\left(\dfrac{\sigma_{a,d_4}}{2} \times d_{O,nom} \times \dfrac{1}{3} d_{O,nom} \right)} = \begin{pmatrix} 2559 \\ 2175.1 \end{pmatrix} \dfrac{kNm}{m}$

total $M_{Ed} = \displaystyle\sum_{i=1}^{4} M_{Ed_i} = \begin{pmatrix} 3204 \\ 3353 \end{pmatrix} \dfrac{kNm}{m}$

Resistance

Partial factor from Sets $\begin{pmatrix} R1 \\ R1 \end{pmatrix}$: $\gamma_{Re} = \begin{pmatrix} 1 \\ 1 \end{pmatrix}$

Horizontal resistance $H_{Rd} = \left[\dfrac{\left(\dfrac{\sigma_{p,d_5} + \sigma_{p,d_6}}{2}\right) \times d_{O,d}}{\gamma_{Re}} \right] = \begin{pmatrix} 1674 \\ 1061 \end{pmatrix} \dfrac{kN}{m}$

Restoring moment about point 'O' (subscripts refer to numbers on diagram)

$M_{Rd_5} = \left(\dfrac{\dfrac{\sigma_{p,d_5}}{2} \times d_{O,d} \times \dfrac{2}{3} d_{O,d}}{\gamma_{Re}} \right) = \begin{pmatrix} 2209 \\ 1169 \end{pmatrix} \dfrac{kNm}{m}$

$M_{Rd_6} = \left(\dfrac{\dfrac{\sigma_{p,d_6}}{2} \times d_{O,d} \times \dfrac{1}{3} d_{O,d}}{\gamma_{Re}} \right) = \begin{pmatrix} 3267 \\ 2187 \end{pmatrix} \dfrac{kNm}{m}$

total $M_{Rd} = \sum\limits_{i\,=\,5}^{6} M_{Rd_i} = \begin{pmatrix} 5476 \\ 3355 \end{pmatrix} \dfrac{kNm}{m}$

Verifications

Rotational equilibrium $M_{Ed} = \begin{pmatrix} 3204 \\ 3353 \end{pmatrix} \dfrac{kNm}{m}$ and $M_{Rd} = \begin{pmatrix} 5476 \\ 3355 \end{pmatrix} \dfrac{kNm}{m}$

Degree of utilization $\boxed{\Lambda_{GEO,1} = \dfrac{M_{Ed}}{M_{Rd}} = \begin{pmatrix} 59 \\ 100 \end{pmatrix} \%}$ ⑧

Design is unacceptable if the degree of utilization is > 100%

Reaction near wall toe $F_{Ed} = H_{Rd} - H_{Ed} = \begin{pmatrix} 668.1 \\ 97.7 \end{pmatrix} \dfrac{kN}{m}$ ⑨

Wall section must be designed for...

Maximum bending moment of $M_{d,max} = 222 \dfrac{kNm}{m}$

Maximum shear force of $V_{d,max} = 59 \dfrac{kN}{m}$

Design Approach 2 (summary)

Verification of rotational stability

Nominal depth of embedment $d_{nom} = 9.8$ m

Rotational equilibrium $M_{Ed} = 3204 \dfrac{kNm}{m}$ and $M_{Rd} = 2898 \dfrac{kNm}{m}$

Degree of utilization $\boxed{\Lambda_{GEO,2} = \dfrac{M_{Ed}}{M_{Rd}} = 111\%}$ **⑩**

Design is unacceptable if the degree of utilization is > 100%

Reaction near wall toe $F_{Ed} = H_{Rd} - H_{Ed} = -120.4 \dfrac{kN}{m}$

Design Approach 3 (summary)

Verification of rotational stability

Nominal depth of embedment $d_{nom} = 9.8$ m

Rotational equilibrium $M_{Ed} = 3353 \dfrac{kNm}{m}$ and $M_{Rd} = 3355 \dfrac{kNm}{m}$

Degree of utilization $\boxed{\Lambda_{GEO,3} = \dfrac{M_{Ed}}{M_{Rd}} = 100\%}$ **⑩**

Design is unacceptable if the degree of utilization is > 100%

Reaction near wall toe $F_{Ed} = H_{Rd} - H_{Ed} = 97.7 \dfrac{kN}{m}$

❷ In traditional practice, the bottom 20% of the wall's embedment is assumed to be below the point of fixity 'O'.[39] This is not specified in Eurocode 7 and would be regarded as part of the calculation model.

❸ The partial factor applied to undrained strength in Design Approach Combination 2 ($\gamma_{cu} = 1.4$) is higher than that applied to the angle of shearing resistance ($\gamma_\varphi = 1.25$), reflecting the greater uncertainty in reliably determining this parameter.

❹ There is debate as to whether the characteristic value of φ_{cv} should be factored to obtain its design value. Proponents of critical state soil mechanics argue that φ_{cv} is effectively a 'worst credible' value already and should therefore not be factored. Others argue that φ_{cv} should be treated no differently to other measures of shearing resistance. In this calculation we illustrate the application of the partial factor γ_φ to tan (φ_{cv}).

❺ The angle of wall friction is estimated from the soil's *constant volume* angle of shearing resistance, φ_{cv}. For steel sheet piling δ must be no more than 2/3 times φ_{cv}.

❻ The earth pressure coefficients are obtained from Annex C of EN 1997-1.

❼ The partial factor on unfavourable permanent actions (γ_G) is applied to effective earth pressures on both the retained and restraining sides of the wall. The passive earth pressures are considered to come from the same 'source' as the active earth pressures and hence they are factored in the same way (involving the Single-Source Principle, discussed in Chapter 3). This is somewhat counter-intuitive to engineers who are used to treating passive earth pressures as a resistance.

❽ Combination 2 governs the wall's required embedded length. Separate calculations to determine the maximum bending moment $M_{d,max}$ and shear force $V_{d,max}$ are given on the book's website (see ❿ below). Combination 2 governs, giving $M_{d,max} = 222$ kNm/m and $V_{d,max} = 59$ kN/m.

❾ It is usual to assume that this horizontal force will be provided by passive resistance below the wall's effective toe, particularly with the assumed extra 20% embedment from ❷ above.

❿ The results for Design Approaches 2 and 3 are presented in summary only. The full calculations are available from the book's website at www.decodingeurocode7.com.

Design Approach 2 applies factors greater than 1.0 to action *effects* (i.e. active earth pressures) and resistance (i.e. passive earth pressures). The degree of utilization (111%) is higher than for DA1.

Design Approach 3 applies factors greater than 1.0 to structural actions and material properties. Earth pressures arising from the self-weight of the ground are treated as a geotechnical action and factored by $\gamma_G = 1.0$; earth pressures arising from the surcharge are treated as structural and factored by $\gamma_Q = 1.5$ (versus 1.3 in DA1-2). Hence the degree of utilization (100%) is the same as in DA1.

12.10.2 Anchored sheet pile wall

Example 12.2 considers the design of the anchored sheet pile retaining wall shown in **Figure 12.16**. The wall supports a sandy soil with a nominal retained height of 6m and is propped 1m below its top by a single row of anchorages. Water level is at ground level on the retained side and maintained at formation level on the restraining side.

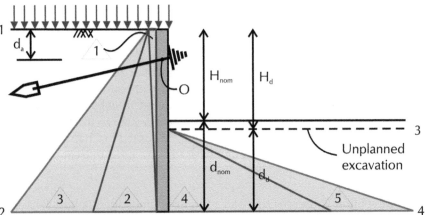

Figure 12.16. Anchored sheet pile wall in sandy soil with groundwater at ground and formation levels

Notes on Example 12.2

❶ Eurocode 7 requires an unplanned 'overdig' to be considered in ultimate limit states. In this example, the upper limit of $\Delta H \ngtr 0.5$m governs.

❷ The hydraulic head is assumed to fall linearly around the wall, resulting in simple triangular distributions of pore water pressure on each side of the wall and the same pore pressure on both sides at the wall toe.

Example 12.2
Anchored sheet pile wall
Verification of strength (limit state GEO)

Design situation

Consider an embedded sheet pile retaining wall which retains $H_{nom} = 6m$ of

medium dense sand with characteristic weight density $\gamma_k = 19\dfrac{kN}{m^3}$, angle of

shearing resistance $\varphi_k = 36°$, and effective cohesion $c'_k = 0kPa$. The soil's

angle of shearing resistance under constant volume conditions is estimated to

be $\varphi_{cv,k} = 32°$. Groundwater is located at ground level on both sides of the

wall. A variable imposed surcharge of $q_{Qk} = 10kPa$ acts at the head of the

wall. The wall is supported by a single row of anchors placed at $d_a = 1m$ below

ground level. The wall toe is at a nominal depth $d_{nom} = 5.85m$ below

formation level. The unit weight of water is $\gamma_w = 9.81\dfrac{kN}{m^3}$

Design Approach 1

Geometry

Unplanned 'overdig' $\Delta H = min\left[10\%(H_{nom} - d_a), 0.5m\right] = 0.5\,m$ ❶

Unplanned height of excavation $H_d = H_{nom} + \Delta H = 6.5\,m$

Reduced depth of embedment $d_d = d_{nom} - \Delta H = 5.35\,m$

Total length of wall $L_d = H_d + d_d = 11.85\,m$

Actions

Vertical total stresses (excluding surcharge) at...

ground level $\sigma_{v,k_1} = 0.kPa$

wall toe (retained side) $\sigma_{v,k_2} = \sigma_{v,k_1} + \gamma_k \times \left(H_d + d_d\right) = 225.2\,kPa$

formation level $\sigma_{v,k_3} = 0kPa$

wall toe (restrained side) $\sigma_{v,k_4} = \sigma_{v,k_3} + \gamma_k \times d_d = 101.7\,kPa$

Difference in hydraulic head $\Delta h = H_d = 6.5\,m$

Distance around wall $x = H_d + 2d_d = 17.2\ \text{m}$

Hydraulic head at wall toe $h_{toe} = \dfrac{\Delta h}{x} \times \left(H_d + d_d\right) = 4.48\ \text{m}$

Pore water pressures at... (assuming head falls linearly around wall)

ground level $u_{k_1} = 0\text{kPa}$

formation level $u_{k_3} = u_{k_1} = 0\ \text{kPa}$

wall toe (retained side) $u_{k_2} = \gamma_w \times \left(H_d + d_d - h_{toe}\right) = 72.3\ \text{kPa}\ \textbf{❷}$

wall toe (restraining side) $u_{k_4} = u_{k_2} = 72.3\ \text{kPa}\ \textbf{❷}$

Vertical effective stresses (excluding surcharge) at...

ground level $\sigma'_{v,k_1} = \sigma_{v,k_1} - u_{k_1} = 0\ \text{kPa}$

wall toe (retained side) $\sigma'_{v,k_2} = \sigma_{v,k_2} - u_{k_2} = 152.9\ \text{kPa}$

formation level $\sigma'_{v,k_3} = \sigma_{v,k_3} - u_{k_3} = 0\ \text{kPa}$

wall toe (restraining side) $\sigma'_{v,k_4} = \sigma_{v,k_4} - u_{k_4} = 29.4\ \text{kPa}$

Material properties

Partial factors from Sets $\begin{pmatrix} M1 \\ M2 \end{pmatrix}$: $\gamma_\varphi = \begin{pmatrix} 1 \\ 1.25 \end{pmatrix}$ and $\gamma_c = \begin{pmatrix} 1 \\ 1.25 \end{pmatrix}$

Design angle of shearing resistance $\varphi_d = \tan^{-1}\left(\dfrac{\tan\left(\varphi_k\right)}{\gamma_\varphi}\right) = \begin{pmatrix} 36 \\ 30.2 \end{pmatrix}^{\circ}$

Design effective cohesion $c'_d = \dfrac{c'_k}{\gamma_c} = \begin{pmatrix} 0 \\ 0 \end{pmatrix} \text{kPa}$

Constant volume angle of shearing resistance (partial factor applied)

$\varphi_{cv,d} = \tan^{-1}\left(\dfrac{\tan\left(\varphi_{cv,k}\right)}{\gamma_\varphi}\right) = \begin{pmatrix} 32 \\ 26.6 \end{pmatrix}^{\circ}$

For soil/steel interface $k = \dfrac{2}{3}$

Design angle of wall friction $\delta_d = k \times \varphi_{cv,d} = \begin{pmatrix} 21.3 \\ 17.7 \end{pmatrix} \text{deg}$

Design friction/shearing ratio $\dfrac{\delta_d}{\varphi_d} = \begin{pmatrix} 0.59 \\ 0.59 \end{pmatrix}$

Effects of actions

Partial factors from Sets $\begin{pmatrix} A1 \\ A2 \end{pmatrix}$: $\gamma_G = \begin{pmatrix} 1.35 \\ 1 \end{pmatrix}$ and $\gamma_Q = \begin{pmatrix} 1.5 \\ 1.3 \end{pmatrix}$

arth pressure coefficients $K_{a\gamma} = \begin{pmatrix} 0.222 \\ 0.287 \end{pmatrix}$ $K_{aq} = \begin{pmatrix} 0.222 \\ 0.287 \end{pmatrix}$ $K_{ac} = \begin{pmatrix} 1.07 \\ 1.226 \end{pmatrix}$,

$K_{p\gamma} = \begin{pmatrix} 6.7 \\ 4.5 \end{pmatrix}$, $K_{pc} = \begin{pmatrix} 7.8 \\ 6.1 \end{pmatrix}$ ❸

Horizontal effective stresses (numbers refer to diagram)

$$\sigma'_{a,d_1} = \overrightarrow{\left[\gamma_G \times \left(K_{a\gamma}\sigma'_{v,k_1} - K_{ac}c'_d \right) + \gamma_Q \times K_{aq}q_{Qk} \right]} = \begin{pmatrix} 3.3 \\ 3.7 \end{pmatrix} \text{kPa}$$

$$\sigma'_{a,d_2} = \overrightarrow{\left[\gamma_G \times \left(K_{a\gamma}\sigma'_{v,k_2} - K_{ac}c'_d \right) + \gamma_Q \times K_{aq}q_{Qk} \right]} = \begin{pmatrix} 49.2 \\ 47.7 \end{pmatrix} \text{kPa}$$

$$\sigma'_{p,d_3} = \overrightarrow{\left[\gamma_G \times \left(K_{p\gamma}\sigma'_{v,k_3} + K_{pc}c'_d \right) \right]} = \begin{pmatrix} 0 \\ 0 \end{pmatrix} \text{kPa} ❹$$

$$\sigma'_{p,d_4} = \overrightarrow{\left[\gamma_G \times \left(K_{p\gamma}\sigma'_{v,k_4} + K_{pc}c'_d \right) \right]} = \begin{pmatrix} 265.6 \\ 132.8 \end{pmatrix} \text{kPa} ❹$$

Water pressures (numbers refer to diagram)

$$u_{a,d_1} = \gamma_G \times u_{k_1} = \begin{pmatrix} 0 \\ 0 \end{pmatrix} \text{kPa}$$

$$u_{a,d_2} = \gamma_G \times u_{k_2} = \begin{pmatrix} 97.6 \\ 72.3 \end{pmatrix} \text{kPa}$$

$$u_{p,d_3} = \gamma_G \times u_{k_3} = \begin{pmatrix} 0 \\ 0 \end{pmatrix} \text{kPa} ❺$$

$$u_{p,d_4} = \gamma_G \times u_{k_4} = \begin{pmatrix} 97.6 \\ 72.3 \end{pmatrix} \text{kPa} ❺$$

Horizontal total stresses (numbers refer to diagram)

$$\sigma_{a,d_1} = \overrightarrow{\left(\sigma'_{a,d_1} + u_{a,d_1} \right)} = \begin{pmatrix} 3.3 \\ 3.7 \end{pmatrix} \text{kPa}$$

$$\sigma_{a,d_2} = \overrightarrow{\left(\sigma'_{a,d_2} + u_{a,d_2} \right)} = \begin{pmatrix} 146.8 \\ 120 \end{pmatrix} \text{kPa}$$

$$\sigma_{p,d_3} = \overrightarrow{\left(\sigma'_{p,d_3} + u_{p,d_3}\right)} = \begin{pmatrix} 0 \\ 0 \end{pmatrix} \text{ kPa}$$

$$\sigma_{p,d_4} = \overrightarrow{\left(\sigma'_{p,d_4} + u_{p,d_4}\right)} = \begin{pmatrix} 363.2 \\ 205.1 \end{pmatrix} \text{ kPa}$$

Horizontal thrust $H_{Ed} = \left(\dfrac{\sigma_{a,d_1} + \sigma_{a,d_2}}{2}\right) \times L_d = \begin{pmatrix} 889.8 \\ 732.9 \end{pmatrix} \dfrac{kN}{m}$

Overturning moment about point 'O'

$$M_{Ed_1} = \left(\dfrac{\sigma_{a,d_1} \times L_d}{2}\right) \times \left(\dfrac{L_d}{3} - d_a\right) = \begin{pmatrix} 58.3 \\ 65.3 \end{pmatrix} \dfrac{kNm}{m}$$

$$M_{Ed_2} = \left(\dfrac{\sigma_{a,d_2} \times L_d}{2}\right) \times \left(\dfrac{2L_d}{3} - d_a\right) = \begin{pmatrix} 6002.9 \\ 4904 \end{pmatrix} \dfrac{kNm}{m}$$

$$\text{sum } M_{Ed} = \sum_{i=1}^{2} M_{Ed_i} = \begin{pmatrix} 6061.2 \\ 4969.3 \end{pmatrix} \dfrac{kNm}{m}$$

Resistance

Partial factor from Sets $\begin{pmatrix} R1 \\ R1 \end{pmatrix}$: $\gamma_{Re} = \begin{pmatrix} 1 \\ 1 \end{pmatrix}$

Horizontal resistance $H_{Rd} = \dfrac{\left(\dfrac{\sigma_{p,d_3} + \sigma_{p,d_4}}{2}\right) \times d_d}{\gamma_{Re}} = \begin{pmatrix} 971.4 \\ 548.6 \end{pmatrix} \dfrac{kN}{m}$

Restoring moment about point 'O'

$$M_{Rd} = \dfrac{\left(\dfrac{\sigma_{p,d_4} d_d}{2}\right) \times \left(\dfrac{2d_d}{3} + H_d - d_a\right)}{\gamma_{Re}} = \begin{pmatrix} 8807.7 \\ 4973.8 \end{pmatrix} \dfrac{kNm}{m}$$

Verifications

Design values $M_{Ed} = \begin{pmatrix} 6061.2 \\ 4969.3 \end{pmatrix} \dfrac{kNm}{m}$ and $M_{Rd} = \begin{pmatrix} 8807.7 \\ 4973.8 \end{pmatrix} \dfrac{kNm}{m}$

Degree of utilization $\Lambda_{GEO,1} = \dfrac{M_{Ed}}{M_{Rd}} = \left(\dfrac{69}{100}\right)\%$ ⑥

Design is unacceptable if the degree of utilization is > 100%

For horizontal equilibrium, anchor must provide design resistance of

$$F_d = H_{Ed} - H_{Rd} = \binom{-81.7}{184.3}\dfrac{kN}{m}$$ ⑦

where $H_{Ed} = \binom{889.8}{732.9}\dfrac{kN}{m}$ and $H_{Rd} = \binom{971.4}{548.6}\dfrac{kN}{m}$

The wall cross-section must now be designed to withstand...

Maximum bending moment in wall $M_{d,max} = -503\dfrac{kNm}{m}$ ⑧

Maximum shear force in wall $V_{d,max} = -176\dfrac{kN}{m}$ ⑧

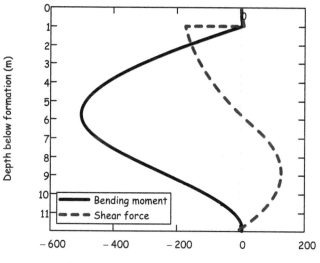

Shear force (kN/m) and bending moment (kNm/m)

Design Approach 2 (summary)

Verification of rotational stability

Nominal depth of embedment $d_{nom} = 7.49 \, m$

Rotational equilibrium $M_{Ed} = 9519 \, \dfrac{kNm}{m}$ and $M_{Rd} = 9514 \, \dfrac{kNm}{m}$

Degree of utilization $\boxed{\Lambda_{GEO,2} = \dfrac{M_{Ed}}{M_{Rd}} = 100 \, \%}$ ⑨

Design is unacceptable if the degree of utilization is > 100%

Anchor must be designed to carry force $F_d = 267 \, \dfrac{kN}{m}$

Wall section must be designed for...

   Maximum bending moment of $M_{d,max} = -819 \, \dfrac{kNm}{m}$

   Maximum shear force of $V_{d,max} = -257 \, \dfrac{kN}{m}$

Design Approach 3 (summary)

Verification of rotational stability

Nominal depth of embedment $d_{nom} = 5.85 \, m$

Rotational equilibrium $M_{Ed} = 4969 \, \dfrac{kNm}{m}$ and $M_{Rd} = 4974 \, \dfrac{kNm}{m}$

Degree of utilization $\boxed{\Lambda_{GEO,3} = \dfrac{M_{Ed}}{M_{Rd}} = 100 \, \%}$ ⑨

Design is unacceptable if the degree of utilization is > 100%

Anchor must be designed to carry force $F_d = 184 \, \dfrac{kN}{m}$

Wall section must be designed for...

   Maximum bending moment of $M_{d,max} = -503 \, \dfrac{kNm}{m}$

   Maximum shear force of $V_{d,max} = -176 \, \dfrac{kN}{m}$

❸ Earth pressure coefficients are taken from Annex C of EN 1997-1.

❹ Applying the 'Single-Source Principle' (see Chapter 3) results in the unfavourable partial factor on actions being applied to both active and passive pressures.

❺ The 'Single-Source Principle' should be applied to pore water pressures as well as earth pressures, resulting in the unfavourable partial factor on actions being applied to the water pressures on both sides of the wall.

❻ Design Approach 1, Combination 2 (DA1-2) gives the most critical result, indicating that the design length of wall is just sufficient to meet the requirements of Eurocode 7.

❼ When considering the forces in the anchor, Design Approach 1 Combination 1 (DA1-1) suggests that a *negative* anchor force is needed to maintain horizontal equilibrium! This arises because passive pressures have been multiplied by $\gamma_G = 1.35$, enhancing their effect and reducing the need for support from the anchor. A more sensible calculation for DA1-1 (given on the book's website – see details below) would be for an embedment d = 4.35m. With this length, the degree of utilization for DA1-1 is 100% and the anchor force needed increases to 181 kN/m.

❽ Once the minimum wall length has been obtained, further analysis is required to establish the maximum bending moment $M_{d,max}$ and shear force $V_{d,max}$ in the wall (to allow selection of an appropriate wall section). Combination 2 gives $M_{d,max}$ = 503 kNm/m and $V_{d,max}$ = 176 kN/m (whereas a similar calculation for Combination 1 gives 450 kNm/m and 172 kN/m, respectively, both of which are smaller than for Combination 2). Furthermore, an additional calculation with all partial factors set to 1.0 should be performed to check that a more onerous anchor force is not obtained under serviceability conditions.

❾ The results for Design Approaches 2 and 3 are presented in summary only. The full calculations are available from the book's website at www.decodingeurocode7.com.

Design Approach 2 applies factors greater than 1.0 to action *effects* (i.e. active earth pressures) and resistance (i.e. passive earth pressures). The depth of embedment needed to obtain a degree of utilization of 100% is 7.49m, which is longer than for DA1. The maximum bending moment and shear force in the wall are 819 kNm/m and 257 kN/m and the anchor force required is

267 kN/m – all of which are considerably greater than for Design Approach 1, because the wall is so much longer.

Design Approach 3 applies factors greater than 1.0 to structural actions and material properties. Earth pressures arising from the self-weight of the ground are treated as a geotechnical action and factored by $\gamma_G = 1.0$; earth pressures arising from the surcharge are treated as structural and factored by $\gamma_Q = 1.5$ (versus 1.3 in DA1-2). The depth of embedment needed to obtain a degree of utilization of 100% is 5.85m, which is identical to DA1-2. The maximum bending moment and shear force in the wall and the required anchor force are also identical to Design Approach 1, Combination 2.

12.11   Notes and references

1. EN 1991-4, Eurocode 1 – Actions on structures, Part 4: Silos and tanks, European Committee for Standardization, Brussels.

2. EN 1993-4-1, Eurocode 3 – Design of steel structures, Part 4-1: Silos, European Committee for Standardization, Brussels.

3. See, for example, Clayton, C. R. I., Milititsky, J., and Woods, R. I. (1993) *Earth pressure and earth-retaining structures* (2nd edition), Glasgow, Blackie Academic & Professional, 398pp.

4. UK National Annex to BS EN 1997-1: 2004, Eurocode 7: Geotechnical design – Part 1: General rules, British Standards Institution.

5. Gaba, A. R., Simpson, B., Powrie, W., and Beadman, D. R. (2003) *Embedded retaining walls - guidance for economic design,* CIRIA Report C580, London: CIRIA, 390pp.

6. BS 5950: 2000, Structural use of steelwork in building, British Standards Institution.

7. BS 8110: 1997, Structural use of concrete, British Standards Institution.

8. Nicholson, D., Tse, C-M., and Penny, C. (1999) *The Observational Method in ground engineering: principles and applications,* CIRIA Report R185, London: CIRIA, 214pp.

9. See §3.2.2.2 of BS 8002: 1994, Code of practice for earth retaining structures, British Standards Institution, with Amendment 2 (dated May 2001).

10. See BS 8102: 1990, Code of practice for protection of structures against water from the ground, British Standards Institution.

11. Terzaghi K. (1954) 'Anchored bulkheads', *Trans. Am. Soc. Civ. Engrs*, 199, pp. 1243-1280.

12. Clayton et al., ibid.

13. Padfield, C. J., and Mair, R. J. (1984) *Design of retaining walls embedded in stiff clays*, CIRIA Report RP104, London: CIRIA, 146pp.

14. EAU (2004), *Recommendations of the committee for Waterfront Structures Harbours and Waterways* (8th edition), Berlin: Ernst & Sohn, 636pp.

15. British Steel (1997), *Piling handbook* (7th edition), Scunthorpe: British Steel plc.

16. BS 8002, ibid.

17. Canadian Geotechnical Society (2006), *Canadian Foundation Engineering Manual* (4th edition), Calgary: Canadian Geotechnical Society, 488pp.

18. Gaba et al., ibid.

19. Padfield and Mair, ibid.

20. Meyerhof, G. G. (1976), 'Bearing capacity and settlement of pile foundations', *J. Geotech. Engng*, Am. Soc. Civ. Engrs, 102(GT3), pp. 197-228.

21. Kezdi, A. (1972), 'Stability of rigid structures', *Proc. 5th European Conf. on Soil Mech. and Found. Engng*, 2, pp. 105-130.

22. UK National Annex, ibid.

23. BS 8002, ibid.

24. Kerisel, J., and Absi, E. (1990), *Active and passive earth pressure tables* (3rd edition), Rotterdam: A.A. Balkema, 220pp.

25. The procedure is not attributed in Annex C, but appears to have been developed by Brinch Hansen. See Christensen, N.H. (1961) *Model tests on plane active earth pressures in sand*, Bulletin No. 10, Copenhagen: Geoteknisk Institut, 19pp. A new derivation is provided by Hansen, B.

(2001) *Advanced theoretical soil mechanics*, Bulletin No. 20, Lyngby: Dansk Geoteknisk Forening, 541pp.

26. Gaba et al., ibid.

27. Simpson, B., and Driscoll, R. (1998) *Eurocode 7 - a commentary*, Garston: BRE.

28. See, for example, Driscoll, R. M. C., Powell, J. J. M., and Scott, P. D. (2008, in preparation) *EC7 – implications for UK practice*, CIRIA Report RP701, London: CIRIA.

29. Padfield and Mair, ibid.

30. Gaba et al., ibid, p. 123.

31. Gaba et al., ibid, p. 160.

32. A similar approach is described in Frank, R., Bauduin, C., Kavvadas, M., Krebs Ovesen, N., Orr, T., and Schuppener, B. (2004), *Designers' guide to EN 1997-1: Eurocode 7: Geotechnical design – General rules*, London: Thomas Telford.

33. See Bauduin, C. (2005), 'Some considerations on the use of finite element methods in ultimate limit state design', *Proc. Int. Workshop on the Evaluation of Eurocode 7* (ed. T. Orr), Dublin, pp. 183-211.

34. Gaba et al., ibid.

35. EN 1992, Eurocode 2 – Design of concrete structures, European Committee for Standardization, Brussels.

36. EN 1993, Eurocode 3 – Design of steel structures, European Committee for Standardization, Brussels.

37. EN 1995, Eurocode 5 – Design of timber structures, European Committee for Standardization, Brussels.

38. EN 1996, Eurocode 6 – Design of masonry structures, European Committee for Standardization, Brussels.

39. See, for example, CIRIA 104, ibid.

Chapter 13

Design of piles

The design of piles is covered by Section 7 of Eurocode 7 Part 1, 'Pile foundations', whose contents are as follows:

§7.1 General (3 paragraphs)
§7.2 Limit states (1)
§7.3 Actions and design situations (18)
§7.4 Design methods and design considerations (8)
§7.5 Pile load tests (20)
§7.6 Axially loaded piles (89)
§7.7 Transversely loaded piles (15)
§7.8 Structural design of piles (5)
§7.9 Supervision of construction (8)

Section 7 of EN 1997-1 applies to end-bearing, friction, tension, and transversely-loaded piles installed by driving, jacking, screwing, and boring (with or without grouting). *[EN 1997-1 §7.1(1)P]*

13.1 Ground investigation for piles

Annex B.3 of Eurocode 7 Part 2 provides outline guidance on the depth of investigation points for pile foundations, as illustrated in **Figure 13.1**. (See Chapter 4 for guidance on the spacing of investigation points.)

The recommended minimum depth of investigation below the base of the deepest pile, z_a, is the greatest of:

$$z_a \geq b_g$$

$$z_a \geq 3D_F$$

$$z_a \geq 5m$$

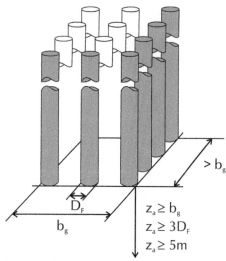

Figure 13.1. Recommended depth of investigation for pile foundations

where b_g is the smaller width of the pile group on plan and D_F is the base diameter of the largest pile.

The depth z_a may be reduced to 2m if the pile foundation is built on competent strata[†] with 'distinct' (i.e. known) geology. With 'indistinct' geology, at least one borehole should go to at least 5m. If bedrock is encountered, it becomes the reference level for z_a.   *[EN 1997-2 §B.3(4)]*

Greater depths of investigation may be needed for very large or highly complex projects or where unfavourable geological conditions are encountered.   *[EN 1997-2 §B.3(2)NOTE and B.3(3)]*

13.2   Design situations and limit states

Possible limit states for pile foundations are illustrated in **Figure 13.2**, from left to right: (top) ground and (middle) structural failure, in compression, tension, and under transverse loading; and (bottom) failure by buckling, shear, and bending.

Pile foundations are used to support buildings and bridges, typically when the upper strata have insufficient bearing capacity to carry the loads or, more commonly, the settlement of a shallow footing exceeds the acceptable limit for the structure.

Eurocode 7 Part 1 contains a long list of design considerations for pile foundations, some of which are mandatory ('shall be taken into account') and some of which are optional ('should receive attention'). For example, the choice of pile type must account for stresses generated during installation and the effect of installation on adjacent structures; while installation-induced vibrations and soil disturbance from boring should be considered.

When either designing a pile group or selecting a pile type these lists provide a useful check to ensure that key factors have not been forgotten. The statements and associated lists do not cover everything that needs to be considered but act as useful aides-memoirs. Whether items are 'shall' or 'should' reflects their relative importance – although 'should' statements are not mandatory, they do reflect good practice, thus the sensible designer will not ignore them.

[†]i.e. weaker strata are unlikely to occur at depth, structural weaknesses such as faults are absent, and solution features and other voids are not expected.

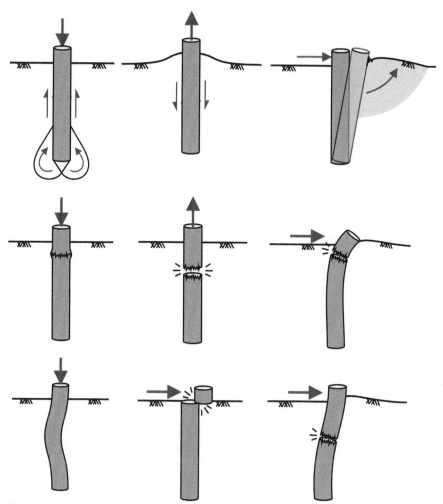

Figure 13.2. Examples of limit states for pile foundations

13.3   Basis of design

Eurocode 7 discusses three methods of designing pile foundations, as summarized in the table below.   *[EN 1997-1 §7.4.1 (1)P]*

The following sub-sections discuss in more detail design by static load tests, calculation, dynamic impact tests, pile driving formulae, and wave equation analysis.

Method	Use	Constraints
Testing	Results of static load tests, provided consistent with relevant experience	Validity must be demonstrated by calculation or other means
	Results of dynamic load tests	Validity must be demonstrated by static load tests in comparable situations
Calculation	Empirical or analytical calculation methods	
Observation	Observed performance of comparable pile foundation	Must be supported by results of site investigation and ground testing

This book does not attempt to provide complete guidance on the design of pile foundations, for which the reader should refer to any well-established text on the subject.[1]

13.3.1 Design by static load tests

As the table above shows, EN 1997-1 places great emphasis on the use of static load tests, either as the primary design method or in providing validity to designs based on dynamic load tests or calculations.

Pile load tests must be performed when there is no comparable experience of the proposed pile type or installation method; the results of previous tests under comparable soil and loading conditions are not available; theory and experience do not provide sufficient confidence in the design for the anticipated loading; or pile behaviour during installation deviates strongly and unfavourably from that anticipated (and additional ground investigations do not explain this deviation).   *[EN 1997-1 §7.5.1(1)P]*

Eurocode 7 distinguishes between static load tests carried out on piles that form part of the permanent works ('working piles') and on piles installed, before the design is finalized, specifically for the purpose of testing ('trial piles' – or what, in the UK, are commonly termed 'preliminary piles').[2] Trial piles must be installed in the same manner and founded in the same stratum as the working piles.   *[EN 1997-1 §7.4.1(3) and 7.6.2.2(2)P]*

The load test procedure must allow the pile's deformation behaviour, creep, rebound, and – for trial piles – the ultimate failure load to be determined. The test load applied to working piles must not be smaller than the foundation's design load; piles tested in tension should be loaded to failure to avoid having to extrapolate the load-displacement curve.

[EN 1997-1 §7.5.2.1(1)P, 7.5.2.1(4), and 7.5.2.3(2)P]

A footnote in Eurocode 7 refers to the International Society of Soil Mechanics and Foundation Engineering's (ISSMFE's) suggested method for axial pile load testing,[3] which will presumably be replaced by reference to EN ISOs 22477-1 to -3[4] when published (see Chapter 4).   *[EN 1997-1 §7.5.2.1(1)P footnote 5]*

Eurocode 7 does not specify how many piles should be tested, leaving this decision to engineering judgement. For trial piles, this must be based on ground conditions and their variability across the site; the structure's Geotechnical Category (see Chapter 3); documented evidence of relevant pile performance in similar ground conditions; and the total number and types of pile in the foundation design. For working piles, that judgement must be based additionally on the piles' installation records.

[EN 1997-1 §7.5.2.2 (1)P and 7.5.2.3 (1)P]

In the absence of detailed rules in Eurocode 7 on the number of test piles, we recommend following the guidance summarized in the table below, which is given by the UK's Institution of Civil Engineers[5] and Federation of Piling Specialists.[6]

Risk	Essential to test ... piles	Number of tests	
		Preliminary	Working
High	Preliminary and working	1 in 250	1 in 100
Medium	Preliminary or working	1 in 500	
Low	(nothing specified)		

Eurocode 7 states that, if only one static load test is performed, it must normally be located where the most adverse ground conditions occur. Failing this, the characteristic compressive resistance must be adjusted accordingly. When two or more tests are performed, one must be located where the most adverse ground conditions occur and the others at locations representative of the pile foundation. In practice, it is often the case that access to the part of the site that has the most adverse conditions is not possible at the time of preliminary testing.

[EN 1997-1 §7.5.1 (4)P and 7.5.1(5)P]

Static load tests must not be performed until the pile material has achieved its desired strength and excess pore pressure generated during installation has fully dissipated. In practice, pore pressures are rarely measured during static load tests, so it is difficult to see this latter requirement being met.

[EN 1997-1 §7.5.1 (6)P]

13.3.2   Design by calculation

Eurocode 7 gives greater emphasis to determining pile resistance from tests (typically static loads, ground, or dynamic impact tests) than by calculation.

Almost as an aside, the standard states that a pile's characteristic base and shaft resistances (R_{bk} and R_{sk}) may be obtained from an 'alternative procedure' involving the equations:

$$R_{bk} = A_b q_{bk}$$

where A_b = the pile's base area and q_{bk} = its unit base resistance; and:

$$R_{sk} = \sum_i A_{s,i} q_{sk,i}$$

where $A_{s,i}$ = the pile's shaft area and $q_{sk,i}$ = its unit shaft resistance in layer 'i'.

In the UK, the majority of pile designs are based on calculations that involve these two equations and Eurocode 7's use of the adjective 'alternative' does not give them sufficient recognition.

13.3.3   Design using dynamic load tests

Eurocode 7 allows the compressive resistance of a pile to be estimated using dynamic load tests, provided the tests are calibrated against static load tests on similar piles, with similar dimensions, installed in similar ground conditions. These requirements limit the applicability of dynamic load tests for design purposes – but they remain useful as an indicator of pile consistency and a detector of weak piles.   [EN 1997-1 §7.5.3.1(1) and 7.5.3.1(3)]

A footnote in Eurocode 7 refers to the American Society of Testing and Materials' (ASTM's) standard test method for high-strain dynamic pile testing,[7] which has no equivalent in the geotechnical investigation and testing standards being prepared by ISO and CEN (see Chapter 4). Draft standards on dynamic load tests and rapid ('Statnamic') testing are expected to be submitted to CEN TC341 in the future.   [EN 1997-1 §7.5.3.1(1) footnote 6]

13.3.4   Design using pile driving formulae or wave equation analysis

Eurocode 7 allows the compressive resistance of a pile to be estimated using pile driving formulae or wave equation analysis, provided the ground's stratification has been determined and the method's validity demonstrated by static load tests on similar piles, with similar dimensions, installed in similar ground conditions.   [EN 1997-1 §7.6.2.5(1)P ,(2)P; 7.6.2.6(1)P,(2)P]

The blow count used in pile driving formulae should be obtained from driving records from at least five piles.   [EN 1997-1 §7.6.2.5(4)]

13.4   Piles subject to compression

For a pile foundation subject to compression, Eurocode 7 requires the design compressive action F_{cd} acting on the pile to be less than or equal to the design bearing resistance R_{cd} of the ground:

$$F_{cd} \leq R_{cd}$$ [EN 1997-1 exp (7.1)]

F_{cd} should include the self-weight of the pile. This equation is merely a restatement of the inequality:

$$E_d \leq R_d$$

discussed at length in Chapter 6.

Figure 13.3 shows a single pile subject to an imposed vertical action P. The total vertical compressive action F_c includes both the imposed action and the pile's self-weight W.

The characteristic value of F_c is given by:

$$F_{ck} = \left(P_{Gk} + \sum_i \psi_i P_{Qk,i}\right) + W_{Gk}$$

where P_{Gk} and $P_{Qk,i}$ are, respectively, characteristic permanent and variable components of P; the symbol W_{Gk} represents the pile's characteristic self-weight (a permanent action); and ψ_i is the combination factor applicable to the *i*th variable action (see Chapter 2).

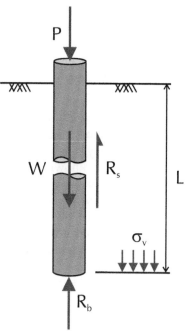

The design value of F_c is given by:

$$F_{cd} = \gamma_G (P_{Gk} + W_{Gk}) + \sum_i \gamma_Q \psi_i P_{Qk,i}$$

where γ_G and γ_Q are partial factors on unfavourable permanent and variable actions.

Figure 13.3. Single pile subject to vertical compression

The total characteristic compressive resistance R_{ck} is given by:

$$R_{ck} = R_{sk} + R_{bk}$$

where R_{sk} is the characteristic shaft resistance and R_{bk} the characteristic base resistance, which should include an allowance for the total overburden pressure at the base of the pile, $\sigma_{v,b}$.   [EN 1997-1 §7.6.2.2(12)]

The design resistance R_{cd} is given by:

$$R_{cd} = \frac{R_{sk}}{\gamma_s} + \frac{R_{bk}}{\gamma_b} \text{ or } R_{cd} = \frac{R_{tk}}{\gamma_t} = \frac{R_{sk} + R_{bk}}{\gamma_t}$$

where γ_s and γ_b are partial factors on the shaft and base resistances, respectively, and γ_t is a partial factor on the total characteristic resistance R_{tk}. The first equation is normally used when designing piles by calculation (i.e. on the basis of ground test results) and the second when the shaft and base components cannot be determined separately (for example, when designing piles using static load or dynamic impact tests).   *[EN 1997-1 §7.6.2.2(14)P]*

The pile's self-weight W is often omitted from traditional pile capacity calculations and the beneficial effect of the overburden pressure at the pile base $\sigma_{v,b}$ is ignored, since they are numerically similar in value:

$$W \approx \sigma_{v,b} A_b$$

where A_b is the area of the pile base. Eurocode 7 allows this simplification to be used if the terms 'cancel approximately'.   *[EN 1997-1 §7.6.2.1(2)]*

13.4.1   Downdrag

As well as carrying loads from the structure, piles can be subject to actions that arise from movement of the ground in which they are installed. This phenomenon is known as 'downdrag' when it involves the ground consolidating, thus bringing additional downwards force onto the pile shaft. Ground movements in other directions (e.g. upwards or horizontal) can induce heaving, stretching, or other displacement of the pile.

Eurocode 7 Part 1 requires such displacements to be treated in one of two ways: either as an indirect action in a soil-structure interaction analysis; or as an equivalent direct action, calculated separately as an upper bound value.

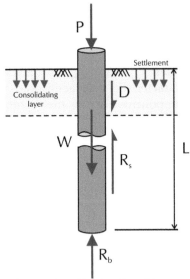

Figure 13.4. Single pile subject to downdrag

Consider a pile installed through a superficial layer as shown in **Figure 13.4**. Consolidation of the layer (owing, for example, to fill being placed upon it) will occur after the pile has been installed, resulting in additional loading being applied to the pile.

Soil-structure interaction analysis allows the 'neutral' depth (where the settlement of the consolidating matches that of the pile under load) to be determined, albeit approximately. In many situations, the effort involved in this type of analysis is outweighed by uncertainties in obtaining suitable ground parameters for use in the analysis.

It is more usual to account for downdrag by inclusion of an appropriate upper-bound action. The characteristic vertical compressive action F_{ck} applied to the pile is then:

$$F_{ck} = P_{Gk} + W_{Gk} + D_{Gk}$$

where P_{Gk} and W_{Gk} are as defined in Section 13.4 and D_{Gk} is the characteristic downdrag acting on the pile (a permanent action). Note that when downdrag is included in this equation, any variable actions may be ignored (hence the absence of P_{Qk}).   [EN 1997-1 §7.3.2.2.(7)]

Typically, the consolidating layer is cohesive and downdrag is calculated from:

$$D_{Gk} = \alpha \times c_{uk} \times A_{s,D}$$

where α = an appropriate adhesion factor, c_{uk} = the clay's characteristic undrained strength, and $A_{s,D}$ = the surface area of the pile shaft in the consolidating layer. In selecting values for α and c_{uk}, it is important to choose *upper* values so as to maximize the value of D_{Gk}.

Some guides[8] recommend calculating the design value of downdrag (D_{Gd}) from:

$$D_{Gd} = \alpha \times c_{ud} \times A_{s,D} = \alpha \times \left(\gamma_{cu} \times c_{uk} \right) \times A_{s,D}$$

where c_{ud} = the clay's design undrained strength and γ_{cu} = a partial factor on strength (= 1.4 in Design Approach 1, Combination 2, and Design Approach 3 – see Chapter 6).

However, care must be taken when using this equation not to select the wrong characteristic value of undrained strength, as illustrated in **Figure 13.5**, which shows the probability of any particular c_u value occurring in the field. In this diagram, c_{uk} is a cautious estimate of the clay's 'inferior' undrained strength[‡] (in this example, c_{uk} = 15 kPa). Multiplying c_{uk} by a partial factor γ_{cu} = 1.4 results in a meaningless value for c_{ud} (= 21 kPa) indicated by the cross. Instead, it is the 'superior' value of the clay's strength $c_{uk,sup}$ (here = 25 kPa) that must be used, resulting in $c_{ud,sup}$ = 35 kPa. In this instance the value of $c_{ud,sup}$ appears excessively large, which may reflect the

---

[‡]See Chapter 5 for discussion of 'inferior' and 'superior' strengths.

fact that the value of $\gamma_{cu} = 1.4$ given in Eurocode 7 is used primarily to determine inferior design strengths (e.g. $c_{ud,inf} = 11$ kPa).

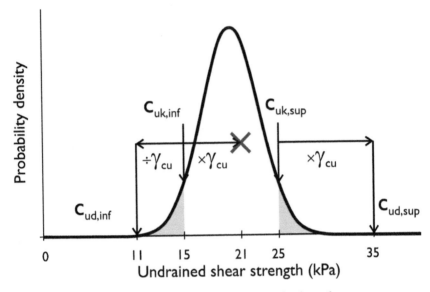

Figure 13.5. *Selection of design undrained strength for downdrag*

In our view, it is more sensible to select the design value of the upper characteristic strength directly (as allowed by Eurocode 7), rather than to calculate it from the characteristic value multiplied by a partial factor.

[EN 1997-1 §2.4.6.2(1)P]

Although EN 1997-1 suggests downdrag should be considered in ultimate limit states, strictly speaking it is only relevant to serviceability limit states. Downdrag results in additional settlement of piles, which needs to be compared with the limiting total and differential settlements defined for the structure. In rare cases when piles are mainly end-bearing, downdrag may result in excessive compressive loads in the piles, leading to end-bearing failure in the ground or structural failure of the pile.

13.5   Piles subject to tension

For a pile foundation subject to tension, Eurocode 7 requires the design tensile action F_{td} acting on the pile to be less than or equal to the design tensile resistance R_{td} of the ground:

$$F_{td} \le R_{td}$$

[EN 1997-1 exp (7.12)]

F_{td} should include the self-weight of the pile. This equation is merely a re-statement of the inequality:

$$E_d \le R_d$$

discussed at length in Chapter 6.

Figure 13.6 shows a single pile subject to an imposed vertical action T, which attempts to pull the pile out of the ground. The uplift force is countered somewhat by the pile's self-weight W and the total vertical tensile action that results is F_t.

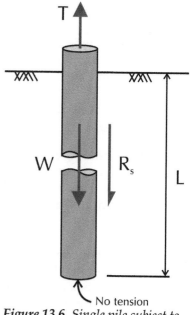

The characteristic value of F_t is given by:

$$F_{tk} = \left(T_{Gk} + \sum_i \psi_i T_{Qk,i}\right) - W_{Gk}$$

where T_{Gk} and T_{Qk} are characteristic permanent and variable components of T, respectively; the symbol W_{Gk} represents the pile's characteristic self-weight (a *favourable* permanent action); and ψ_i is the combination factor applicable to the *i*th variable action (see Chapter 2).

No tension

Figure 13.6. Single pile subject to tension

The design value of F_t is given by:

$$F_{td} = \left(\gamma_G T_{Gk} - \gamma_{G,fav} W_{Gk}\right) + \sum_i \gamma_Q \psi_i T_{Qk,i}$$

where γ_G and γ_Q are partial factors on *unfavourable* permanent and variable actions and $\gamma_{G,fav}$ (= 1.0) is a partial factor on *favourable* permanent actions. Since it is conservative to do so, the pile's self-weight is often omitted from traditional calculations of pile pullout.

The characteristic tensile resistance R_{tk} of the pile is given by:

$$R_{tk} = R_{stk}$$

where R_{stk} is the characteristic tensile shaft resistance. It is assumed that the base offers no resistance to pullout.

The design value of R_t is given by:

$$R_{td} = \frac{R_{stk}}{\gamma_{st}}$$

where γ_{st} is a partial factor on shaft resistance in tension. Values of the partial factor γ_{st} are greater than γ_s to reflect the potential consequences of failure.

[EN 1997-1 §7.6.3.2(2)P]

13.6   Piles subject to transverse actions

Figure 13.7 shows a pile subject to transverse loading. This may be due to horizontal loads and moments applied at the head of the pile or due to lateral ground movements.

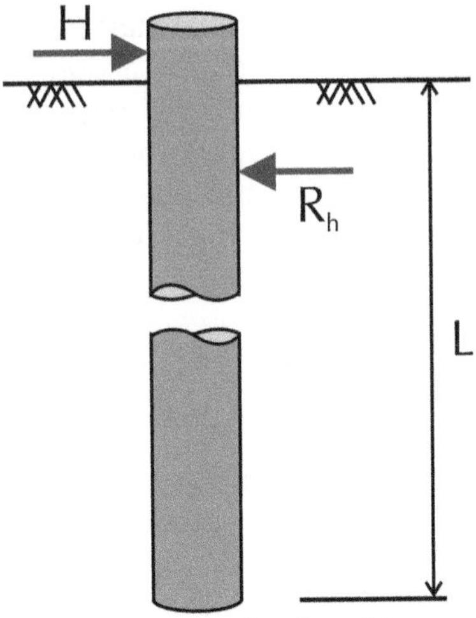

The characteristic horizontal action is:

$$H_k = H_{Gk} + \sum_i \psi_i H_{Qk,i}$$

where H_{Gk} and H_{Qk} are characteristic permanent and variable components of H; and ψ_i is the combination factor applicable to the ith variable action (see Chapter 2).

Figure 13.7. Single pile subject to transverse actions

Whether the horizontal resistance is governed by the lateral resistance of the ground or a combination of ground and pile strength depends on the pile's stiffness and length and the strength-to-stiffness ratio of the ground. Furthermore, the horizontal resistance is also governed by any fixity of the pile head. 'Short' piles are those where the horizontal resistance is governed by the ground strength alone:

$$H_{Rk} = R\left\{X_{k,ground}\right\}$$

whereas 'long' piles are those where the resistance is governed by both pile and ground strength:

$$H_{Rk} = R\left\{X_{k,ground}, M_{Rk,pile}, V_{Rk,pile}\right\}$$

where H_{Rk} is the characteristic horizontal resistance of the pile.

13.7   Introducing reliability into the design of piles

Figure 13.8 illustrates the verification of strength for pile foundations according to Eurocode 7.

The diagram contains significant differences from the equivalent flow-chart discussed in Chapter 6. Chief amongst these is the inclusion of an additional 'channel' (on the right-hand side) dealing with piles designed by testing. This topic is discussed at length in Section 13.9.

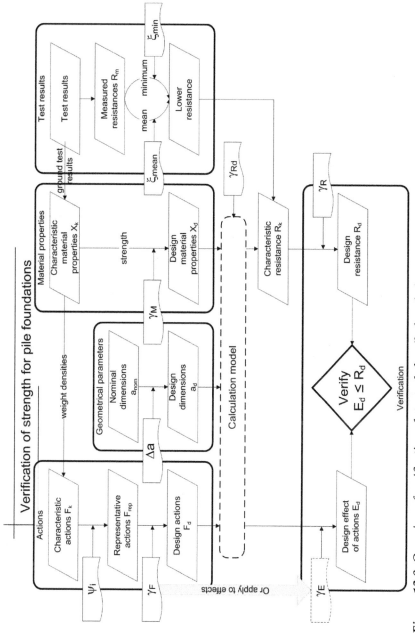

Figure 13.8. Overview of verification of strength for pile foundations

A further change from the flow-chart of Chapter 6 is the introduction of an explicit 'model factor' (γ_{Rd}, applied to the calculation model) when designing piles by calculation. This topic is discussed at length in Section 13.8.

13.7.1   Partial factors

Partial factors for verification of the resistance of pile foundations are specified in Tables A.6–8 of EN 1997-1 and summarized in the table below. Most of these values are amended in the UK National Annex to BS EN 1997-1. Partial factors for actions are the same as for general foundations (see Chapter 6).

Parameter	Model factor	Resistance factors			
		R1	R2	R3	R4
Base resistance (R_b)	γ_b		1.1	1.0	
... driven pile		1.0			1.3
... bored pile		1.25			1.6
... CFA pile		1.1			1.45
Shaft resistance (R_s)	γ_s	1.0	1.1	1.0	1.3
Total resistance (R_c)	γ_t		1.1	1.0	
... driven pile		1.0			1.3
... bored pile		1.15			1.5
... CFA pile		1.1			1.4
Tensile resistance (R_{st})	γ_{st}	1.25	1.15	1.1	1.6
Calculation model	γ_{Rd}	†			

†Value not given in EN 1997-1, but may be specified in National Annex

13.7.2 Design Approach 1

As discussed in Chapter 6, the philosophy of Design Approach 1 is to check reliability with two different combinations of partial factors.

In Combination 1 for pile foundations, partial factors are applied to actions and small factors to resistances, while ground strengths (when used) are left unfactored. This is achieved by employing factors from Sets A1, M1, and R1, as illustrated in **Figure 13.9**. The crosses on the diagram indicate that the factors in Set M1 are all 1.0 (and hence strengths are, in effect, unfactored) and that tolerances Δa are not routinely applied to dimensions.

In Combination 2, partial factors are applied to resistances and to variable actions, while permanent actions and ground strengths (when used) are left unfactored. This is achieved by employing factors from Sets A2, M1, and R4, as illustrated in **Figure 13.10**. Once again, the crosses on the diagram indicate that the factors in Sets A2 and M1 are all 1.0 (except those applied to variable actions) and hence non-variable actions and strengths are unfactored.

Numerical values of the partial factors for Design Approach 1 are given below. See Section 13.11.1 for UK National Annex values for Sets R1 and R4.

Design Approach 1			Combination 1			Combination 2		
			A1	M1	R1	A2	M1	R4
Permanent actions (G)	Unfavourable †	γ_G	1.35			1.0		
	Favourable	$\gamma_{G,fav}$	1.0			1.0		
Variable actions (Q)	Unfavourable †	γ_Q	1.5			1.3		
	Favourable	$\gamma_{Q,fav}$	0			0		
Material properties (X)		γ_M		1.0			1.0	
Base resistance (R_b)	Driven pile	γ_b			1.0			1.3
	Bored pile				1.25			1.6
	CFA pile				1.1			1.45
Shaft resistance (R_s)		γ_s			1.0			1.3
Total resistance (R_c)	Driven pile	γ_t			1.0			1.3
	Bored pile				1.15			1.5
	CFA pile				1.1			1.4
Tensile resistance (R_{st})		γ_{st}			1.3			1.6

†Partial factors for accidental design situations are 1.0

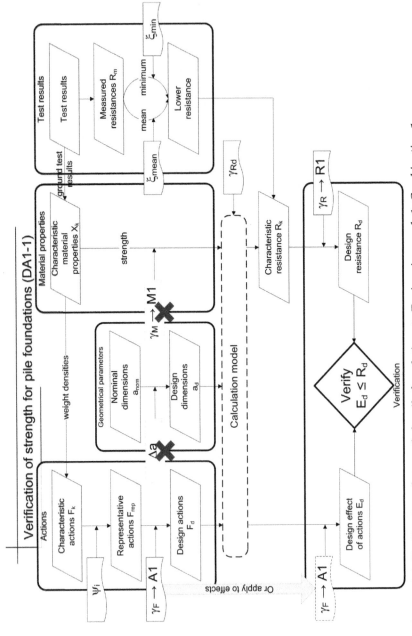

Figure 13.9. Verification of strength for pile foundations to Design Approach 1, Combination 1

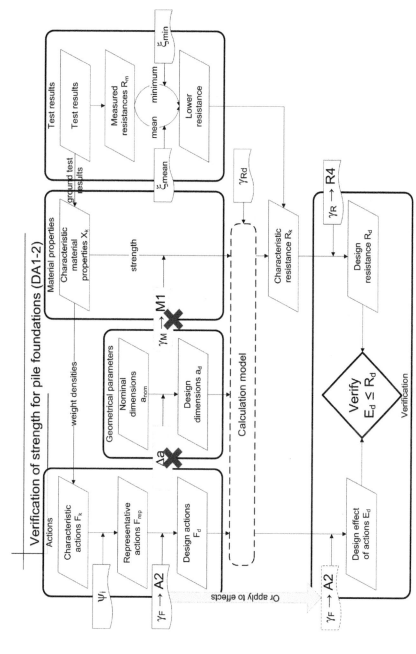

Figure 13.10. Verification of strength for pile foundations to Design Approach 1, Combination 2

13.7.3   Design Approach 2

As discussed in Chapter 6, the philosophy of Design Approach 2 is to check reliability by applying partial factors to actions or action effects and to resistance, while ground strengths (when used) are left unfactored. This philosophy requires no amendment when used to design pile foundations.

Design Approach 2 employs factors from Sets A1, M1, and R2, as illustrated in **Figure 13.11**. The factors in Set M1 are all 1.0 (and hence strengths are unfactored) and tolerances Δa are not routinely applied to dimensions.

Numerical values of the partial factors for Design Approach 2 are:

Design Approach 2			A1	M1	R2
Permanent actions (G)	Unfavourable†	γ_G	1.35		
	Favourable	$\gamma_{G,fav}$	1.0		
Variable actions (Q)	Unfavourable†	γ_Q	1.5		
	Favourable	$\gamma_{Q,fav}$	0		
Material properties (X)		γ_M		1.0	
Base resistance (R$_b$)		γ_b			1.1
Shaft resistance (R$_s$)		γ_s			
Total resistance (R$_c$)		γ_t			
Tensile resistance (R$_{st}$)		γ_{st}			1.15

†Partial factors for accidental design situations are 1.0

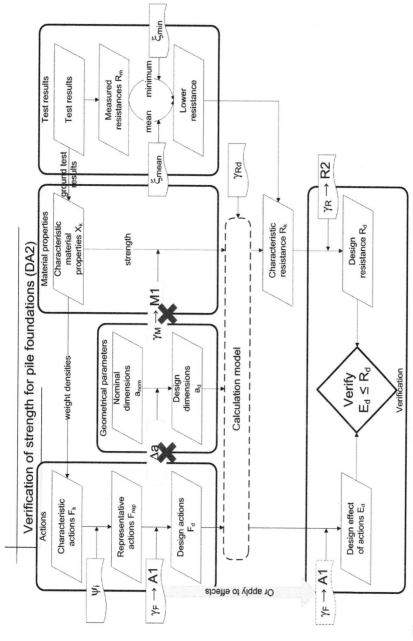

Figure 13.11. Verification of strength for pile foundations to Design Approach 2

13.7.4   Design Approach 3

As discussed in Chapter 6, the philosophy of Design Approach 3 is to check reliability by applying partial factors to actions and to material properties (when used), while resistances are left unfactored. This philosophy remains unchanged when used in the design of pile foundations.

Design Approach 3 employs factors from Sets A1 or A2 (on structural and geotechnical actions respectively), M2, and R3, as illustrated in **Figure 13.12**. The factors in Set R3 are all 1.0 (except those applied to tensile pile resistance) and hence resistance is mostly unfactored. Tolerances, Δa, are not routinely applied to dimensions and the model factor γ_{Rd} is usually not needed.

Numerical values of the partial factors for Design Approach 3 are:

Design Approach 3			A1	A2	M2	R3
Permanent actions (G)	Unfavourable†	γ_G	1.35	1.0		
	Favourable	$\gamma_{G,fav}$	1.0	1.0		
Variable actions (Q)	Unfavourable†	γ_Q	1.5	1.3		
	Favourable	$\gamma_{Q,fav}$	0	0		
Coefficient of shearing resistance (tan φ)		γ_φ			1.25	
Effective cohesion (c′)		$\gamma_{c'}$			1.25	
Undrained strength (c_u)		γ_{cu}			1.4	
Unconfined compressive strength (q_u)		γ_{qu}			1.4	
Weight density (γ)		γ_γ			1.0	
Base resistance (R_b)		γ_b				1.0
Shaft resistance (R_s)		γ_s				
Total resistance (R_t)		γ_t				
Shaft resistance in tension (R_{st})		γ_{st}				1.1

†Partial factors for accidental design situations are 1.0

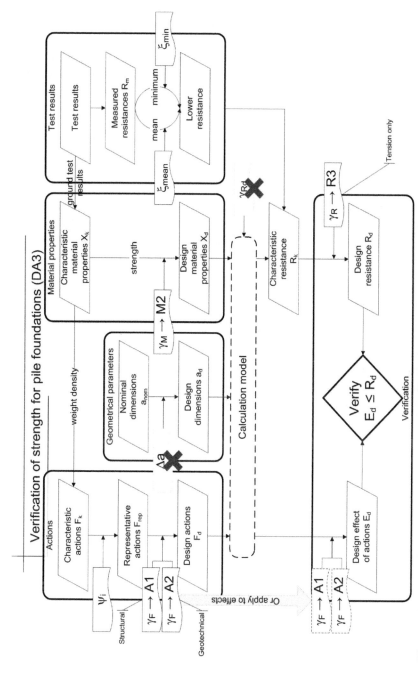

Figure 13.12. Verification of strength for pile foundations to Design Approach 3

13.8   Design by calculation

Design by calculation involves the use of equations that relate soil and rock parameters to shaft friction and end bearing. Unfortunately, available calculation models do not reliably predict the ultimate capacity of piles, owing to the complex interaction between pile type, construction processes, workmanship, and group effects. Consequently, relatively large factors of safety (in the range 2.0–3.0) are applied in such calculations.

The partial factors given in Eurocode 7 for pile resistance are numerically much lower than have traditionally been used in the design of pile foundations, i.e. 1.0–1.6 on base resistance and 1.0–1.3 on shaft (see Section 13.7.1). These factors were chosen for the purpose of design by testing, in which additional 'correlation factors' – numerically between 1.0 and 1.6 – are used to ensure the reliability of the foundation (see Section 13.9). For this reason, Eurocode 7 Part 1 allows a 'model factor' $\gamma_{Rd} > 1$ to be applied in design by calculation to 'correct' the resistance factors:

> *the values of the partial factors γ_b and γ_s recommended in Annex A may need*
> *to be corrected by a model factor larger than [1.0]. The value of the model*
> *factor may be set by the National annex.*   *[EN 1997-1 §7.6.2.3(8) NOTE]*

Some guides[9] on Eurocode 7 suggest combining γ_{Rd} with γ_b and γ_s to produce a single set of 'enhanced' resistance factors, but we do not recommend this. Instead, we prefer to include the model factor in the formulation of the pile's characteristic base and shaft resistances (R_{bk} and R_{sk}), as follows:

$$R_{bk} = \frac{A_b q_{bk}}{\gamma_{Rd}} \text{ and } R_{sk} = \frac{\sum_i A_{s,i} q_{sk,i}}{\gamma_{Rd}}$$

where the symbols are as defined in Section 13.3.2. The design resistance in compression is then given by:

$$R_{cd} = \frac{R_{sk}}{\gamma_s} + \frac{R_{bk}}{\gamma_b} = \frac{\sum_i A_{s,i} q_{sk,i}}{\gamma_{Rd} \times \gamma_s} + \frac{A_b q_{bk}}{\gamma_{Rd} \times \gamma_b}$$

Note that **Figure 13.9** to **Figure 13.11** omit the box for 'Design material properties X_d' that appears in the flow-charts in Chapter 6. The reason for this is to maintain a distinction between characteristic values, which have not had partial factors applied to them, and design values, which have. Since all the factors in Set M1 are unity, we consider the material strengths that go into the calculation model to be characteristic rather than design values.

13.9 Design by testing

Design by testing involves using the results of static load, dynamic impact, or ground tests to define the total pile resistance. This approach can only work where trial piles are installed and the results of tests on these piles are used to design the working piles. Traditionally, on smaller contracts, pile testing is avoided by using a large factor of safety on the calculated capacity.

Although EN 1997-1 emphasizes design by static load testing, for most contracts this is generally impractical as there is insufficient lead time between the main piling works and the test programme. Preliminary tests are rarely performed on piles with similar diameters and lengths, making it difficult to derive a sensible mean test result. In many cases, the ultimate load from a test is obtained by extrapolation of the load-displacement curve, adding further to the uncertainty in any calculated mean.

Verification of strength for pile foundations is illustrated in **Figure 13.8**, where the right-hand channel ('Test results') deals with piles designed by testing. **Figure 13.13** expands this channel to show the specific correlation factors (ξ_1 to ξ_6) that are applied to measured or calculated resistances (R_m or R_{cal}) to determine the characteristic resistance R_k. Factors ξ_1, ξ_3, and ξ_5 are applied to the mean values and ξ_2, ξ_4, and ξ_6 to the minimum values of R_m or R_{cal}. The characteristic resistance R_k is given by the smaller of the results, i.e.:

$$R_k = \min\left(\frac{R_{mean}}{\xi_{mean}}, \frac{R_{min}}{\xi_{min}} \right)$$

where $\xi_{mean} = \xi_1$, ξ_3, or ξ_5 and $\xi_{min} = \xi_2$, ξ_4, or ξ_6.

13.9.1 Correlation factors

Correlation factors for calculating pile resistance are specified in Annex A of EN 1997-1, in Tables A.9 for static load tests, A.10 for ground tests, and A.11 for dynamic impact tests. The table on page 467 summarizes their values.

Eurocode 7 allows the correlation factors for static load tests (ξ_1 and ξ_2) and for ground tests (ξ_3 and ξ_4) to be reduced by 10% when designing piles in groups, provided the structure has sufficient stiffness and strength to transfer loads from weak to strong piles (but ξ_1 and ξ_3 must not be taken as less than 1.0).   *[EN 1997-1 §7.6.2.2(9) and 7.6.2.3(7)]*

The correlation factors for dynamic impact tests (ξ_5 and ξ_6) may be reduced by 15% if signal matching is used. When using pile driving formulae, they should be increased by 10% if the quasi-elastic pile head displacement during impact is measured; and by 20% if not.   *[EN 1997-1 §A.3.3.3]*

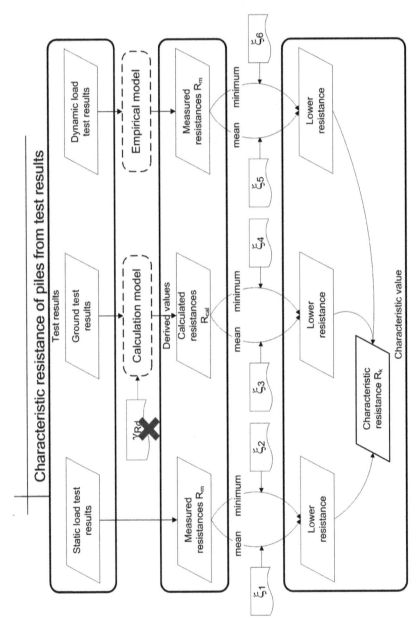

Figure 13.13. Characteristic resistance of pile foundations from test results

No of tests	Static load tests		Ground tests		No of tests	Dynamic impact	
	ξ_1 (on mean)	ξ_2 (on min.)	ξ_3 (on mean)	ξ_4 (on min.)		ξ_5 (on mean)	ξ_6 (on min.)
1	1.4		1.4		-	-	-
2	1.3	1.2	1.35	1.27	2–4	1.6	1.5
3	1.2	1.05	1.33	1.23			
4	1.1	1.0	1.31	1.20			
5	1.0	1.0	1.29	1.15	5–9	1.5	1.35
7			1.27	1.12			
10			1.25	1.08	10–14	1.45	1.3
					15–19	1.42	1.25
					≥20	1.4	1.25

Figure 13.14 shows the variation in correlation factors specified in EN 1997-1.

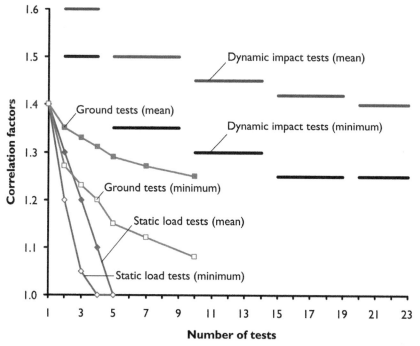

Figure 13.14. Correlation factors specified in EN 1997-1

13.10   Traditional design

Traditional design based on geotechnical calculations is often verified by static pile load tests. Wherever possible preliminary pile load tests are carried out to failure, thereby establishing with some certainty the ultimate capacity of the piles in the particular ground conditions at the site.

> *'A factor of safety of 2.0 is often deemed sufficient when test piles have been loaded to failure. However ... 2.5 is recommended where only proof loads are applied to working piles.'*[10]

When no pile tests are carried out, it is usual to apply a larger factor of safety normally no greater than 3.0.[11]

There are two common methods of introducing safety factors into calculations of the 'allowable' (or 'safe working') load Q_a. Either a single factor F is applied to the pile's ultimate capacity Q_{ult}, or two factors F_s and F_b are applied separately to the shaft and base capacities $Q_{s,ult}$ and $Q_{b,ult}$, i.e.:

$$Q_a = \frac{Q_{ult}}{F} = \frac{Q_{s,ult} + Q_{b,ult}}{F} \text{ or } Q_a = \frac{Q_{s,ult}}{F_s} + \frac{Q_{b,ult}}{F_b}$$

Q_a is often taken as the lower of the values calculated by the two methods.

Various researchers and local and national authorities have recommended values for these factors of safety – for bored and continuous flight auger (CFA) piles – and some of these recommendations are summarized below.

Reference		Factor of safety for bored/CFA piles		
		Global (F)	Shaft (F_s)	Base (F_b)
Burland et al.[12]		2.0	1.0	3.0
Tomlinson[13]		2.5	1.5	3.0
Bowles[14]		2.0–4.0*	-	-
Lord et al. (chalk)[15]	s < 10mm†	2.5	1.5	3.5
	s ≥ 10mm†		1.0	
No base‡		-	1.2–1.5	∞

*'depending on designer uncertainties'
†s = pile head settlement
‡commonly used when base capacity is uncertain[16]

Figure 13.15 compares the equivalent global factor of safety F* obtained from the various recommendations given above, for different amounts of shaft capacity (Q_s) expressed as a percentage of the total ultimate capacity (Q_{ult}). The left-hand end of the horizontal scale represents mainly end-bearing piles and the right-hand end mainly friction piles.

For each curve, the allowable capacity Q_a is calculated as the lower of the values obtained from the application of a single factor F and the application of twin factors F_s and F_b described above. The equivalent global factor is then the ratio of the ultimate capacity to the allowable (i.e. $F^* = Q_{ult}/Q_a$).

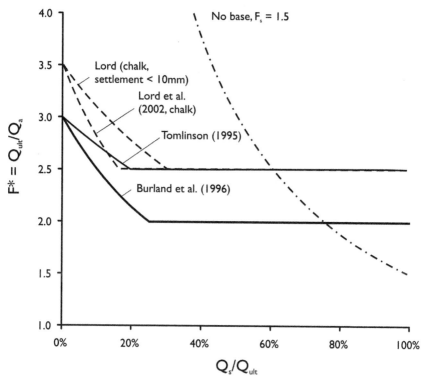

Figure 13.15. Equivalent global factors of safety recommended in the literature

The following table summarizes guidance issued by the London District Surveyors' Association (LDSA)[17] for bored piles in London Clay. The LDSA's advice is to decrease the factor of safety F as more testing is undertaken. This guidance is currently being revised to bring it into line with Eurocode 7.

Adequate SI?	Pile load tests		LDSA recommended		F for α = 0.6
	Preliminary	Working	F	α	
No	None	None	3.0	0.5	3.6
Yes				0.6	3.0
		1% of piles	2.5		2.5
	CRP†		2.25		2.25
	ML‡		2.0	0.5	2.4

†Constant rate of penetration test, ‡Maintained load test

13.11 Changes made in the UK National Annex

The UK National Annex (NA) has reconsidered the partial factor system for piling recommended in EN 1997-1 Annex A in the light of concerns from the UK piling industry that the system does not reflect current UK practice. There are worries that Eurocode 7 might result in overly conservative or, in some instances, unsafe designs. The aim of the UK NA is to provide an equivalent level of safety to traditional good practice, taking into account the benefits of pile testing.

13.11.1 Changes to resistance factors

The table below shows the values of the partial factors in Set R4 that are recommended in the UK National Annex and compares these with values from EN 1997-1 and its earlier draft (ENV 1997-1). In addition, the National Annex sets all the resistance factors in set R1 to 1.0 for piles.

A key feature of the Nation Annex recommendations is the link between the values of the model (γ_{Rd}) and resistance (γ_R) factors and the type and amount of pile testing performed on the site.

With no testing ('no explicit SLS check'), a design relies solely on calculation to guard against the occurrence of ultimate or serviceability limit states – there is no independent corroboration of either. Hence a high level of reliability is needed in the calculations.

With tests on working piles, verification of the ultimate limit state still relies on calculation alone. However, the piles' performance under serviceability conditions will be examined by the static load tests and, if found wanting, would lead to re-design of the piles. Hence lower partial factors can be tolerated than when no tests are performed.

Standard/type and amount of static load testing (where applicable)	MF* γ_{Rd}	Partial resistance factors for Set R4§				γ_t	γ_{st}
		Bored/CFA†		Driven‡			
		γ_b	γ_s	γ_b	γ_s		
ENV 1997-1	1.5	1.45–1.6	1.3	1.3		1.3–1.5	2.0
EN 1997-1	?						
UK NA — No explicit SLS check	1.4	2.0	1.6	1.7	1.5	Use γ_b	2.0
UK NA — > 1% working piles to 1.5x characteristic load		1.7	1.4	1.5	1.3		1.7
UK NA — Preliminary load tests	1.2						

*MF = model factor; ? = value not given in EN 1997-1
γ_{Rd} = model, γ_b = base, γ_s = shaft, γ_t = total, γ_{st} = shaft tension factors
†Replacement piles, ‡displacement piles
§Partial factors for Set R1 are 1.0

Finally, with tests on preliminary piles, verification of both ultimate and serviceability limit states will be corroborated by static load tests – and, again, if found wanting, would lead to re-design of the piles. However, if the results of the static load tests confirm the assumptions made in the calculations, then even lower partial factors can be tolerated than when just working tests are performed.

A common practice in current pile design is to ensure the working load is less than the shaft resistance (with $F_s = 1.2$–1.5) to control settlements where the base performance is uncertain, e.g. in sands. Although the UK NA factors may appear to deal with serviceability implicitly, there will be cases where settlement at characteristic load could still be excessive. Although this can be checked by testing, there will be cases where no preliminary testing is specified. In these cases an assessment of the likely settlement should be carried out.

Figure 13.16 illustrates these ideas by plotting the equivalent global factor of safety F* (defined in Section 13.10) for the three levels of testing discussed above. The diagram also compares the values of F* with those obtained using traditional safety factors for bored piles (taken from **Figure 13.15**). As can be seen, the National Annex ensures roughly equivalent levels of reliability to traditional practice. Note that, because displacement piles increase the stresses in the ground during installation – and are, in effect, tested during installation – they are inherently more reliable than replacement piles and hence require a lower equivalent factor of safety.

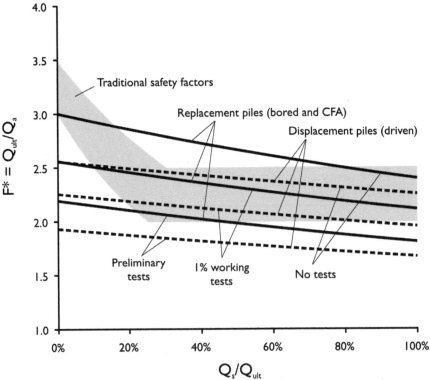

Figure 13.16. Comparison between pile design according to UK National Annex to EN 1997-1 and traditional factors of safety for bored piles

13.11.2 Changes to correlation factors

The table below shows the values of the correlation factors recommended in the UK National Annex to BS EN 1997-1. The values have been changed from those given in EN 1997-1 to ensure compatibility with the alterations made to resistance factors, discussed in Section 13.11.1.

No of tests	Static load tests		Ground tests		No of tests	Dynamic impact tests	
	Mean	Min.	Mean	Min.		Mean	Min.
	ξ_1	ξ_2	ξ_3	ξ_4		ξ_5	ξ_6
1	1.55		1.55		-	-	-
2	1.47	1.35	1.47	1.39	2–4	1.94	1.90
3	1.42	1.23	1.42	1.33			
4	1.38	1.15	1.38	1.29			
5	1.35	1.08	1.36	1.26	5–9	1.85	1.76
6			(1.34)	(1.23)			
7			1.33	1.20			
8			(1.32)	(1.18)			
9			(1.31)	(1.17)			
10			1.30	1.15	10–14	1.83	1.70
					15–19	1.82	1.67
					≥20	1.81	1.66

Values in (brackets) interpolated

The Eurocode 7 rules that allow the correlation factors for static load tests and ground tests to be reduced by 10% when designing piles in groups still apply, as do the rules for adjusting the factors for dynamic impact tests if signal-matching or pile-driving formulae are used (see Section 13.9.1).

Note that, when designing piles by testing on the basis of static load tests, the relevant partial resistance factors to use are those for preliminary tests (i.e. γ_b = 1.7 and γ_s = 1.4 for replacement piles or γ_b = 1.5 and γ_s = 1.3 for displacement piles – see Section 13.11.1).

Figure 13.17 shows the variation in the correlation factors specified in the UK National Annex with the number of tests performed. Not only are the values larger than the ones in EN 1997-1, they also follow a smoother progression with the number of tests.

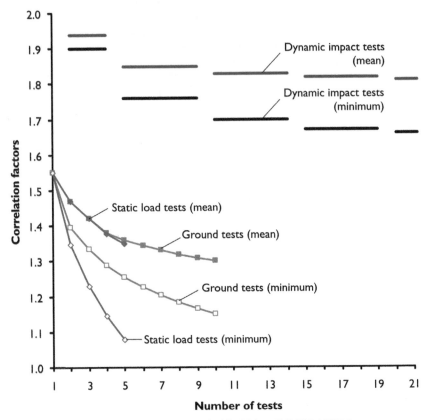

Figure 13.17. Correlation factors specified in NA to BS EN 1997-1

13.12 Supervision, monitoring, and maintenance

Section 7 of EN 1997-1 requires a pile installation plan to form the basis of the piling works. Piles should be monitored and records made as they are installed. These records should be kept for at least five years after completion of the works. Section 7 refers the reader to the following execution standards (which are discussed in Chapter 15) for detailed guidance on pile installation:

EN 1536	Bored piles
EN 12063	Sheet pile walls
EN 12699	Displacement piles
EN 14199	Micropiles

13.13   Summary of key points

The design of piles to Eurocode 7 involves checking that the ground surrounding the piles has sufficient resistance to withstand compression, tension, and transverse actions at the ultimate limit state. Piles may be designed by static load testing (validated by calculation), by dynamic impact testing (validated by static load tests), or by calculation (validated by static load tests).

Verification of ultimate limit states of compression and tension is demonstrated by satisfying the inequalities:

$F_{cd} \leq R_{cd}$ and $F_{td} \leq R_{td}$

(where the symbols are as defined earlier in this chapter). These equations are merely specific forms of:

$E_d \leq R_d$

which is discussed at length in Chapter 6.

Verification of serviceability limit states (SLSs) is normally demonstrated implicitly by verification of the ultimate limit state.

The UK National Annex changes the correlation and partial resistance factors recommended in EN 1997-1 and links the latter to the type and amount of static load testing performed.

13.14   Worked examples

The worked examples in this chapter look at a concrete pile driven into clay and sand (Example 13.1); the same pile designed to the UK National Annex to EN 1997-1 (Example 13.2); the analysis of static load tests for the Emirates Stadium in London (Example 13.3); the use of cone penetration tests to design a continuous flight auger pile in gravel (Example 13.4); and the design of driven piles to a specified set using a pile driving formula (Example 13.5).

Specific parts of the calculations are marked ❶, ❷, ❸, etc., where the numbers refer to the notes that accompany each example.

13.14.1   Concrete pile driven into clay and sand

Example 13.1 looks at the design of a concrete pile that is to be driven through 8m of medium strength sandy CLAY into a medium dense gravelly SAND, as shown in **Figure 13.18**.

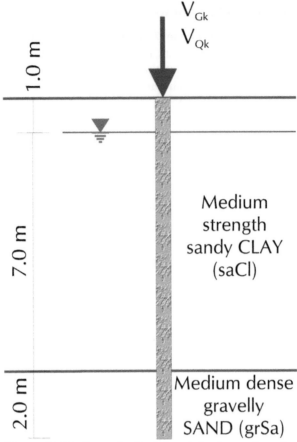

Figure 13.18. Concrete pile driven into clay and sand

The design in this example compares the three Design Approaches discussed in Chapter 6, using the partial factors specified in Annex A of EN 1997-1. Eurocode 7 emphasises design of piles via testing and gives little guidance to design by calculation. This example uses standard UK calculation procedures and highlights some areas where there is debate how Eurocode 7 should be applied.

Notes on Example 13.1

❶ Unlike other geotechnical structures, for DA1 the partial factors applied to material properties are always 1.0 and therefore characteristic material properties are used to derive characteristic shaft resistance and end bearing.

Example 13.1
Concrete pile driven into clay and sand
Verification of strength (limit state GEO)

Design situation

Consider a square concrete pile of breadth b = 400mm, which is driven to a depth L = 10.0m into a medium strength sandy CLAY overlying medium dense gravelly SAND, and carries imposed loads V_{Gk} = 650kN (permanent) and V_{Qk} = 250kN (variable). The weight density of reinforced concrete is

γ_{ck} = 25 $\dfrac{kN}{m^3}$ (as per EN 1991-1-1 Table A.1). The clay, which is t_1 = 8m thick, has characteristic undrained strength c_{uk} = 45kPa and weight density

γ_{k_1} = 18.5 $\dfrac{kN}{m^3}$. The sand has drained strength parameters φ_k = 36° and

c'_k = 0kPa and weight density γ_{k_2} = 20 $\dfrac{kN}{m^3}$. The sand's constant-volume angle of shearing resistance is $\varphi_{cv,k}$ = 33°. Groundwater has been found at a depth d_w = 1m and skin friction on the pile will be ignored above that depth (to allow for whipping during installation, etc.).

Design Approach 1

Actions and effects

Characteristic self-weight of pile is $W_{Gk} = \gamma_{ck} \times b^2 \times L$ = 40 kN

Partial factors from Sets $\begin{pmatrix} A1 \\ A2 \end{pmatrix}$: $\gamma_G = \begin{pmatrix} 1.35 \\ 1 \end{pmatrix}$ and $\gamma_Q = \begin{pmatrix} 1.5 \\ 1.3 \end{pmatrix}$

Design vertical action is $V_d = \overrightarrow{\left[\gamma_G \times \left(W_{Gk} + V_{Gk} \right) + \gamma_Q \times V_{Qk} \right]} = \begin{pmatrix} 1307 \\ 1015 \end{pmatrix}$ kN

Material properties

Partial factors, Sets $\begin{pmatrix} M1 \\ M1 \end{pmatrix}$: $\gamma_\varphi = \begin{pmatrix} 1 \\ 1 \end{pmatrix}$, $\gamma_c = \begin{pmatrix} 1 \\ 1 \end{pmatrix}$, and $\gamma_{cu} = \begin{pmatrix} 1 \\ 1 \end{pmatrix}$ ❶

Hence design values of material strengths are identical to their characteristic values, i.e. undrained strength of clay is c_{uk} = 45 kPa, shearing resistance of sand is φ_k = 36°, and effective cohesion of sand is c'_k = 0 kPa.

Resistance

Partial factors from Sets $\begin{pmatrix} R1 \\ R4 \end{pmatrix}$: $\gamma_b = \begin{pmatrix} 1.0 \\ 1.3 \end{pmatrix}$ and $\gamma_s = \begin{pmatrix} 1.0 \\ 1.3 \end{pmatrix}$ ❶

Model factor (from Irish NA to EN 1997-1): $\gamma_{Rd} = 1.5$ ❷

Base in sand

Area of base is $A_b = b \times b = 0.16 \, m^2$

Length of pile penetrating sand is $t_2 = L - t_1 = 2 \, m$

Slenderness ratio (in sand) $\lambda = \dfrac{t_2}{b} = 5.0$

Bearing capacity factor (from Berezantzev): basic value $B_k = 95.5$,

slenderness correction $\alpha_t = 0.83$, giving bearing capacity factor

$N_q = B_k \times \alpha_t = 79.3$ ❸

Stresses at pile base

Vertical total stress $\sigma_{vk,b} = \left(\gamma_{k_1} \times t_1 \right) + \left(\gamma_{k_2} \times t_2 \right) = 188 \, kPa$

Pore pressure $u_b = \gamma_w \left(L - d_w \right) = 88.3 \, kPa$

Vertical effective stress $\sigma'_{vk,b} = \sigma_{vk,b} - u_b = 99.7 \, kPa$

Characteristic base resistance

$$R_{bk} = \dfrac{\left[\left(N_q + 1 \right) \times \sigma'_{vk,b} + u_b \right] \times A_b}{\gamma_{Rd}} = 863.6 \, kN$$ ❹

Shaft in clay

Area of shaft in clay below water table is $A_{s_1} = 4b \times \left(t_1 - d_w \right) = 11.2 \, m^2$

Undrained adhesion factor (US Army Corps of Engineers) $\alpha = 0.8$ ❸

Characteristic shaft resistance in clay $R_{sk_1} = \dfrac{\alpha \times c_{uk} \times A_{s_1}}{\gamma_{Rd}} = 268.8 \, kN$ ❹

Shaft in sand

Area of shaft in sand is $A_{s_2} = 4b \times \left(L - t_1 \right) = 3.2 \, m^2$

Adopting the method given by Fleming et al. for calculating skin friction...

Earth pressure coefficient $K_s = \dfrac{N_q}{50} = 1.59$

Shaft friction $\delta = \varphi_{cv,k} = 33°$

Equivalent beta factor $\beta = K_s \tan(\delta) = 1.03$

Stresses at top of sand

Vertical total stress $\sigma_{vk} = \gamma_{k_1} \times t_1 = 148\ kPa$

Pore pressure $u = \gamma_w \times \left(t_1 - d_w\right) = 68.6\ kPa$

Vertical effective stress $\sigma'_{vk} = \sigma_{vk} - u = 79.4\ kPa$

Average vertical stress along shaft $\sigma'_{vk,av} = \dfrac{\sigma'_{vk} + \sigma'_{vk,b}}{2} = 89.5\ kPa$

Average skin friction in sand $\tau_{av} = \beta \times \sigma'_{vk.av} = 92.2\ kPa$ ❸

Characteristic shaft resistance in sand $R_{sk_2} = \dfrac{\tau_{av} \times A_{s_2}}{\gamma_{Rd}} = 196.7\ kN$ ❹

Totals

Total characteristic shaft resistance $R_{sk} = \displaystyle\sum_{i=1}^{2} R_{sk_i} = 465.5\ kN$

Total design shaft resistance $R_{sd} = \dfrac{R_{sk}}{\gamma_s} = \begin{pmatrix} 465.5 \\ 358.1 \end{pmatrix} kN$ ❺

Design base resistance $R_{bd} = \dfrac{R_{bk}}{\gamma_b} = \begin{pmatrix} 863.6 \\ 664.3 \end{pmatrix} kN$ ❺

Total design resistance $\qquad\qquad R_d = R_{bd} + R_{sd} = \begin{pmatrix} 1329 \\ 1022 \end{pmatrix} kN$

Verification of compressive resistance

Design values $V_d = \begin{pmatrix} 1307 \\ 1015 \end{pmatrix} kN$ and $R_d = \begin{pmatrix} 1329 \\ 1022 \end{pmatrix} kN$

Degree of utilization $\Lambda_{GEO,1} = \dfrac{V_d}{R_d} = \begin{pmatrix} 98 \\ 99 \end{pmatrix} \%$ ❻

Design is unacceptable if degree of utilization is > 100%

Design Approach 2

Actions and effects

Partial factors from set A1: $\gamma_G = 1.35$ and $\gamma_Q = 1.5$

Design vertical action is $V_d = \gamma_G \times \left(W_{Gk} + V_{Gk}\right) + \gamma_Q \times V_{Qk} = 1307 \text{ kN}$

Material properties

Partial factors from set M1: $\gamma_\varphi = 1$, $\gamma_c = 1$, and $\gamma_{cu} = 1$ ❼

Material properties are therefore identical to Design Approach 1.

Resistance

Partial factors from set R2: $\gamma_b = 1.1$ and $\gamma_s = 1.1$ ❼

Model factor (from Irish NA to EN 1997-1): $\gamma_{Rd} = 1.5$ ❷

Characteristic resistances are unchanged from Design Approach 1

Total design shaft resistance is $R_{sd} = \dfrac{R_{sk}}{\gamma_s} = 423.2 \text{ kN}$ ❺

Design base resistance $R_{bd} = \dfrac{R_{bk}}{\gamma_b} = 785.1 \text{ kN}$ ❺

Total resistance is $R_d = R_{bd} + R_{sd} = 1208 \text{ kN}$

Verification of compressive resistance

Design values $V_d = 1307 \text{ kN}$ and $R_d = 1208 \text{ kN}$

Degree of utilization $\Lambda_{GEO,2} = \dfrac{V_d}{R_d} = 108\ \%$ ❻

Design is unacceptable if degree of utilization is > 100%

Design Approach 3

Actions and effects

Partial factors from set A1: $\gamma_G = 1.35$ and $\gamma_Q = 1.5$

Design vertical action is $V_d = \gamma_G \times \left(W_{Gk} + V_{Gk} \right) + \gamma_Q \times V_{Qk} = 1307$ kN

Material properties

Partial factors from set M2: $\gamma_\varphi = 1.25$, $\gamma_c = 1.25$, and $\gamma_{cu} = 1.4$ ⑧

Design undrained strength of clay is $c_{ud} = \dfrac{c_{uk}}{\gamma_{cu}} = 32.1$ kPa

Design shearing resistance of sand is $\varphi_d = \tan^{-1}\left(\dfrac{\tan\left(\varphi_k\right)}{\gamma_\varphi} \right) = 30.2\,°$

Design constant volume angle of shearing resistance of sand is

$\varphi_{cv.d} = \min\left(\varphi_{cv.k}, \varphi_d \right) = 30.2\,°$ ⑨

Design effective cohesion of sand is $c'_d = \dfrac{c'_k}{\gamma_c} = 0$ kPa

Resistance

Partial factors from set R3: $\gamma_b = 1.0$ and $\gamma_s = 1.0$ ⑧

Bearing capacity factor (Berezantzev): basic value $B_k = 35.4$, slenderness

correction $\alpha_t = 0.78$, giving $N_q = B_k \times \alpha_t = 27.8$ ④

Characteristic base resistance is:

$$R_{bk} = \dfrac{\left[\left(N_q + 1 \right) \times \sigma'_{vk,b} + u_b \right] \times A_b}{\gamma_{Rd}} = 315.6 \text{ kN} ③$$

Characteristic shaft resistance in clay $R_{sk_1} = \dfrac{\alpha \times c_{ud} \times A_{s_1}}{\gamma_{Rd}} = 192 \text{ kN} ③ ④$

Earth pressure coefficient for sand $K_s = \dfrac{N_q}{50} = 0.56$

Shaft friction in sand $\delta = \varphi_{cv,d} = 30.2\,°$

Equivalent beta factor $\beta = K_s \tan(\delta) = 0.32$

Average skin friction in sand is $\tau_{av} = \beta \times \sigma'_{vk.av} = 28.9\ kPa$ ❸

Characteristic shaft resistance in sand $R_{sk_2} = \dfrac{\tau_{av} \times A_{s_2}}{\gamma_{Rd}} = 61.7\ kN$ ❹

Totals

Total characteristic shaft resistance $R_{sk} = \sum\limits_{i=1}^{2} R_{sk_i} = 253.7\ kN$

Total design shaft resistance $R_{sd} = \dfrac{R_{sk}}{\gamma_s} = 253.7\ kN$ ❺

Design base resistance is $R_{bd} = \dfrac{R_{bk}}{\gamma_b} = 315.6\ kN$ ❺

Total resistance is $R_d = R_{bd} + R_{sd} = 569.3\ kN$

Verification of compressive resistance
Design values $V_d = 1307\ kN$ and $R_d = 569\ kN$

Degree of utilization $\boxed{\Lambda_{GEO,3} = \dfrac{V_d}{R_d} = 230\%}$ ❻

Design is unacceptable if degree of utilization is > 100%

If the model factor $\gamma_{Rd} = 1.5$ had been omitted, then

$R_d = R_d \times \gamma_{Rd} = 853.9\ kN$

Degree of utilization $\boxed{\Lambda_{GEO,3} = \dfrac{V_d}{R_d} = 153\%}$ ❿

Design is unacceptable if degree of utilization is > 100%

❷ No value for the model factor is given in EN 1997-1 so we have arbitrarily chosen to use the one given in the Irish National Annex.

❸ The methods used to calculate shaft resistance in clay and sand and the end bearing resistance are standard methods taken from the literature.

❹ The characteristic resistance values are calculated by applying the model factor to the calculated resistance.

❺ To obtain design values for the shaft and end bearing resistance the characteristic values are divided by the appropriate partial factor.

❻ The resultant design is shown to be just satisfactory for DA1 but under-designed according to DA2 and DA3.

❼ DA2 applies partial factors greater than 1.0 to actions and resistance but not to material properties. The values of partial factors for shaft and base resistance are lower than those used in DA1-2 but greater than those used in DA1-1.

❽ DA3 applies partial factors to both the actions and the material properties but not to the resultant shaft and base resistance.

❾ It could be argued that the material partial factor of 1.25 should be applied to φ_{cv} as well as to φ_k, however φ_{cv} already represents the lowest likely value for the material and for the purposes of assessing the shaft resistance through the sand it is acceptable to use the lower of φ_{cv} and φ_d.

❿ DA3 gives a much more conservative design than the other two approaches due to the large effect on N_q when $\varphi_d = \tan^{-1}(\tan \varphi_k)/1.25$ is used. Excluding the model factor improves the situation and it could be argued that it should be taken as 1.0 for DA3. Further, if actions were considered as geotechnical rather than structural, a more economical design would result.

13.14.2   Concrete pile from Example 13.1 to UK National Annex

Example 13.2 revisits the design of the concrete driven pile of Example 13.1 in view of the specific requirements of the UK National Annex to EN 1997-1.

The UK National Annex uses the model factor to account for increased confidence in the calculation model provided by static load testing. The resistance factors given in the National Annex are larger than those in Annex A to EN 1997-1. The partial factors aim to provide similar levels of reliability for Eurocode 7 and traditional UK designs, while allowing potential savings if a significant programme of pile testing is put in place.

Notes on Example 13.2

❶ For no pile tests the values of partial factors for DA1-2 are $\gamma_b = 1.7$, $\gamma_s = 1.5$ and $\gamma_{Rd} = 1.4$.

❷ The resulting design is shown to be unsatisfactory.

❸ For proof load pile tests on 1% of working piles the values of partial factors for DA1-2 are $\gamma_b = 1.5$, $\gamma_s = 1.3$ and $\gamma_{Rd} = 1.4$.

❹ The resulting design is shown to be unsatisfactory but only just and it is likely that a judgement would be made to accept the pile design provided proof load tests were carried out.

❺ For proof load pile tests on 1% of working piles the values of partial factors for DA1-2 are $\gamma_b = 1.5$, $\gamma_s = 1.3$ and $\gamma_{Rd} = 1.2$.

❻ The resulting design is shown to be satisfactory provided preliminary pile tests are carried out. It could be argued that the design is too conservative and some economy could be made if preliminary tests to failure were carried out. It would be a matter of judgement to decide whether it was more economic to be less conservative in the pile design and carry out preliminary pile load tests, or limit the testing to proof loading of working piles.

Example 13.2
Concrete pile from Example 13.1 to UK National Annex
Verification of strength (limit state GEO)

Design situation

Re-consider the design of the square concrete pile from the previous example, this time using the UK National Annex to BS EN 1997-1. Consider what size pile is needed if (a) no static pile load tests are scheduled, (b) tests on 1% of the working piles are scheduled to 1.5 times the characteristic imposed load, and (c) preliminary static pile load tests to the calculated ultimate imposed load are performed.

UK National Annex (Design Approach 1)

Actions, effects, and material properties

From previous calcuation: $V_d = \begin{pmatrix} 1307 \\ 1015 \end{pmatrix}$ kN

Material properties are unchanged from the previous calculation.

Resistance if no explicit SLS check is undertaken

Partial factors from Sets $\begin{pmatrix} R1 \\ R4 \end{pmatrix}$: $\gamma_b = \begin{pmatrix} 1 \\ 1.7 \end{pmatrix}$ and $\gamma_s = \begin{pmatrix} 1 \\ 1.5 \end{pmatrix}$ ❶

Model factor from UK NA to EN 1997-1: $\gamma_{Rd} = 1.4$ ❶

From previous calculation: $A_b = 0.16\,m^2$, $N_q = 79.3$, $\sigma'_{vk,b} = 99.7\,kPa$

Characteristic base resistance:

$$R_{bk} = \frac{\left[\left(N_q + 1\right) \times \sigma'_{vk,b} + u_b\right] \times A_b}{\gamma_{Rd}} = 925.3\,kN$$

From previous calculation: $A_{s_1} = 11.2\,m^2$, $\alpha = 0.8$, $c_{uk} = 45\,kPa$

Characteristic shaft resistance in clay: $R_{sk_1} = \dfrac{\alpha \times c_{uk} \times A_{s_1}}{\gamma_{Rd}} = 288\,kN$

From previous calculation: $A_{s_2} = 3.2\,m^2$, $\sigma'_{vk,av} = 90\,kPa$, $\beta = 1.03$

Characteristic shaft resistance in sand: $R_{sk_2} = \dfrac{\beta \times \sigma'_{vk,av} \times A_{s_2}}{\gamma_{Rd}} = 210.8\,kN$

Characteristic shaft resistance $R_{sk} = \sum\limits_{i=1}^{2} R_{sk_i} = 498.8\,kN$

Design shaft resistance $R_{sd} = \dfrac{R_{sk}}{\gamma_s} = \begin{pmatrix} 498.8 \\ 332.5 \end{pmatrix} kN$

Design base resistance $R_{bd} = \dfrac{R_{bk}}{\gamma_b} = \begin{pmatrix} 925.3 \\ 544.3 \end{pmatrix} kN$

Total resistance is $R_d = R_{bd} + R_{sd} = \begin{pmatrix} 1424 \\ 877 \end{pmatrix} kN$

Verification of compressive resistance

Degree of utilization $\Lambda_{GEO,1} = \dfrac{V_d}{R_d} = \begin{pmatrix} 92 \\ 116 \end{pmatrix} \%$ ❷

Design is unacceptable if the degree of utilization is > 100%

Resistance if SLS checked by static pile load tests on 1% working piles

Partial factors from Sets $\begin{pmatrix} R1 \\ R4 \end{pmatrix}$: $\gamma_b = \begin{pmatrix} 1 \\ 1.5 \end{pmatrix}$ and $\gamma_s = \begin{pmatrix} 1 \\ 1.3 \end{pmatrix}$ ❸

Model factor from UK NA to EN 1997-1: $\gamma_{Rd} = 1.4$ ❸

Total design shaft resistance is $R_{sd} = \dfrac{R_{sk}}{\gamma_s} = \begin{pmatrix} 498.8 \\ 383.7 \end{pmatrix} kN$

Design base resistance is $R_{bd} = \dfrac{R_{bk}}{\gamma_b} = \begin{pmatrix} 925.3 \\ 616.9 \end{pmatrix} kN$

Total resistance is $R_d = R_{bd} + R_{sd} = \begin{pmatrix} 1424 \\ 1001 \end{pmatrix} kN$

Verification of compressive resistance

Degree of utilization $\Lambda_{GEO,1} = \dfrac{V_d}{R_d} = \begin{pmatrix} 92 \\ 101 \end{pmatrix} \%$ ❹

Design is unacceptable if the degree of utilization is > 100%

Resistance if SLS & ULS checked by static pile load tests to ultimate load

Partial factors from Sets $\begin{pmatrix} R1 \\ R4 \end{pmatrix}$: $\gamma_b = \begin{pmatrix} 1 \\ 1.5 \end{pmatrix}$ and $\gamma_s = \begin{pmatrix} 1 \\ 1.3 \end{pmatrix}$ **⑤**

Model factor from UK NA to EN 1997-1: $\gamma_{Rd} = 1.2$ **⑤**

Characteristic base resistance $R_{bk} = \dfrac{\left[(N_q + 1) \times \sigma'_{vk,b} + u_b \right] \times A_b}{\gamma_{Rd}} = 1080 \text{ kN}$

Characteristic shaft resistance in clay: $R_{sk_1} = \dfrac{\alpha \times c_{uk} \times A_{s_1}}{\gamma_{Rd}} = 336 \text{ kN}$

Characteristic shaft resistance in sand: $R_{sk_2} = \dfrac{\beta \times \sigma'_{vk,av} \times A_{s_2}}{\gamma_{Rd}} = 245.9 \text{ kN}$

Total characteristic shaft resistance: $R_{sk} = \displaystyle\sum_{i=1}^{2} R_{sk_i} = 581.9 \text{ kN}$

Total design shaft resistance: $R_{sd} = \dfrac{R_{sk}}{\gamma_s} = \begin{pmatrix} 581.9 \\ 447.6 \end{pmatrix} \text{ kN}$

Design base resistance is $R_{bd} = \dfrac{R_{bk}}{\gamma_b} = \begin{pmatrix} 1079.5 \\ 719.7 \end{pmatrix} \text{ kN}$

Total resistance is $R_d = R_{bd} + R_{sd} = \begin{pmatrix} 1661 \\ 1167 \end{pmatrix} \text{ kN}$

Verification of compressive resistance

Degree of utilization $\Lambda_{GEO,1} = \dfrac{V_d}{R_d} = \begin{pmatrix} 79 \\ 87 \end{pmatrix} \%$ **⑥**

Design is unacceptable if the degree of utilization is > 100%

13.14.3   Static load tests for the Emirates Stadium in London

Example 13.3 looks at the design of bored piles for the Emirates Stadium in London, the new home of Arsenal Football Club.[18]

Seven preliminary pile tests were carried out on piles of the same diameter, but with slightly different depths of penetration. Six of the piles (P1–3 and P5–7) were of similar length, between 23.5m and 26.3m, and one pile (P4) was significantly shorter, at 16.9m. All the piles were of bored construction. Ground conditions at the pile toes were similar for all piles except two (P1 and P5), which were founded on a stiffer calcareous layer.

For design based on pile testing, it is essential that the pile tests are all carried out on similarly constructed piles in similar ground conditions. For this reason, it is arguable that not all the pile tests at the Emirates Stadium should be included in the design calculation. In this example, we have chosen to consider only those piles that are clearly comparable (i.e. P2–4 and P6–7).

The design is checked for the partial and correlation factors recommended in both Annex A to EN 1997-1 and the UK National Annex to BS EN 1997-1 (to highlight the effects of the National Annex on design).

Notes on Example 13.3

❶ Since the load is measured at the head of the pile, it is appropriate to ignore the weight of the pile in the analysis.

❷ In order to apply the correlation factors given in EN 1997-1, it is essential that the pile tests should be considered to be representative of the same data set. Therefore those tests that cannot be treated as being essentially of similar design, construction, and founded in similar materials should be excluded. It could be debated whether the two piles of similar length to the others but with significantly lower settlement do or do not form part of the same data set. It is highly unlikely that one could ever get a sufficiently large data set of similar tests to take advantage of this approach and yet this is Eurocode 7's preferred method.

❸ The factor of 1.1 is specified in §7.6.2.2 of EN 1997-1.

❹ Adopting the EN 1997-1 values for partial factors results in a design that meets the requirements of the code.

Example 13.3
Static load tests for the Emirates Stadium load tests in London
Verification of strength (limit state GEO)

Design situation

Consider the design of piles for the Emirates Stadium in London (the new home of Arsenal Football Club) where a large number of bored piles of different diameters are to be installed. Ground conditions at the site comprise 2.9m Made Ground, 2.1m Terrace Gravels, 31.0m London Clay, and at least 5m Lambeth Clay. Assume the top $L_0 = 5m$ of the ground profile provides negligible skin friction to the piles.

Load tests have been perfomed on seven piles, all with the same diameter $D_m = 600mm$ but with different total lengths L_m (Pile test data courtesy Stent Foundations Ltd.) The peak applied load P and the corresponding settlement at that load is given below for each pile:

$$L_m = \begin{pmatrix} 25.4m \\ 24.9m \\ 23.5m \\ 16.9m \\ 26.3m \\ 24.3m \\ 24.4m \end{pmatrix} \quad P_m = \begin{pmatrix} 6000kN \\ 4956kN \\ 4000kN \\ 2310kN \\ 4300kN \\ 4200kN \\ 4200kN \end{pmatrix} \quad S_m = \begin{pmatrix} 26.2mm \\ 61.9mm \\ 60.5mm \\ 60.7mm \\ 23.2mm \\ 61.5mm \\ 45mm \end{pmatrix}$$

The two piles with less than 30mm settlement at peak load were founded on siltstone within the London Clay, whereas the other piles were founded on London Clay. The settlement reading for the last pile is an underestimate, owing to instrument failure.

A group of piles, with diameter D = 600mm and length L = 25m, are each required to carry a permanent action $F_{Gk} = 1500kN$ together with a variable action $F_{Qk} = 700kN$. The weight density of reinforced concrete is $\gamma_{ck} = 25\dfrac{kN}{m^3}$ (as per EN 1991-1-1 Table A.1).

Measured resistance
Assume that the majority of the measured resistance from the pile load tests comes from shaft resistance. Regard the results of the first and fifth load tests as unrepresentative because the pile toes are founded on a siltstone layer. The results from the remaining tests on piles with different embedded lengths can then be 'normalized' to the same length as the piles being designed, as follows: ❷

Length of shaft ignored is $L_0 = 5\,m$

Normalized measured resistance $R_m = \left[P_m \times \left[\dfrac{D \times (L - L_0)}{D_m \times (L_m - L_0)} \right] \right] = \begin{pmatrix} 0 \\ 4981 \\ 4324 \\ 3882 \\ 0 \\ 4352 \\ 4330 \end{pmatrix}$ kN

No of tests considered is $n = 5$

Mean resistance is $R_{m,mean} = \dfrac{\sum R_m}{n} = 4374\ kN$

Minimum resistance is $R_{m,min} = min\left(R_{m_2}, R_{m_3}, R_{m_4}, R_{m_6}, R_{m_7} \right) = 3882\ kN$

Design Approach 1

Actions and effects

Ignore the self-weight of pile

Characteristic total action $F_{ck} = F_{Gk} + F_{Qk} = 2200$ kN ❶

Partial factors from Sets $\begin{pmatrix} A1 \\ A2 \end{pmatrix}$: $\gamma_G = \begin{pmatrix} 1.35 \\ 1 \end{pmatrix}$ and $\gamma_Q = \begin{pmatrix} 1.5 \\ 1.3 \end{pmatrix}$

Design total action is $F_{cd} = \gamma_G F_{Gk} + \gamma_Q F_{Qk} = \begin{pmatrix} 3075 \\ 2410 \end{pmatrix}$ kN

Characteristic resistance

Correlation factor on mean measured resistance $\xi_1 = 1.0$

Correlation factor on minimum measured resistance $\xi_2 = 1.0$

For a pile group that can transfer load from weak to strong piles

(§7.6.2.2.(9)), ξ may be divided by 1.1 (but ξ_1 cannot fall beneath 1.0). ❸

Thus $\xi_1 = \max\left(\dfrac{\xi_1}{1.1}, 1.0\right) = 1.0$ and $\xi_2 = \dfrac{\xi_2}{1.1} = 0.91$

Mean/min measured resistance $\dfrac{R_{m,mean}}{\xi_1} = 4374$ kN and $\dfrac{R_{m,min}}{\xi_2} = 4271$ kN

Characteristic resistance is $R_{ck} = \min\left(\dfrac{R_{m,mean}}{\xi_1}, \dfrac{R_{m,min}}{\xi_2}\right) = 4271$ kN

Design resistance

Partial factors from Sets $\begin{pmatrix} R1 \\ R4 \end{pmatrix}$: $\gamma_t = \begin{pmatrix} 1.15 \\ 1.5 \end{pmatrix}$

Design resistance is $R_{cd} = \dfrac{R_{ck}}{\gamma_t} = \begin{pmatrix} 3714 \\ 2847 \end{pmatrix}$ kN

Verification of compression resistance

Degree of utilization $\Lambda_{GEO,1} = \dfrac{F_{cd}}{R_{cd}} = \begin{pmatrix} 83 \\ 85 \end{pmatrix}$ % ❹

Design is unacceptable if the degree of utilization is > 100%

Design Approach 2

Actions and effects

Ignore the self-weight of pile

Characteristic total action $F_{ck} = F_{Gk} + F_{Qk} = 2200$ kN

Partial factors from set A1: $\gamma_G = 1.35$ and $\gamma_Q = 1.5$

Design total action is $F_{cd} = \gamma_G F_{Gk} + \gamma_Q F_{Qk} = 3075$ kN

Characteristic resistance

Characteristic resistance is unchanged: $R_{ck} = 4271$ kN

Design resistance

Partial factors from sets R2: $\gamma_t = 1.1$

Design resistance is $R_{cd} = \dfrac{R_{ck}}{\gamma_t} = 3882$ kN

Verification of compression resistance

Degree of utilization $\boxed{\Lambda_{GEO,2} = \dfrac{F_{cd}}{R_{cd}} = 79\ \%}$

Design is unacceptable if the degree of utilization is > 100%

Design Approach 3

Is not suitable for pile design

Design to UK National Annex to BS EN 1997-1

The UK National Annex changes both the resistance factors and the correlation factors that must be used to verify pile resistance. All other factors remain unchanged from their EN values.

Characteristic resistance

Correlation factor on mean measured resistance is $\xi_1 = 1.35$

Correlation factor on minimum measured resistance is $\xi_2 = 1.08$

For a pile group that can transfer load from weak to strong piles

(§7.6.2.2.(9)): $\xi_1 = \max\left(\dfrac{\xi_1}{1.1}, 1.0\right) = 1.23$ and $\xi_2 = \dfrac{\xi_2}{1.1} = 0.98$ ❺

Mean/min measured resistance $\dfrac{R_{m,mean}}{\xi_1} = 3564 \text{ kN}$ and $\dfrac{R_{m,min}}{\xi_2} = 3954 \text{ kN}$

Characteristic resistance is $R_{ck} = \min\left(\dfrac{R_{m,mean}}{\xi_1}, \dfrac{R_{m,min}}{\xi_2}\right) = 3564 \text{ kN}$ ❻

Design resistance

Partial factors, Sets $\begin{pmatrix} R1 \\ R4 \end{pmatrix}$: $\gamma_t = \begin{pmatrix} 1 \\ 1.7 \end{pmatrix}$, $\gamma_s = \begin{pmatrix} 1 \\ 1.4 \end{pmatrix}$, $\gamma_b = \begin{pmatrix} 1 \\ 1.7 \end{pmatrix}$ ❼

Design resistance is $R_{cd} = \dfrac{R_{ck}}{\gamma_t} = \begin{pmatrix} 3564 \\ 2096 \end{pmatrix} \text{ kN}$ ❽

Verification of compression resistance

Degree of utilization $\Lambda_{GEO,1} = \dfrac{F_{cd}}{R_{cd}} = \begin{pmatrix} 86 \\ 115 \end{pmatrix} \%$ ❾

Design is unacceptable if the degree of utilization is > 100%

Assuming $\chi = 85\%$ of the characteristic resistance comes from shaft

friction, design resistance $R_{cd} = \chi\left(\dfrac{R_{ck}}{\gamma_s}\right) + (1-\chi)\left(\dfrac{R_{ck}}{\gamma_b}\right) = \begin{pmatrix} 3564 \\ 2478 \end{pmatrix} \text{ kN}$

Degree of utilization $\Lambda_{GEO,1} = \dfrac{F_{cd}}{R_{cd}} = \begin{pmatrix} 86 \\ 97 \end{pmatrix} \%$ ❿

Design is unacceptable if the degree of utilization is > 100%

❺ The correlation factors given in the UK NA to BS EN 1997-1 are higher because existing UK practice suggests the correlation factors in EN 1997-1 may lead to unsafe designs.

❻ The resulting characteristic resistance is significantly reduced compared with the EN 1997-1 calculation.

❼ In addition, the R4 resistance factor is higher than that given in Annex A of EN 1997-1, leading to a much-reduced design resistance in DA1-2.

❽ The partial factor on the total resistance must be used because no attempt has been made to separate the base and shaft components of resistance.

❾ A design to BS EN 1997-1 suggests DA1-2 is the most critical and that the requirements of the standard are not met.

❿ If we separate the base and shaft resistances, then a lower resistance factor may be applied to the shaft component (since $\gamma_s < \gamma_t$), but not to the base (since $\gamma_b = \gamma_t$). Assuming a relatively small proportion of load taken by the base (15%), which is not untypical for floating piles, implies that the piles would be satisfactory to DA1-2. This emphasizes the importance of representing the pile behaviour as accurately as possible if unnecessary conservatism is to be avoided.

13.14.4 Design of continuous flight auger piles from cone tests

Example 13.4 demonstrates how continuous flight auger (CFA) piles may be designed from the results of field testing.

An investigation comprising both cable percussion boreholes and cone penetration tests (CPTs) was carried out at a site in Richmond, West London.[19] A series of CPTs was carried out and **Figure 13.19** shows a typical result from the site. The CPTs confirm the visual identification of the strata from the boreholes and provide data for direct pile design based on cone resistance. The sequence of strata is approximately 8m of Kempton Park Gravel overlying London Clay.

The design of a pile group comprising six, 400mm-diameter, 6m-long, continuous flight auger (CFA) piles is to be based on four CPT ground profiles. All three Design Approaches are considered. Design Approach 3 is not suitable for designing piles based on ground tests, as it requires partial factors to be applied to material properties that are not derived directly from the test.

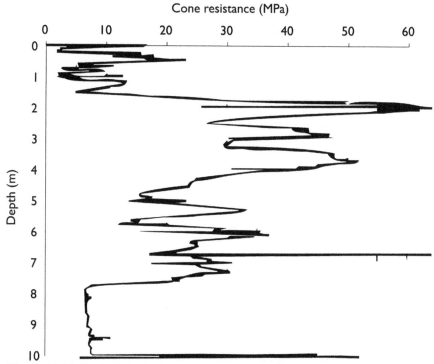

Figure 13.19. *Typical cone penetration test profile for site in Twickenham, through Kempton Park Gravel*

The calculation for Design Approach 1 is also re-worked using the correlation and partial factors recommended in the UK National Annex to BS EN 1997-1.

Notes on Example 13.4

❶ Four cone penetration (CPT) tests have been used to derive average shaft and base resistance, using standard procedures to obtain derived values of shaft and base resistance.

❷ Partial factor sets to be used for permanent and variable actions are defined in Annex A to EN 1997-1.

❸ Correlation factors are defined in Annex A to EN 1997-1 for four ground profiles. Correlation factors are applied to the mean and minimum of the data set.

Example 13.4
Design of continuous flight auger piles from cone tests
Verification of strength (limit state GEO)

Design situation

Consider the design of continuous flight auger (CFA) piles for a site in Twickenham, London. Ground conditions at the site comprise dense, becoming loose gravelly, SAND. Cone penetration tests have been perfomed at the site to a depth of 8m. (Data courtesy CL Associates.) The limiting average unit shaft resistance p_s and limiting unit base resistance p_b at each cone location are estimated to be:

$$P_s = \begin{pmatrix} 120\text{kPa} \\ 120\text{kPa} \\ 100\text{kPa} \\ 120\text{kPa} \end{pmatrix} \quad P_b = \begin{pmatrix} 2800\text{kPa} \\ 3000\text{kPa} \\ 2000\text{kPa} \\ 3000\text{kPa} \end{pmatrix} \quad \mathbf{①}$$

A group of N = 6 piles with diameter D = 400mm and length L = 6m are required to carry between them a permanent action F_{Gk} = 2100kN together with a variable action F_{Qk} = 750kN. The weight density of reinforced concrete is $\gamma_{ck} = 25\dfrac{kN}{m^3}$ (as per EN 1991-1-1 Table A.1).

Design Approach 1

Actions and effects

The self-weight of pile is $W_{Gk} = \left(\dfrac{\pi \times D^2}{4} \right) \times L \times \gamma_{ck} = 18.8$ kN

Partial factors from Sets $\begin{pmatrix} A1 \\ A2 \end{pmatrix} : \gamma_G = \begin{pmatrix} 1.35 \\ 1 \end{pmatrix}$ and $\gamma_Q = \begin{pmatrix} 1.5 \\ 1.3 \end{pmatrix}$ $\mathbf{②}$

Design total action per pile is:

$$F_{cd} = \frac{\gamma_G \times \left(F_{Gk} + W_{Gk} \right) + \gamma_Q \times F_{Qk}}{N} = \begin{pmatrix} 664 \\ 516 \end{pmatrix} \text{ kN}$$

Calculated shaft resistance

Number of cone penetration tests n = 4

Calculated shaft resistance $R_s = \pi \times D \times L \times P_s = \begin{pmatrix} 905 \\ 905 \\ 754 \\ 905 \end{pmatrix}$ kN

Mean calculated shaft resistance $R_{s,mean} = \dfrac{\sum R_s}{n} = 867$ kN

Minimum calculated shaft resistance $R_{s,min} = \min(R_s) = 754$ kN

Calculated base resistance

Calculated base resistance $R_b = \left(\dfrac{\pi \times D^2}{4}\right) \times P_b = \begin{pmatrix} 352 \\ 377 \\ 251 \\ 377 \end{pmatrix}$ kN

Mean calculated base resistance $R_{b,mean} = \dfrac{\sum R_b}{n} = 339$ kN

Minimum calculated base resistance $R_{b,min} = \min(R_b) = 251$ kN

Calculated total resistance

Mean calculated total resistance $R_{t,mean} = R_{s,mean} + R_{b,mean} = 1206$ kN

Minimum calculated total resistance $R_{t,min} = R_{s,min} + R_{b,min} = 1005$ kN

Characteristic resistance

Correlation factor on mean measured resistance $\xi_3 = 1.31$ ❸

Correlation factor on minimum measured resistance $\xi_4 = 1.20$ ❸
For a pile group that can transfer load from weak to strong piles
(§7.6.2.2.(9)), ξ may be divided by 1.1 (but ξ_3 cannot fall beneath 1.0).

Thus $\xi_3 = \max\left(\dfrac{\xi_3}{1.1}, 1.0\right) = 1.19$ and $\xi_4 = \dfrac{\xi_4}{1.1} = 1.09$

Calculated resistances $\dfrac{R_{t,mean}}{\xi_3} = 1013$ kN and $\dfrac{R_{t,min}}{\xi_4} = 922$ kN❹

Characteristic resistance should therefore be based on the minimum value.

Characteristic shaft resistance is $R_{sk} = \dfrac{R_{s,min}}{\xi_4} = 691$ kN **⑤**

Characteristic base resistance is $R_{bk} = \dfrac{R_{b,min}}{\xi_4} = 230$ kN **⑤**

Design resistance

Partial factors from Sets $\begin{pmatrix} R1 \\ R4 \end{pmatrix}$: $\gamma_s = \begin{pmatrix} 1 \\ 1.3 \end{pmatrix}$ and $\gamma_b = \begin{pmatrix} 1.1 \\ 1.45 \end{pmatrix}$ **⑤**

Design resistance is $R_{cd} = \dfrac{R_{sk}}{\gamma_s} + \dfrac{R_{bk}}{\gamma_b} = \begin{pmatrix} 901 \\ 691 \end{pmatrix}$ kN

Verification of compression resistance

Degree of utilization $\boxed{\Lambda_{GEO,1} = \dfrac{F_{cd}}{R_{cd}} = \begin{pmatrix} 74 \\ 75 \end{pmatrix} \%}$ **⑥**

Design is unacceptable if degree of utilization is > 100%

Design Approach 2

Actions and effects

Partial factors from set A1: $\gamma_G = 1.35$ and $\gamma_Q = 1.5$ **②**

Design total action per pile is $F_{cd} = \dfrac{\gamma_G \times \left(F_{Gk} + W_{Gk}\right) + \gamma_Q \times F_{Qk}}{N} = 664$ kN

Design resistance

Characteristic shaft and base resistances are unchanged from DA1

Partial factors from set R2: $\gamma_s = 1.1$ and $\gamma_b = 1.1$ **⑦**

Design resistance is $R_{cd} = \dfrac{R_{sk}}{\gamma_s} + \dfrac{R_{bk}}{\gamma_b} = 838$ kN

Verification of compression resistance

Degree of utilization $\left| \Lambda_{GEO,2} = \dfrac{F_{cd}}{R_{cd}} = 79\ \% \right|$ **⑥**

Design is unacceptable if degree of utilization is > 100%

Design Approach 3

Actions and effects

Partial factors from set A1: $\gamma_G = 1.35$ and $\gamma_Q = 1.5$ **❷**

Design total action per pile is $F_{cd} = \dfrac{\gamma_G \times \left(F_{Gk} + W_{Gk}\right) + \gamma_Q \times F_{Qk}}{N} = 664\ kN$

Characteristic resistance

Partial factors from set M2 should be applied to material properties... but since there are no material properties to factor, we will factor the resistances instead using $\gamma_\varphi = 1.25$. Since resistances are governed by the minimum calculated resistance (as per DAs 1 and 2)...

Characteristic shaft resistance is $R_{sk} = \dfrac{R_{s,min}}{\xi_4 \times \gamma_\varphi} = 553\ kN$

Characteristic base resistance is $R_{bk} = \dfrac{R_{b,min}}{\xi_4 \times \gamma_\varphi} = 184\ kN$

Design resistance

Partial factors from set R3: $\gamma_s = 1$ and $\gamma_b = 1$

Design resistance is $R_{cd} = \dfrac{R_{sk}}{\gamma_s} + \dfrac{R_{bk}}{\gamma_b} = 737\ kN$

Verification of compression resistance

Degree of utilization $\left| \Lambda_{GEO,3} = \dfrac{F_{cd}}{R_{cd}} = 90\ \% \right|$ **⑥**

Design is unacceptable if degree of utilization is > 100%

Design to UK National Annex to BS EN 1997-1

Characteristic resistance

Correlation factor on mean measured resistance $\xi_3 = 1.38$ **⑧**

Correlation factor on minimum measured resistance $\xi_4 = 1.29$ **⑧**
For a pile group that can transfer load from weak to strong piles
(§7.6.2.2.(9)), ξ may be divided by 1.1 (but ξ_3 cannot fall beneath 1.0).

Thus $\xi_3 = \max\left(\dfrac{\xi_3}{1.1}, 1.0\right) = 1.25$ and $\xi_4 = \dfrac{\xi_4}{1.1} = 1.17$

Calculated resistances $\dfrac{R_{t,mean}}{\xi_3} = 961.6$ kN and $\dfrac{R_{t,min}}{\xi_4} = 857$ kN **④**

Characteristic resistance should therefore be based on the minimum value,
so...

Characteristic shaft resistance is $R_{sk} = \dfrac{R_{s,min}}{\xi_4} = 643$ kN

Characteristic base resistance is $R_{bk} = \dfrac{R_{b,min}}{\xi_4} = 214$ kN

Design resistance

Partial factors from Sets $\begin{pmatrix} R1 \\ R4 \end{pmatrix}$: $\gamma_s = \begin{pmatrix} 1 \\ 1.6 \end{pmatrix}$ and $\gamma_b = \begin{pmatrix} 1 \\ 2 \end{pmatrix}$ **⑨**

Design resistance is $R_{cd} = \dfrac{R_{sk}}{\gamma_s} + \dfrac{R_{bk}}{\gamma_b} = \begin{pmatrix} 857 \\ 509 \end{pmatrix}$ kN

Verification of compression resistance

Degree of utilization $\Lambda_{GEO,1} = \dfrac{F_{cd}}{R_{cd}} = \begin{pmatrix} 77 \\ 101 \end{pmatrix}$ % **⑩**

Design is unacceptable if degree of utilization is > 100%

❹ Correlation factors are applied to the total derived pile resistance to identify whether the minimum or the mean gives the most critical result. In this case the minimum value governs.

❺ Once it has been established that the minimum is critical, the base and shaft components need to be identified so that the different shaft and base partial factors for CFA piles may be applied.

❻ For DA1, DA1-2 governs and suggests that the piles may be slightly over-designed. DA2 and DA3 yield similar levels of utilization.

❼ DA2 provides small partial factors to both the shaft and the base at the same time as applying partial factors to the actions. The net result is to produce a slightly more onerous condition than DA1-1.

❽ The UK National Annex (NA) provides more onerous correlation factors than Annex A of EN 1997-1. The minimum derived value of resistance still governs the design.

❾ The UK NA values for partial resistance factors for CFA piles are larger than those given in Annex A of EN 1997-1.

❿ Based on the UK NA, the design just exceeds the requirements of BS EN 1997-1 but at 101% utilization would probably be considered satisfactory.

13.14.5 Designing to a set with a pile driving formula

Example 13.5 looks at how driven concrete piles may be designed according to Eurocode 7 to a specified 'set' with a commonly-used pile driving formula.

A new five-storey teaching block is needed to improve facilities at an existing school. Details of the design of driven piles for an existing school building (including the results of static load tests) are available. It is proposed to found the new teaching block on similar driven piles. Measurements of pile set will be used to demonstrate that the piles have adequate resistance and settlement performance.

The Hiley formula[20] was used for the previous construction and suitably calibrated against the pile tests.[21] In this example, 300mm-square x12m-long concrete piles will form a 16-pile group. Design Approaches 1 and 2 have been considered (Design Approach 3 is not suitable for use with driving formulae). The effects of using the revised correlation factors given in the UK National Annex to BS EN 1997-1 are also investigated.

Example 13.5
Designing to a set with a pile driving formula
Verification of strength (limit state GEO)

Design situation

Consider the design of piles for a new 5-storey teaching block at a school. Ground conditions have been established from a ground investigation and are similar to those under the existing school buildings. These buildings are founded on driven concrete piles which were originally designed using the Hiley pile driving formula, and whose capacity was confirmed by static load tests.

The new building will be founded on $B = 300$mm square piles, $L = 12$m long. A total of $N = 16$ piles will be used to carry characteristic permanent and variable actions $F_{Gk} = 600$kN and $F_{Qk} = 200$kN.

A drop hammer of weight $W = 50$kN, drop height $h = 500$mm, and efficiency $e = 90\%$ will be used to drive the concrete piles to a maximum set of $S = 6$mm per blow. The total temporary compression of the pile and ground is estimated to be $C = 12$mm. ❶

Design Approach 1

Actions and effects

Ignore the self-weight of pile

Characteristic total action $F_{ck} = F_{Gk} + F_{Qk} = 800$ kN

Partial factors from Sets $\begin{pmatrix} A1 \\ A2 \end{pmatrix}$: $\gamma_G = \begin{pmatrix} 1.35 \\ 1 \end{pmatrix}$ and $\gamma_Q = \begin{pmatrix} 1.5 \\ 1.3 \end{pmatrix}$

Design total action per pile is $F_{cd} = \gamma_G F_{Gk} + \gamma_Q F_{Qk} = \begin{pmatrix} 1110 \\ 860 \end{pmatrix}$ kN

Measured resistance

The energy transferred from the drop hammer to each pile is:

$$E = W \times h \times e = 22500 \text{ kN mm } ❷$$

Using Hiley's formula to calculate the resistance of the pile:

$$R_{cm} = \frac{E}{\left(S + \dfrac{C}{2} \right)} = 1875 \text{ kN } ❸$$

Assuming all the piles are installed to the same maximum set $S = 6$ mm

Mean resistance is $R_{cm,mean} = R_{cm} = 1875$ kN

Minimum resistance is $R_{cm,min} = R_{cm} = 1875$ kN

Characteristic resistance

Correlation factor on mean measured resistance $\xi_5 = 1.42$ ❹

Correlation factor on minimum measured resistance $\xi_6 = 1.25$ ❹

These values must be increased by a model factor $\gamma_{Rd} = 1.2$ when using a pile driving formula without measurement of quasi-elastic pile head displacement during the impact. Thus $\xi_5 = \xi_5 \times \gamma_{Rd} = 1.70$ and $\xi_6 = \xi_6 \times \gamma_{Rd} = 1.50$

Based on mean measured resistance $\dfrac{R_{cm,mean}}{\xi_5} = 1100$ kN

Based on minimum measured resistance $\dfrac{R_{cm,min}}{\xi_6} = 1250$ kN

Characteristic resistance is $R_{ck} = \min\left(\dfrac{R_{cm,mean}}{\xi_5}, \dfrac{R_{cm,min}}{\xi_6}\right) = 1100$ kN

Design resistance

Partial factors from Sets $\begin{pmatrix} R1 \\ R4 \end{pmatrix}$: $\gamma_t = \begin{pmatrix} 1 \\ 1.3 \end{pmatrix}$

Design resistance is $R_{cd} = \dfrac{R_{ck}}{\gamma_t} = \begin{pmatrix} 1100 \\ 846 \end{pmatrix}$ kN

Verification of compression resistance

Degree of utilization $\Lambda_{GEO,1} = \dfrac{F_{cd}}{R_{cd}} = \begin{pmatrix} 101 \\ 102 \end{pmatrix}$ % ❺

Design is unacceptable if degree of utilization is > 100%

Design Approach 2

Actions and effects

Characteristic total action same as Design Approach 1

Partial factors from set A1: $\gamma_G = 1.35$ and $\gamma_Q = 1.5$

Design total action per pile is $F_{cd} = \gamma_G\, F_{Gk} + \gamma_Q\, F_{Qk} = 1110$ kN **❻**

Characteristic resistance

Characteristic resistance is same as Design Approach 1

Design resistance

Partial factors from set R2: $\gamma_t = 1.1$

Design resistance is $R_{cd} = \dfrac{R_{ck}}{\gamma_t} = 1000$ kN

Verification of compression resistance

Design values $F_{cd} = 1110$ kN and $R_{cd} = 1000$ kN

$$\text{Degree of utilization}\ \boxed{\Lambda_{GEO,2} = \dfrac{F_{cd}}{R_{cd}} = 111\ \%}\ \text{❼}$$

Design is unacceptable if degree of utilization is > 100%

Design Approach 3

Cannot use Design Approach 3 for pile design with pile driving formulae

Design to UK National Annex to BS EN 1997-1

The UK National Annex changes both the resistance factors and the correlation factors that must be used to verify pile resistance. All other factors remain unchanged from their EN values.

Measured resistance

Measured resistance is unchanged from previous calculation

Correlation factor on mean measured resistance $\xi_5 = 1.57$ **(8)**

Correlation factor on minimum measured resistance $\xi_6 = 1.44$ **(8)**

These values must be increased by a model factor $\gamma_{Rd} = 1.2$ when using a pile driving formula without measurement of quasi-elastic pile head displacement during the impact. Thus $\xi_5 = \xi_5 \times \gamma_{Rd} = 1.88$ and $\xi_6 = \xi_6 \times \gamma_{Rd} = 1.73$

Based on mean measured resistance $\dfrac{R_{cm,mean}}{\xi_5} = 995$ kN

Based on minimum measured resistance $\dfrac{R_{cm,min}}{\xi_6} = 1085$ kN

Characteristic resistance is $R_{ck} = \min\left(\dfrac{R_{cm,mean}}{\xi_5}, \dfrac{R_{cm,min}}{\xi_6}\right) = 995$ kN

Design resistance

Partial factors from Sets $\begin{pmatrix} R1 \\ R4 \end{pmatrix}$: $\gamma_t = \begin{pmatrix} 1 \\ 1.5 \end{pmatrix}$

Design resistance is $R_{cd} = \dfrac{R_{ck}}{\gamma_t} = \begin{pmatrix} 995 \\ 663 \end{pmatrix}$ kN

Verification of compression resistance

Degree of utilization $\Lambda_{GEO,1} = \dfrac{F_{cd}}{R_{cd}} = \begin{pmatrix} 112 \\ 130 \end{pmatrix}$ % **(9)**

Design is unacceptable if degree of utilization is > 100%

Assuming $\chi = 50\%$ of the characteristic resistance comes from shaft friction, design resistance $R_{cd} = \chi\left(\dfrac{R_{ck}}{\gamma_s}\right) + (1-\chi)\left(\dfrac{R_{ck}}{\gamma_b}\right) = \begin{pmatrix} 995 \\ 715 \end{pmatrix}$ kN

Degree of utilization $\Lambda_{GEO,1} = \dfrac{F_{cd}}{R_{cd}} = \begin{pmatrix} 112 \\ 120 \end{pmatrix}$ % **(10)**

Design is unacceptable if degree of utilization is > 100%

Notes on Example 13.5

❶ The maximum set of 6mm per blow has been arbitrarily assumed for this example.

❷ The formula for energy transfer is a standard used by industry.

❸ This form of the Hiley formula is one of many variants that are used in the industry. It represents a simplified form of the equation with fewer constants required to specify the hammer performance than may be required for other variants.

❹ Correlation factors are those specified in Annex A of EN 1997-1 for between 15 and 20 piles tested.

❺ On the basis of the requirements of EN 1997-1 the pile tests indicate that the piles do not fully comply with DA1 but with a utilization factor of 102% are likely to be regarded as adequate. Note: traditionally a factor of safety of 2.0 is often applied to provide safe capacities from driving formulae, an equivalent global factor of safety of 2.34 is implied by the DA1 calculation.

❻ The same design action is calculated for DA2 as for DA1-1.

❼ DA2 indicates that the design is unacceptable with a utilization factor of 111%.

❽ The UK NA to BS EN 1997-1 provides larger correlation factors to those given in Annex A of EN 19997-1 and therefore both the characteristic and design resistances are reduced.

❾ The utilization factor for DA1 to the UK NA is much greater than 100% and would suggest that the piles would be unsatisfactory if the design set was 6mm per blow. This would imply that a lower set would be required giving an inferred global factor of safety much higher than 2.0. However, it could be argued that the reliability of pile driving formulae to properly predict pile performance is low and therefore the current practice by some contractors of adopting a global factor of safety of 2.0 may be too optimistic.

❿ By splitting the base and shaft components lower levels of utilization can be calculated but these are still higher than is acceptable. Also it is not technically possible to identify from pile driving formulae the proportion of capacity derived from the base and the shaft, thus in practice advantage cannot be taken of the reduced factors.

13.15   Notes and references

1. See, for example, Fleming, W. G. K., Weltman, A. J., Randolph, M. F., and Elson, W. K. (1992) *Piling Engineering* (2nd edition), Glasgow: Blackie & Son Ltd., 390pp.

2. Institution of Civil Engineers (2007) *ICE specification for piling and embedded retaining walls* (known as SPERW).

3. ISSMFE Subcommittee on Field and Laboratory Testing (1985), 'Axial Pile Loading Test, Suggested Method', *ASTM Journal*, pp. 79–90.

4. BS EN ISO 22477, Geotechnical investigation and testing — Testing of geotechnical structures, British Standards Institution.
    Part 1: Pile load test by static axially loaded compression.
    Part 2: Pile load test by static axially loaded tension.
    Part 3: Pile load test by static transversely loaded tension.

5. Institution of Civil Engineers (2007), ibid.

6. Federation of Piling Specialists (2006), *Handbook on pile load testing*, Beckenham, Kent: Federation of Piling Specialists.

7. ASTM Designation D 4945, Standard Test Method for High-Strain Dynamic Testing of Piles.

8. See, for example, Frank, R., Bauduin, C., Kavvadas, M., Krebs Ovesen, N., Orr, T., and Schuppener,B. (2004) *Designers' guide to EN 1997-1: Eurocode 7: Geotechnical design — General rules*, London: Thomas Telford.

9. See, for example, Driscoll, R.M.C., Powell , J.J.M., and Scott, P.D. (2008, in preparation) *EC7 — implications for UK practice*, CIRIA RP701.

10. Fleming et al., ibid., p. 212.

11. BS 8004: 1986, Code of practice for foundations, British Standards Institution.

12. Burland, J. B., Broms, B. B., and de Mello, V. F. (1977) 'Behaviour of foundations and structures', *9th Int. Conf. on Soil Mechanics and Fdn Engng*, Tokyo, 2, pp. 495–547.

13. Tomlinson, M. J. (1994) *Pile design and construction practice*, E & FN Spon.

14. Bowles, J. E. (1997) *Foundation analysis and design*, McGraw-Hill.

15. Lord, J. A., Clayton C. R. I., and Mortimore, R. N. (2002) *Engineering in chalk*, CIRIA Report C574.

16. Viv Troughton (pers. comm., 2008).

17. London District Surveyors' Association (1999), *Guidance notes for the design of straight shafted bored piles in London Clay.*

18. Data kindly provided by Stent Foundations (pers. comm., 2007).

19. Data kindly provided by CL Associates (pers. comm., 2007).

20. Hiley, A. (1930), 'Pile-driving calculations with notes on driving forces and ground resistances', *The Structural Engineer*, 8, pp. 246–259 and 278–288.

21. Data kindly provided by by Aarslef Piling (pers. comm., 2008).

Chapter 14

Design of anchorages

The design of anchorages is covered by Section 8 of Eurocode 7 Part 1, 'Anchorages', whose contents are as follows:

§8.1   General (12 paragraphs)
§8.2   Limit states (1)
§8.3   Design situations and actions (2)
§8.4   Design and construction considerations (15)
§8.5   Ultimate limit state design (10)
§8.6   Serviceability limit state design (6)
§8.7   Suitability tests (4)
§8.8   Acceptance tests (3)
§8.9   Supervision and monitoring (1)

Section 8 of EN 1997-1 applies to prestressed and non-prestressed anchorages (both temporary and permanent), used to support retaining structures, to stabilize slopes, cuts, and tunnels, and to resist uplift of structures. Section 8 does *not* apply to tension piles used to anchor structures.

[EN 1997-1 §8.1.1(1)P]

An anchor is a structure consisting of a tendon free length (designed to satisfy overall structural stability) and a restraint (designed to transmit tensile forces to the surrounding ground). In prestressed anchorages, the restraint is provided by a tendon bond length bonded to the ground by grout; in non-prestressed anchorages, the restraint is provided by a deadman anchorage, a screw anchor, or a rock bolt.

An anchorage is a structure consisting of an anchor head and an anchor. The anchor head (typically, a nut, flat washer, plate, stud, or bolt head) transmits restraining forces to the structure.

The Principles and Application Rules given in §8 are written with grouted anchorages in mind – deadman anchorages, screw anchors, and rock bolts are given scant attention. EN 1997-1 explicitly states that §8 does not cover the design of soil nails.

14.1   Ground investigation for anchorages

Annex B.3 of Eurocode 7 Part 2 provides outline guidance on the depth of investigation points for the principal types of geotechnical structure, but does not give direct guidance on the spacing or depth of investigation points for anchorages.

Since anchorages are used in conjunction with other structures, e.g. slopes, retaining walls, holding down basements, etc., the scope of investigation is governed by the structure that the anchorages help to stabilize. The properties of the ground should be investigated over the full length of the anchor, thereby enabling pull-out resistance to be adequately assessed.

EN 1537[1] provides some guidance on ground investigations for anchorage design but adds little to the advice given in Eurocode 7 Part 2.[2] The corrosion potential of the ground and the existence of stray electric currents are given particular prominence since the long-term performance of an anchorage is largely governed by corrosion. In addition, because anchors are normally installed at inclinations close to horizontal, lateral variation in ground conditions as well as ground conditions outside the site boundary may be of concern.

14.2   Design situations and limit states

Situations where the performance of anchorages is crucial to the design are illustrated in **Figure 14.1**. From left to right, these include: (top) anchorage support to a retaining wall and for a potential slip in soil; (middle) anchorages helping to resist uplift and supporting a wind turbine; and (bottom) anchoring a potential wedge failure in rock.

Limit states for anchorages typically involve structural failure of the anchor tendon or anchor head; corrosion or distortion of the anchor head; failure at the interface of the grout with the tendon or the surrounding ground; instability of the ground containing the anchor and associated structures; or loss of anchorage force owing to creep, relaxation, or excessive displacement of the anchor head.   *[EN 1997-1 §8.3(1)P]*

14.3   Basis of design

Guidance on the design of anchorages is currently split between EN 1997-1 and EN 1537[3] (the execution standard for ground anchors – see Chapter 15). Unfortunately some of this guidance is contradictory. Furthermore, EN 1537's testing requirements for ground anchors overlap the planned scope of ISO EN 22477-5[4] (the geotechnical investigation and testing standard for anchorages – see Chapter 4). The relevant CEN Technical Committees are considering moving design-related material from EN 1537 into EN 1997-1 and testing-related material into ISO EN 22477-5, leaving EN 1537 to deal solely with execution matters. However, UK practitioners argue[5] that separating testing from execution is not a good idea and prefer the integration of content employed in BS 8081.[6]

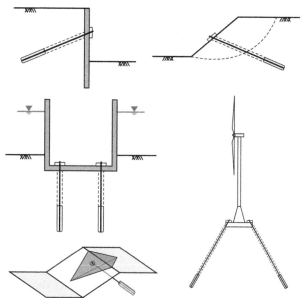

Figure 14.1. Examples of limit states for anchorages

In view of the above, the following discussion provides our tentative understanding of the way ground anchorages should be designed to Eurocode 7, highlighting areas where contradictions exist and need to be resolved in future editions of the codes. This book does not attempt to provide complete guidance on the design of anchorages, for which the reader should refer to any well-established text on the subject.[7]

The design pull-out resistance of an anchorage $R_{a,d}$ must satisfy the following inequality:

$$P_d \leq R_{a,d}$$

where P_d is the design load in the anchorage, i.e. the larger of the anchor force derived from an ultimate limit state verification of the retained structure (P_{ULS}) and that derived from a serviceability limit state verification (P_{SLS}), i.e.:   [EN 1997-1 §8.5.5(1)P]

$$P_d = \max(P_{ULS}, P_{SLS})$$

There is some debate as to whether the anchor force that is derived from an ultimate limit state (ULS) verification of an anchored retaining wall (i.e. P_{ULS}) is the largest force the anchor will have to withstand. At the ULS, earth pressures on the back of the wall approach limiting active (K_a) values and those on the front limiting passive (K_p) values.

In some circumstances, it is possible for the serviceability limit state (SLS) force P_{SLS} to be of similar magnitude to – or even larger than – P_{ULS}. At the SLS, earth pressures on the back of the wall may remain close to their in situ (i.e. K_0) values, particularly for inflexible walls in stiff soils. Since values of K_0 can be considerably larger than K_a, it is possible for P_{ULS} and P_{SLS} to be of similar magnitude.

The value of P_{ULS} obtained from ultimate limit state calculations of wall stability implicitly includes a load factor γ_G, which is equal to 1.35 in Design Approaches 1[t] and 2 and 1.0 in Design Approach 3. Because of this, anchorage designs based on Design Approach 3 may not be sufficiently reliable, particularly if the ratio P_{ULS}/P_{SLS} is 1.1–1.2 (which can often be the case when a soil-structure interaction model is used). Frank et al.[8] recommend multiplying P_{SLS} by a model factor $\gamma_{Rd} = \gamma_G = 1.35$ to rectify this. The design load in the anchorage P_d is then:

$$P_d = \max(P_{ULS}, \gamma_{Rd} P_{SLS})$$

Eurocode 7 Part 1 discusses two ways of determining the design pull-out resistance $R_{a,d}$ from the results of tests (as discussed in Section 14.5 below) and by calculation (discussed in Section 14.6).

Eurocode 7 Part 1 has specific requirements regarding anchorage tests. Unless their performance and durability can be demonstrated by (documented) successful comparable experience, anchorage systems must be verified by *investigation* tests (see Section 14.4.1); the characteristic pull-out resistance $R_{a,k}$ of grouted and screw anchorages must be determined from *suitability* tests (Section 14.4.2); and all grouted anchorages must undergo *acceptance* tests (Section 14.4.3).   *[EN 1997-1 §8.4(8)P, (10)P, and 8.8(1)P]*

Suitability tests are not intended to determine the characteristic pull-out resistance. Their purpose is to prove that the anchorages are suitable for the conditions on site. The important distinction is that the loading in a suitability test does not exceed the proof load. Usually the anchorage is subjected to a more rigorous number of load cycles. The test is not designed

---

[t]γ_G = 1.35 in Combination 1 (which governs) and 1.0 in Combination 2.

to ascertain information on the ultimate or failure condition when establishing pull-out resistance.

Factors that must be considered in the design of anchorage systems include: tolerances on angular deviations of the anchor force and accommodation for other deformations; compatibility between the deformation performance of different materials used in the anchorage; and adverse effects of tensile stresses[†] transmitted to ground beyond the anchorage.

[EN 1997-1 §8.4(1), (4)P, and (5)P]

Factors that must be considered in the construction of anchorage systems include: applying proof and lock-off loads to pre-stressed tendons; allowing pre-stressed anchorages to be de-stressed and re-stressed, as required; and avoiding adverse effects caused by the retained ground or existing foundations being too close to the ground that provides the resisting force.

[EN 1997-1 §8.4(3)P and (6)P]

Corrosion protection must comply with the requirements of EN 1537.[9]

14.4   Anchorage tests

Eurocode 7 discusses three types of anchorage test: investigation, suitability, and acceptance.

14.4.1   Investigation tests

An investigation test is a 'load test to establish the ultimate resistance of an anchor at the grout/ground interface and to determine the characteristics of the anchorage in the working load range'. (This definition is identical to the one given in EN 1537.)

[EN 1997-1 §8.1.2.5]

Investigation tests are performed, before working anchorages are installed, to establish the anchorage's ultimate pull-out resistance in the ground conditions at the site, to prove the contractor's competence, and to prove novel types of ground anchorage. Investigation tests should be carried out when anchorages have not previously been tested in similar ground conditions or if higher working loads than previously tested are anticipated.

Eurocode 7 does not give any recommendations concerning the number of investigation tests that should be performed. However, the draft version of the standard[10] suggested that at least 1% of temporary and 2% of permanent

---

[†]which, due to differences in relative elasticity between the tendon and the grouted ground, produces progressive de-bonding within the bonded length.

anchorages should undergo 'assessment' tests, a term that embraces both investigation and suitability tests.

Anchorages used in investigation tests must not be incorporated into the permanent works if they have been loaded to failure. Since the test anchorages are subject to higher loads than working anchorages, their tendon size may need to be increased above that of the working anchorages. Alternatively, investigation tests may incorporate anchorages with shorter fixed lengths. Correction factors may also be applied to account for progressive de-bonding within the fixed anchor length.

In investigation tests, the anchorage should either be loaded to its failure load R_a or to a proof load P_p, which must be limited to (for Test Methods 1 and 2) the smaller of:

$$P_p \leq 0.8 P_{t,k} \text{ and } P_p \leq 0.95 P_{t0.1,k}$$

or (for Test Method 3) the smaller of:

$$P_p \leq 0.8 P_{t,k} \text{ and } P_p \leq 0.9 P_{t0.1,k}$$

where $P_{t,k}$ is the tendon's characteristic tensile load capacity and $P_{t0.1,k}$ is the characteristic tensile load at 0.1% strain. See Chapter 15 for a summary of the differences between Test Methods 1-3.   [EN 1537 §9.5]

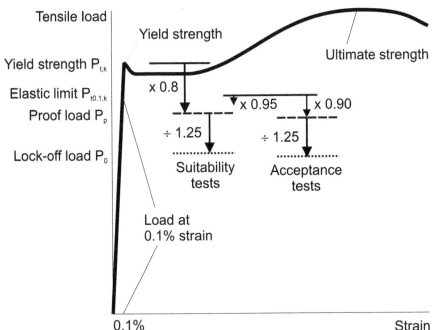

Figure 14.2. Determination of proof and lock-off loads from load vs strain curve for pre-stressed anchor

Figure 14.2 illustrates the relative magnitude of the tendon's tensile yield strength $P_{t,k}$, its elastic limit load $P_{t0.1,k}$, the maximum proof load P_p that should be applied, and the lock-off load P_0.

14.4.2   Suitability tests

A suitability test is a 'load test on site to confirm that a particular anchor design will be adequate in particular ground conditions'. (This definition is identical to the one given in EN 1537.)   *[EN 1997-1 §8.1.2.4]*

Suitability tests are normally carried out on a selected number of anchorages to confirm that a particular anchor design is adequate. Their intent is to examine creep characteristics, elastic extension behaviour, and load loss with time. Anchorages subjected to suitability tests may be used as working anchorages.

At least three suitability tests should[†] be performed on anchorages constructed under identical conditions to the working anchorages.
                                                                   [EN 1997-1 §8.7(2)]

The proof load P_p applied in suitability tests should be the greater of:
$$P_p \geq 1.25P_0 \text{ and } P_p \geq R_d$$
where P_0 is the lock-off load and R_d is the required design resistance of the anchor; and must be limited to (for Test Methods 1 and 2):
$$P_p \leq 0.95P_{t0.1,k}$$
or (for Test Method 3):
$$P_p \leq 0.9P_{t0.1,k}$$
where $P_{t0.1,k}$ is the tendon's characteristic tensile load at 0.1% strain. See Chapter 15 for a summary of the differences between Test Methods 1–3.
                                                    [EN 1537 §E.2.2, E.3.2, and E.4.2]

The requirement to limit the proof load P_p to 90–95% of the tendon's characteristic tensile load $P_{t0.1k}$ conflicts with the requirement that P_p should be greater than or equal to the design resistance of the anchor R_d. This conflict arises because R_d is obtained by dividing the tendon's tensile strength P_{tk} by a partial factor γ_a, which Eurocode 7 specifies as 1.1. Hence R_d is likely to be greater than 90–95% of $P_{t0.1k}$.

Work is underway within CEN Technical Committees 250/SC7 and 288 to rectify this conflict. In the meantime, in order to prevent overstress of the

---

[†]'shall' in EN 1537.

anchor in suitability tests, we recommend that the proof load be limited to 90–95% of $P_{t0.1k}$ and is not based on R_d.

Figure 14.2 illustrates the relative magnitude of the tendon's tensile yield strength $P_{t,k}$, its elastic limit load $P_{t0.1,k}$, the maximum proof load P_p that should be applied, and the lock-off load P_0.

14.4.3   Acceptance tests

An acceptance test is a 'load test on site to confirm that each anchorage meets the design requirements'. (This definition differs slightly from the one given in EN 1537.)   *[EN 1997-1 §8.1.2.3]*

Acceptance tests must be carried out on all working anchorages to demonstrate that a proof load P_p can be sustained; to determine the apparent tendon free length; to ensure the lock-off load is at its design level; and to determine creep or load loss characteristics under serviceability conditions.

The proof load P_p applied in acceptance tests should be (for Test Methods 1 and 2):

$$1.25P_0 \le P_p \le 0.9P_{t0.1,k}$$

or (for Test Method 3):

$$P_p = 1.25P_0 \text{ or } P_p = R_d$$

where P_0 is the lock-off load, $P_{t0.1,k}$ is the tendon's characteristic tensile load at 0.1% strain, and R_d is the required design resistance of the anchor. See Chapter 15 for a summary of the differences between Test Methods 1–3.

[EN 1537 §E.2.3, E.3.3, and E.4.3]

Figure 14.2 illustrates the relative magnitude of the tendon's tensile yield strength $P_{t,k}$, its elastic limit load $P_{t0.1,k}$, the maximum proof load P_p that should be applied, and the lock-off load P_0.

With the resistance factors recommended in EN 1997-1 (see Section 14.5.2), the proof load in an acceptance test using Test Method 3 can approach the anchor's ultimate pull-out resistance, which could lead to unacceptable creep in the anchors. To avoid this, it has been recommended[11] that P_p be further limited to:

$$P_p \le 1.15P_{k,SLS}$$

This is likely to conflict with the requirement for Test Method 3 that the proof load P_p be at least equal to design resistance of the anchor R_d, since the latter is likely to be greater than 1.15 $P_{k,sls}$. Until this conflict is rectified, we recommend ignoring the requirement for the proof load to exceed the anchor's design resistance.

14.5 Pull-out resistance from tests

14.5.1 Characteristic pull-out resistance

The characteristic pull-out resistance of a grouted anchorage is the lowest of the following (see **Figure 14.3**):
- bond resistance between the grout and ground (external resistance, $R_{a,k}$)
- bond resistance between the grout and tendon (internal resistance, $R_{i,k}$)
- tensile capacity of the tendon ($P_{t,k}$)
- capacity of the anchor head.

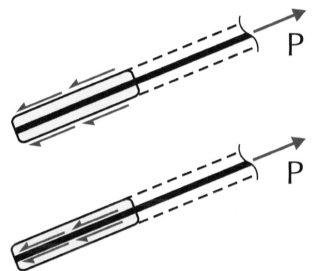

Figure 14.3. *Anchorage resistance: (top) external, (bottom) internal*

For anchorages not formed by grouting, e.g. screw anchorages and rock bolts, it is not necessary to consider the internal resistance as this is covered by the tensile cap.

During an anchorage test, failure will occur in the weakest element. Design based on testing does not identify which failure mode is involved.

When the characteristic pull-out resistance $R_{a,k}$ is obtained from investigation tests (see Section 14.4.1), its value is given by:

$$R_{a,k} = \min(R_a, P_p)$$

where R_a is the measured failure load and P_p the maximum proof load applied in the test.

When the characteristic pull-out resistance $R_{a,k}$ is obtained from suitability tests (see Section 14.4.2), its value is given by:

$$R_{a,k} = \frac{P_p}{\xi_a}$$

where P_p is the measured proof load and ξ_a is a correlation factor that accounts for the number of suitability tests performed. Eurocode 7 provides no recommended values for ξ_a and neither does the UK National Annex to EN 1997-1. Previous suggestions for the values of ξ_a are summarized in the table below:

Reference	ξ_a on measured resistance (R_{am})	No of anchorage tests		
		1	2	> 2
ENV 1997-1[12]	mean ($R_{am,mean}$)	1.5	1.35	1.3
	minimum ($R_{am,min}$)	1.5	1.25	1.1
Designers' Guide[13]	mean ($R_{am,mean}$)	1.2	-	1.1
	minimum ($R_{am,min}$)	1.2	-	1.05
EN 1997-1		No values given		

It is impossible to design anchorages on the basis of suitability tests until suitable values are made available in a future revision of the standard.

When the characteristic pull-out resistance $R_{a,k}$ is obtained from acceptance tests (see Section 14.4.3), its value is given by:

$$R_{a,k} = P_p$$

where P_p is the measured proof load.

14.5.2   Design pull-out resistance

The design pull-out resistance of an anchorage $R_{a,d}$ is given by:

$$R_{a,d} = \frac{R_{a,k}}{\gamma_a}$$

where $R_{a,k}$ is the anchor's characteristic pull-out resistance and γ_a is a partial factor.   [EN 1997-1 §8.5.2]

Figure 14.4 shows the relative magnitude of actions and resistances used in anchorage design. In this diagram, we assume that the anchor's characteristic external resistance $R_{a,k}$ is greater than its internal resistance $R_{i,k}$, given by the tendon's tensile strength P_{tk}.

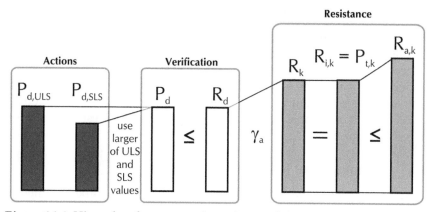

Figure 14.4. Hierarchy of parameters in anchorage design

The recommended values of γ_a for pre-stressed anchorages are given in Annex A of EN 1997-1 and repeated in the table below. In Design Approaches 1 and 2, $\gamma_a = 1.1$; and in Design Approach 3, $\gamma_a = 1.0$ (see Chapter 6 for discussion of the Design Approaches).

Reference	Anchorage		Partial factor			
ENV 1997-1[14]	Temporary	γ_m	1.25			
	Permanent	γ_m	1.5			
EN 1537[15]	(all)	γ_R	1.35			
EN 1997-1	*Partial factor set*		*R1*	*R2*	*R3*	*R4*
	Temporary	$\gamma_{a,t}$	1.1	1.1	1.0	1.1
	Permanent	$\gamma_{a,p}$	1.1	1.1	1.0	1.1
	Design Approach		*1-1*	*2*	*3*	*1-2*

The recommended value of γ_a (termed γ_R) given in EN 1537 is 1.35, which contradicts Eurocode 7. Since the latter takes precedence, $\gamma_a = 1.1$ or 1.0 should be applied instead, according to the Design Approach adopted.

The UK National Annex to EN 1997-1 confirms $\gamma_a = 1.1$ for anchorage design in the UK, but also states that 'larger values ... should be used for non-prestressed anchorages to make their designs consistent with those of tension piles ... or retaining structures.'

14.6   Pull-out resistance by calculation

No guidance is given in Eurocode 7 on how to design anchorages by calculation, despite the standard stating that it is an acceptable way to proceed. Presumably, calculations would need to consider all possible failure modes and compare these with the design anchor load derived from the ultimate and serviceability limit state verifications of the retained structure. In the absence of suitable guidance, we recommend that designs are based on testing instead.

14.7   Summary of key points

There is some confusion about the way anchorages should be designed to Eurocode 7, owing no doubt to the design rules being developed partly by the committee responsible for EN 1997-1 and partly by the committee for the execution standard EN 1537.

A further complication is the planned introduction (discussed in Chapter 4) of a separate testing standard for anchorages, ISO EN 22477-5.[16] Practitioners are nervous that the close relationship between design and testing of anchors will be at best diluted and at worst lost in the new arrangement.

The failure of Eurocode 7 Part 1 to recommend values for the correlation factors to be used to process the results of suitability tests means that, for the time being at least, ground anchorages can be designed solely on the basis of investigation and acceptance tests.

Revision of ENs 1997-1 and 1537 is already being discussed within their respective CEN technical committees and the reader is advised to wait for those revisions before designing anchorages to Eurocode 7.

14.8   Worked example

The worked example in this chapter looks at the design of grouted prestressed anchorages to support an embedded wall (Example 14.1).

Specific parts of the calculation are marked ❶, ❷, ❸, etc., where the numbers refer to the notes that accompany the example.

14.8.1   Grouted anchorages supporting an embedded retaining wall

Example 14.1 considers the design of anchorages to support an embedded retaining wall, as shown in **Figure 14.5**. We will assume that the design resistance the anchorages need to provide to prevent an ultimate limit state is known already from a separate calculation of the wall's stability (see Chapter 12 for details about how to obtain this).

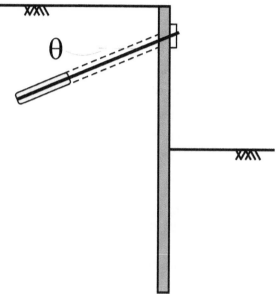

Figure 14.5. Row of anchorages used to support an embedded retaining wall

The design of the anchorages will rely on the results of investigation and suitability tests. The final design resistance will be confirmed by acceptance tests. The proof loads and lock-off loads required in these tests must be calculated.

Notes on Example 14.1

❶ The value of characteristic stress at 0.1% tensile strain is one of the main criteria for assessing the capacity of the anchor based on the tendon strength as required in EN 1537.

❷ Equation for establishing anchorage resistance from the grout/ground interface taken from BS 8081.[17] In a traditional design, it would be usual to have a factor of safety of at least 3.0 applied to the anchor's design load. In this case, it is approximately 3.03.

Example 14.1
Grouted anchor
Verification of strength (GEO)

Design situation

Consider a grouted anchorage that is required to support an embedded retaining wall. Separate calculations of wall stability indicate that the anchorage system must provide horizontal resistance of at least $R_{d,SLS} = 121\text{kN/m}$ to avoid a serviceability limit state and

$R_{d,ULS} = 133\text{kN/m}$ to avoid an ultimate limit state.

Working anchorages, with bond length $L = 4\text{m}$, will be installed into sand with characteristic angle of shearing resistance $\varphi_k = 35°$. The anchorages will be installed at an angle $\theta = 30°$ to the horizontal and at a horizontal spacing $s = 1.2\text{m}$. The anchor bore diameter $D = 133\text{mm}$. Factor from BS 8081 for Type B anchorages to obtain shear stress along anchor bond length

$n_a = 150\dfrac{\text{kN}}{\text{m}}$.

Preliminary design of working anchorages
Design Approach 1

Actions and effects

The design axial load that each anchorage must resist is:

$$P_d = \max\left(R_{d,ULS}, R_{d,SLS}\right) \times \frac{s}{\cos(\theta)} = 184.3\text{ kN}$$

Material properties and resistance

The tendon chosen for the working anchorages is a multi-unit comprising seven drawn-wire strands of nominal diameter $d = 15.2\text{mm}$, with sectional area $A_t = 165\text{mm}^2$, characteristic tensile strength $f_{pk} = 1820\dfrac{\text{N}}{\text{mm}^2}$, and

characteristic stress at 0.1% tensile strain $f_{p0.1k} = 1547\dfrac{\text{N}}{\text{mm}^2}$.

The characteristic tensile capacity of each tendon: $P_{tk} = f_{pk} \times A_t = 300\text{ kN}$

Characteristic tensile stress at 0.1% strain: $P_{t0.1k} = f_{p0.1k} \times A_t = 255\text{ kN}$ ❶

Characteristic internal resistance of anchorage is $R_{i,k} = P_{tk} = 300\text{ kN}$

Characteristic external resistance at grout/ground interface (based on BS

8081 §6.2.4.2 for a $D_{ref} = 0.1m$ borehole) is:

$$R_{e,k} = L \times n_a \times \tan(\varphi_k) \times \left(\frac{D}{D_{ref}}\right) = 559 \, kN \;❷$$

Characteristic resistance of anchorage is $R_{a,k} = \min(R_{e,k}, R_{i,k}) = 300 \, kN \;❸$
Partial factor from set R4: $\gamma_a = 1.1$

Design resistance is: $R_{a,d} = \dfrac{R_{a,k}}{\gamma_a} = 273 \, kN \;❹$

However, if we use partial factor from EN 1537: $\gamma_R = 1.35$

Design resistance is: $R_{a,d,EN1537} = \dfrac{R_{a,k}}{\gamma_R} = 222 \, kN \;❹$

Verification of strength

Degree of utilization is $\Lambda_{GEO,1} = \dfrac{P_d}{R_{a,d}} = 68 \, \%$ based on EN 1997-1

Degree of utilization is $\Lambda_{GEO,1} = \dfrac{P_d}{R_{a,d,EN1537}} = 83 \, \%$ based on EN 1537

The design is unacceptable if the degree of utilization is > 100%

Design of anchorages for investigation tests
Design Approach 1

Proof load for investigation tests (Test Method 2)
The purpose of the investigation test is to determine the anchorage's ultimate tensile resistance, without reaching the ultimate tensile resistance of the tendon.

The test will therefore be performed using $n_s = 3$ strands ❺
Proof load should be at least equal to $P_d = 184.3 \, kN$

Proof load must not exceed $n_s \times 0.8 \times P_{tk} = 720.7 \, kN \;❻$
Proof load must also not exceed $n_s \times 0.95 \times P_{t0.1k} = 727.5 \, kN \;❻$
Select a proof load $P_p = 700kN$

The maximum load that the anchorage was able to carry in the investigation test was $R_{a,m} = 580$kN ➐

Lock-off and proof loads for suitability tests (Test Method 2)
The lock-off load should not exceed $0.6P_{tk} = 180.2$ kN

Select a lock-off load $P_0 = 150$kN➑

The proof load should be at least equal to $1.25 \times P_0 = 187.5$ kN, but must not

exceed $0.95 \times P_{t0.1k} = 242.5$ kN

Select a proof load $P_p = 225$kN

Assume that the ultimate resistances measured in the suitability tests were

$$R_m = \begin{pmatrix} 225 \\ 220 \\ 225 \end{pmatrix} \text{ kN}$$

Hence the minimum resistance is $R_{m,min} = min\left(R_{m_1}, R_{m_2}, R_{m_3}\right) = 220$ kN and

the mean is $R_{m,mean} = \dfrac{\sum R_m}{n} = 223$ kN

Final design of working anchorages
Design Approach 1

Material properties and resistance
The characteristic resistance of the anchorage can be re-calculated from the suitability tests as:

$$R_{ak,mean} = \frac{R_{m,mean}}{\xi_{a,mean}} = 203 \text{ kN}$$

$$R_{ak,min} = \frac{R_{m,min}}{\xi_{a,min}} = 210 \text{ kN}$$

where $\xi_{a,mean} = 1.1$ and $\xi_{a,min} = 1.05$ ➒
and hence the characteristic resistance of the anchorage is:

$$R_{a,k} = min\left(R_{ak,mean}, R_{ak,min}\right) = 203 \text{ kN}$$

and its design resistance is:

$$R_{a,d} = \frac{R_{a,k}}{\gamma_a} = 184.6 \text{ kN according to EN 1997-1, or}$$

$$R_{a,d,EN1537} = \frac{R_{a,k}}{\gamma_R} = 150.4 \text{ kN according to EN 1537}$$

Verification of strength

Degree of utilization is $\Lambda_{GEO,1} = \dfrac{P_d}{R_{a,d}} = 100\,\%$ to EN 1997-1

Degree of utilization is $\Lambda_{EN1537} = \dfrac{P_d}{R_{a,d,EN1537}} = 123\,\%$ to EN 1537 ⑩

The design is unacceptable if the degree of utilization is > 100%

Lock-off and proof loads for acceptance tests (Test Method 2)

Select the same lock-off load as before, i.e. $P_0 = 150 \text{ kN}$

The proof load in acceptance tests must not exceed $0.9 \times P_{t0.1k} = 229.7 \text{ kN}$

The proof load must not be less than $1.25 P_0 = 187.5 \text{ kN}$

However, it is better to limit the proof load to the smaller of

$1.25 P_0 = 187.5 \text{ kN}$ and $1.15 R_{d,SLS} \times \dfrac{s}{\cos(\theta)} = 192.8 \text{ kN}$ ①

Select a proof $P_p = 187.5 \text{kN}$

Provided each anchorage performs satisfactorily up to the proof load, the anchorage is deemed to meet the requirements of Eurocode 7

❸ EN 1537 states that the characteristic anchorage capacity shall be the minimum of the capacity from a geotechnical calculation of the grout/ground interface resistance and the tensile capacity of the tendon. EN 1537 implies that the tendon capacity should govern the design.

❹ ENs 1997-1 and 1537 provide different values for γ_a. Work is underway in the relevant CEN technical committees to resolve these differences.

❺ It is normal to use a larger-diameter tendon in investigation tests than for the working tendon to ensure that where failure occurs it is between the grout and the tendon or the grout and the ground, not the tendon itself.

❻ EN 1537 provides upper limits for the test load based on the tendon capacity.

❼ The investigation test failed at 580kN (less than the proof load), which confirms that the capacity of the grout/ground interface is at least as large as the calculated value of 559kN.

❽ EN 1537 recommends that the lock-off load P_0 be a maximum of $0.6P_{tk}$. Typically, P_0 is between 40 and 60% of P_{tk}.

❾ EN 1537 recommends these limits for suitability tests coupled with correlation factors for three or more tests. Both ENs 1997-1 and 1537 recommend that at least three suitability tests are performed.

⓿ If EN 1537's value of γ_a is used, the suitability results indicate that the anchor does not meet requirements.

① The recommendation of Frank et al.[18] — that the proof load for acceptance tests should not exceed $1.15 R_{SLS}$, where R_{SLS} is the anchor load derived from a serviceability limit state calculation of the retaining wall — has not been adopted here.

14.9   Notes and references

1. BS EN 1537: 2000 Execution of special geotechnical work — Ground anchors, British Standards Institution.

2. BS EN 1997-2: 2007 Eurocode 7: Geotechnical design — Part 2: Ground investigation and testing, British Standards Institution.

3. BS EN 1537, ibid.

4. BS EN ISO 22477-5, Geotechnical investigation and testing — Testing of geotechnical structures, Part 5: Testing of anchorages, British Standards Institution.

5. Devon Mothersille (pers. comm., 2008).

6. BS 8081: 1989, Code of practice for ground anchorages, British Standards Institution.

7. See, for example, Ostermayer, H., and Barley, T. (2003) 'Ground anchors', *Geotechnical engineering handbook, Vol 2: Procedures* (ed. Ulrich Smoltczyk), Berlin: Ernst & Sohn, pp.169–219; and Merrifield, C. M., Barley, A. D., and Von Matt, U. (1997) 'The execution of ground anchor works: the European standard EN1537', *ICE Conference on Ground Anchors and Anchored Structures*, London.

8. See Frank, R., Bauduin, C., Kavvadas, M., Krebs Ovesen, N., Orr, T., and Schuppener, B. (2004) *Designers' guide to EN 1997-1: Eurocode 7: Geotechnical design — General rules*, London: Thomas Telford.

9. BS EN 1537, ibid.

10. DD ENV 1997-1: 1995 Eurocode 7: Geotechnical design — Part 1: General rules, British Standards Institution.

11. Frank et al., ibid.

12. DD ENV 1997-1, ibid.

13. Frank et. al., ibid.

14. DD ENV 1997-1, ibid.

15. BS EN 1537, ibid.

16. BS EN ISO 22477-5, ibid.

17. BS 8081, ibid.

18. Frank et al., ibid.

Chapter 15

Execution of geotechnical works

'The remit of CEN/TC 288 is the standardization of the execution procedures for geotechnical works (including testing and control methods) and of the required material properties ... [these documents have] been prepared to stand alongside [Eurocode 7] ... [they provide] full coverage of the construction and supervision requirements.'[1]

15.1   The work of CEN TC 288

The 'execution standards' comprise a suite of twelve European standards, published over the period 1999 to 2007, which provide detailed guidance about construction and supervision of 'special geotechnical works'[2] (see

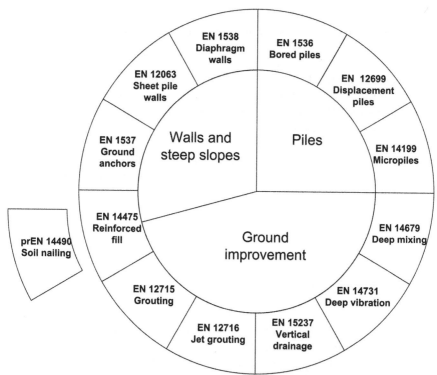

Figure 15.1. Overview of the geotechnical execution standards

Figure 15.1). Several of these standards (ENs 1536, 1537, 12063, 12699, and 14199) are referenced explicitly in EN 1997-1: 2004, the others were in preparation at the time Eurocode 7 was written and hence are not mentioned explicitly.

All of the execution standards follow a common list of contents, illustrated in **Figure 15.2**. The most significant sections amongst these are the ones dealing with materials and products; execution; supervision, testing, and monitoring; and records.

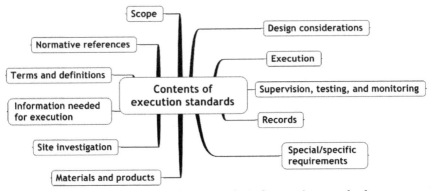

Figure 15.2. Common contents of the geotechnical execution standards

Whether testing should remain within the scope of the execution standards following the publication of EN 22477 (which covers testing of geotechnical structures – see Chapter 4) is open to debate. There may, therefore, be some revision to the execution standards when they are next up for review.

The remainder of this chapter looks at the main features of the execution standards, as far as they concern the implementation of Eurocode 7.

15.2   Piles

As **Figure 15.1** shows, there are three execution standards concerned with piles, as discussed in the following sub-sections.

15.2.1   Bored piles

Figure 15.3 summarizes the scope of EN 1536,[3] which deals with the execution of bored piles, including bored piles constructed of plain concrete, concrete with bar reinforcement, concrete with special reinforcement (e.g. a steel section or tube), pre-cast concrete elements (with temporary casing or uncased), and steel tubes.

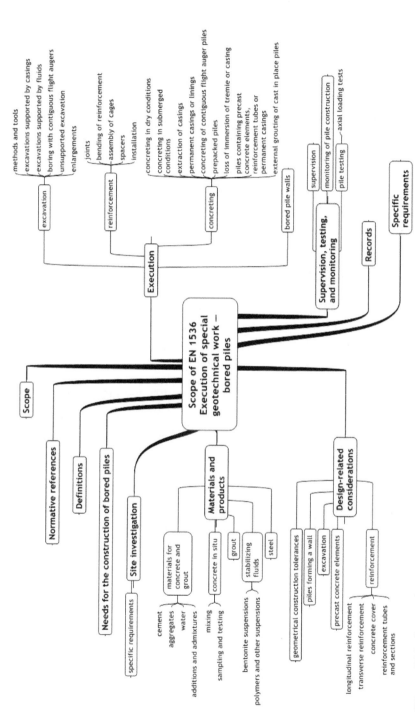

Figure 15.3. Scope of EN 1536, Execution of ... bored piles

EN 1536 restricts itself to bored piles with shaft diameters between 0.3m and 3m. Smaller diameter piles are classified as micropiles (see Section 15.2.3).

The structural design of bored piles is governed by European Standards ENs 1990, 1992-1-1, and 1994-1-1 and no further guidance on this subject is provided in EN 1536. Bored piles may be designed as unreinforced concrete elements provided the various actions upon them produce only compressive stresses in the pile (and the foundation is not in a seismic area).

The recommended design strength of concrete for bored piles is between grades C20/25 and C30/37 and the cement content should not be less than $325kg/m^3$ using dry placement and $375kg/m^3$ using submerged. Specific requirements for water/cement ratio, fines content, and concrete slump are also given.

The plan deviation of a bored pile from its required position should be ≤ 100mm for piles < 1000mm in diameter and ≤ 150mm for piles > 1500mm in diameter. Linear interpolation may be used for diameters between these limits. The inclination of piles raked less than 1:15 from the vertical should deviate no more than 2% from that specified and that of piles raked between 1:15 and 1:4 no more than 4%.

EN 1536 gives detailed recommendations for monitoring bored pile construction, including subjects in the following general areas: setting out (11 subjects), stabilizing fluid (3), reinforcement (8), fresh concrete (6), concrete placement in dry or submerged conditions (11), continuous flight auger piles (5), pre-packed piles (6), external grouting and shaft-base grouting (3), and cutting-off (11).

For example, the eight subjects covered for reinforcement include delivery of materials, dimensions, fabrication of cages, spacers, and installation of cases. For each subject, guidance is given on control (what should be monitored?), purpose (why should it be monitored?), and frequency (when should it be monitored?).

EN 1536 was published by CEN in 1999, confirmed in 2002 for a further five years, and in 2007 was scheduled for revision by TC 288 following a systematic review.

15.2.2   Displacement piles

Figure 15.4 summarizes the scope of EN 12699,[4] which deals with the execution of displacement piles including driven cast-in-place, screwed cast-in-place, prefabricated concrete (round or square), steel (round or H), and prefabricated concrete conical (round or square) piles with or without enlarged bases.

EN 12699 restricts itelf to displacement piles with diameters greater than 150mm. Smaller diameter piles are classified as micropiles (see Section 15.2.3).

Displacement piles are classified according to how they are fabricated (pre-fabricated or cast-in-place) and then according to material type (concrete, steel, or timber) for prefabricated piles or casing type (temporarily cased concrete pile or permanently concrete- or steel-cased) for cast-in-place piles.

The structural design of displacement piles is governed by European Standards ENs 1991-1, 1992-3, 1993-5, 1994-1-1, and 1995-1-1 and no further guidance is provided on this subject in EN 12699.

The plan deviation of a displacement pile from its required position must be ≤ 100mm on land but may be higher over water. The inclination of vertical and raked piles must deviate no more than 4% from the specified inclination.

During driving the maximum calculated stress in the pile must not exceed 80% of the characteristic compressive strength of the concrete or timber and 90% of the characteristic yield stress for steel piles. Particular requirements are also given to govern the reinforcement in driven cast-in-place piles, including: 0.5% minimum longitudinal reinforcement; 100mm clear distance between bars; 5mm minimum diameter for transverse reinforcement; and 50mm cover for piles with temporary casing, rising to 75mm for high exposure risks.

For all displacement pile types useful lists are provided for factors that need to be considered as part of the construction process. Specific guidance is given for driven cast-in-place piles to control the driving of tubes adjacent to piles already constructed. This states that piles may not be installed within six pile diameters until the concrete has reached the desired resistance. Further limits are also given depending on the strength of the soil.

Amongst the special requirements discussed for displacement piles are site safety (security of the site, operational safety of driving and auxiliary equipment and tools, and safety of the working practices), noise and vibration hindrance, environmental damage (noise pollution), and impact on surrounding structures and slopes.

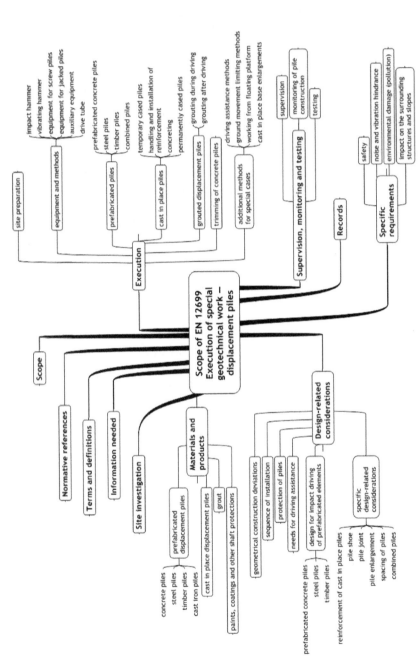

Figure 15.4. Scope of EN 12699, Execution of ... displacement piles

EN 12699 was published by CEN in 2000 and re-confirmed for a further five years in 2005.

15.2.3 Micropiles

Figure 15.5 summarizes the scope of EN 14199,[5] which deals with the execution of bored micropiles with diameters not exceeding 300mm and driven micropiles with diameters not exceeding 150mm. Micropiles can be constructed of steel, reinforced concrete, grout, or mortar and are particularly useful where there is restricted access or difficult drilling conditions (obstructions, rock, etc.), but may also be used for standard foundation applications.

The structural design of micropiles is governed by European Standards ENs 1991-1, 1992-3, 1993, and 1994-1-1.

EN 14199 provides guidance on the information required for execution of the works and associated geotechnical investigations. In particular, it requires additional investigations to be carried out if the available information is deemed insufficient. Since the installation of micropiles may require special tools to overcome obstructions, the geotechnical investigation report must record their presence if encountered.

EN 14199 makes no specific requirements relating to the steel used either for reinforcement or as bearing elements, but instead refers to related European standards. For cement grouts, mortars, and concrete used for micropiles, the following specific requirements are stated: minimum unconfined compressive strength of 25MPa at 28 days; a water cement ratio less than 0.55 for grouts and 0.6 for mortar/concrete; and cement content at least 375kg/m^3.

EN 14199 contains no details on the structural design of micropiles but refers to other standards, particularly the codes dealing with structural materials and EN 1997-1.

A load testing regime incorporating preliminary piles is recommended where conditions are not well known. For micropiles working in compression, at least 2% of the first 100 piles should be tested and then 1% of every additional 100 piles. For micropiles working in tension, at least 8% of the first 100 piles should be tested and then 4% of every additional 100 piles.

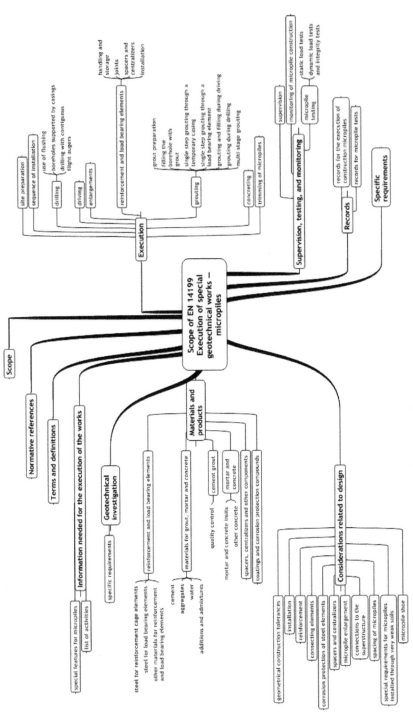

Figure 15.5. Scope of EN 14199, Execution of ... micropiles

The use of dynamic load testing and integrity testing of micropiles should only be used where there is relevant experience or comparison with static load tests to demonstrate that these methods are suitable.

EN 14199 was published by CEN in 2005 and is due for systematic review by TC 288 in 2010.

15.3   Walls and steep slopes

As **Figure 15.1** shows, there are four published execution standards and one draft that are concerned with the execution of walls and steep slopes, as discussed in the following sub-sections.

15.3.1   Sheet pile walls

Figure 15.6 summarizes the scope of EN 12063,[6] which deals with the execution of sheet pile walls including tubes and sheet piles, U-box and U-sheet piles, Z-box and Z-sheet piles, H-beams, and timber piles.

No design methods are provided for establishing the length or modulus requirements for sheet piles: these should be established from other standards and acceptable design methods.

Guidance on the handling of sheet piles is provided in Annex A and on suitable driving methods in Annexes C and D.

Welding requirements for sheet piling are discussed in detail and depend on the type of welded assembly (lengthened, strengthened, junction, or box piles). The type (visual or ultra-sonic) and extent (10, 50, or 100%) of testing of various joints (butt, lap, corner, T, or oblique T) are indicated.

Particular consideration is given to the permeability of sheet piles and where there is a need to reduce the overall permeability of the wall interlock sealing. Annex E provides an analytical method for assessing discharge through a sheet pile wall.

EN 12063 was published by CEN in 1999 and confirmed in 2005 for a further five years.

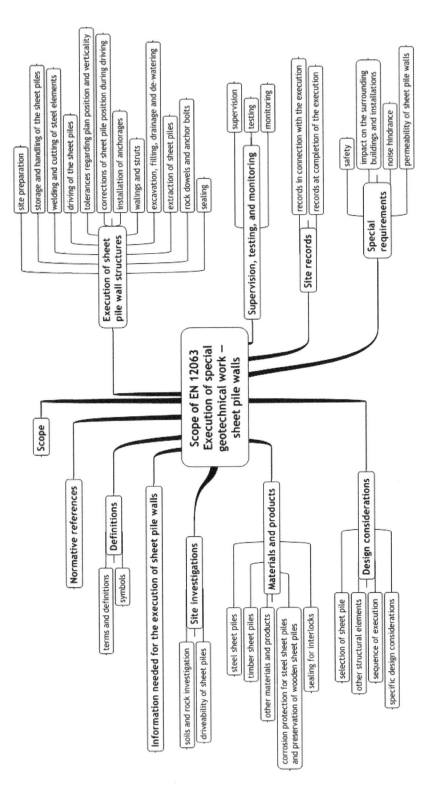

Figure 15.6. Scope of EN 12063, Execution of ... sheet pile walls

15.3.2   Diaphragm walls

Figure 15.7 summarizes the scope of EN 1538,[7] which deals with the execution of diaphragm walls including cast-in-situ, precast concrete, and reinforced slurry walls (for retention) and slurry and plastic concrete walls (for cut-off).

Guidance is provided on the suitable characteristics for bentonite support fluid. No specific characteristics are given for polymer support fluids. Design minimum cement content of concrete for various maximum aggregate sizes is provided ranging from $350kg/m^3$ to $400kg/m^3$ for aggregate sizes of 32mm to 16mm. The water cement/ratio must not exceed 0.6. Mix design guidance is given for plastic concretes for use in cut-offs in Annex A.

A series of tables is provided which details the supervision of execution and monitoring. Tables 3 to 7 cover the requirements for: cast-in-situ concrete diaphragm walls; precast concrete diaphragm walls; reinforced slurry walls; slurry cut-off walls; and plastic concrete cut-off walls.

EN 1538 provides a key diagram that defines the main parts of a diaphragm wall (panel, guide wall, and reinforcement cage) and their dimensions (panel length and thickness, depth of excavation, cut-off level, horizontal and vertical cage lengths, cage width, platform level, and casting level).

EN 1538 was published by CEN in 2000, confirmed in 2002 for a further five years, and scheduled in 2007 for revision by TC 288 following a systematic review.

15.3.3   Ground anchors

Figure 15.8 summarizes the scope of EN 1537,[8] which deals with the execution of ground anchors.

The standard provides a key diagram that defines the main parts of a ground anchor, including: the anchorage point at the jack during stressing and at the anchor head in service; the bearing plate, load transfer block, and structural element; and de-bonding sleeve, tendon, and grout body.

EN 1537 defines three types of test relevant to ground anchors, the purposes of which are summarized below.

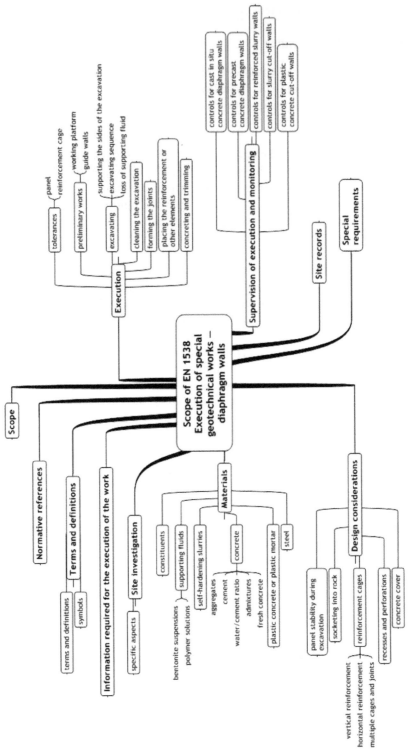

Figure 15.7. Scope of EN 1538, Execution of ... diaphragm walls

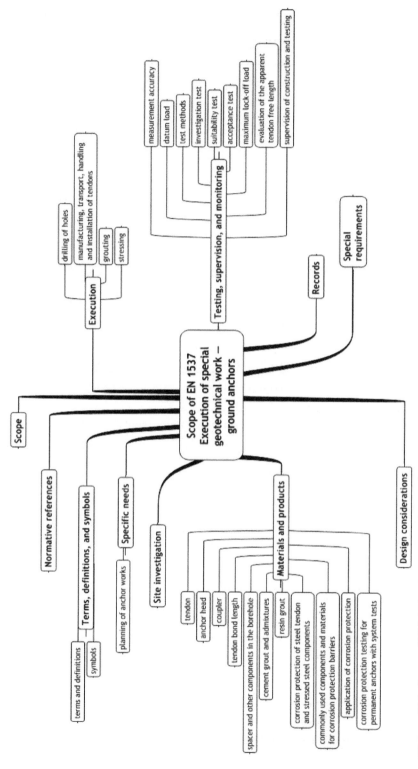

Figure 15.8. Scope of EN 1537, Execution of … ground anchors

Test type	Purpose is to...	
Investigation	establish	resistance at grout/ground interface critical creep load of anchor system, or creep characteristics up to failure load, or load loss characteristic at SLS apparent tendon free length
Suitability	confirm for design situation	ability to sustain the proof load creep or load loss up to proof load apparent tendon free length
Acceptance	confirm for each anchor	ability to sustain the proof load creep or load loss at SLS, where necessary apparent tendon free length

EN 1537 describes three different test methods for ground anchors, summarized below.

Test method	Loads applied in incremental	Loaded to	Measurements of
1	cycles	maximum test load	Anchor head displacement
2	cycles	maximum test load or failure	Loss of load at anchor head at lock-off load
3	steps	maximum test load	Anchor head displacement under maintained load

There is extensive guidance, which was debated at length during drafting of EN 1537,[9] on corrosion protection of steel tendons and stressed steel components, including discussion of common components and materials for corrosion protection barriers such as plastic sheaths and ducts, heat shrink sleeves, seals, cement grout, and resins.

EN 1537 provides far more design guidance than other execution standards, including a method for designing ground anchors which is discussed in Chapter 14. Work is already underway within CEN's Technical Committees TC 250/SC7 and TC 288 to move this design guidance from EN 1537 to Eurocode 7. The specification of the various anchor tests is also likely to be moved (into EN 22477-5), leaving EN 1537 to concentrate on matters relating solely to execution.

The national foreword to BS EN 1537 states that it supersedes those parts of BS 8081:1989[10] that deal with the construction of ground anchorages. BS 8081 is currently being revised in order to remove this conflicting material.

EN 1537 was published by CEN in 1999, confirmed in 2005 for two years, and scheduled in 2007 for revision by TC 288 following a systematic review. A corrigendum was issued in 2000.

15.3.4   Reinforced fill

Figure 15.9 summarizes the scope of EN 14475,[11] which deals with the execution of reinforced fill.

The standard provides a review of various facing systems, including: partial- and full-height facing panels; sloping panels; planter units; segmental concrete blocks; king post systems; semi-elliptical steel faces; steel wire grids; gabion baskets; wrap-arounds (with or without formwork and bagged); and in situ concrete facings. For each facing system, the standard reviews its main application; any reinforcement needed; the technology involved; its longitudinal and transverse flexibility; fill material; and tolerances in alignment, differential settlement, and compressibility.

EN 14475 helpfully details the design output necessary for reinforced fills. This includes things such as drainage, construction phases, level of control, frost susceptibility, type and configuration of reinforcement, steel grade and type of coating, geosynthetic creep behaviour, aesthetic requirements, top soils for greened faces, and a further twenty or so topics not listed here.

The foreword to EN 144475 clarifies its relationship with Eurocode 7:

> 'Eurocode 7 ... does not currently cover the detailed design of reinforced fill structures. The values of partial factors and load factors given in EN 1997-1 have not been calibrated for [them].
>
> ... a two stage approach has been adopted ... first producing an EN giving guidance on the Execution of reinforced fill, before working towards a common method of design. This standard represents the implementation of the first part ...'[12]

The national foreword to BS EN 144475 further clarifies its relationship with BS 8006:[13]

> 'This [standard] partially supersedes BS 8006:1995 which is currently being revised in order to remove conflicting material. In the meantime, where conflict arises between the two documents the provisions of BS EN 14475 should take precedence'[14]

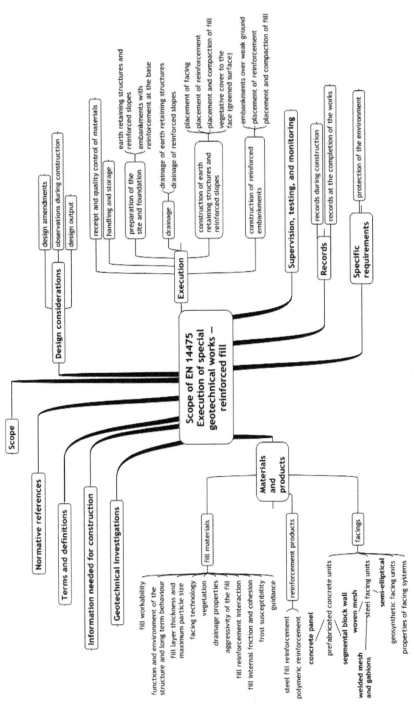

Figure 15.9. Scope of EN 14475, Execution of ... reinforced fill

And finally, the UK National Annex to EN 1997-1 states:

> 'In the UK, the design and execution of reinforced fill structures and soil
> nailing should be carried out in accordance with BS 8006, BS EN 14475 and
> prEN 14490.'[15]

EN 14475 was published by CEN in 2006 and is due for systematic review by TC 288 in 2011. A corrigendum was issued in 2006.

15.3.5 Soil nailing

At the time of writing (early 2008), development of EN 14490, which will deal with the execution of soil nailing, has not reached completion and the document is only available as pre-standard prEN 14490.[16] Publication of EN 14490 is expected before 2010.

15.4 Ground improvement

As **Figure 15.1** shows, there are five execution standards concerned with the execution of ground improvement, as discussed in the following sub-sections.

15.4.1 Grouting

Figure 15.10 summarizes the scope of EN 12715,[17] which deals with the execution of non-displacement and displacement grouting but excluding jet grouting, which is covered by EN 12716 (see Section 15.4.2). The standard identifies three types of grout (solution, suspension, and mortar) and outlines their characteristics and applicability.

The purpose of grouting is either to reduce the mass permeability (hydraulic conductivity) and/or to improve the stiffness and density of the ground. This can be achieved by a number of different combinations of grout type, grout design, and injection processes. Emphasis is placed on various properties of grouts, monitoring of their properties and the injection process, and measurement of ground properties before and after grouting.

The importance of an appropriate site investigation is emphasized with checklists of information that an investigation should provide in order to facilitate the grouting operation and, in particular, to enable the design of the grout and injection process. Coupled with an adequate site investigation, EN 12715 emphasizes the need for grouting trials to confirm the design.

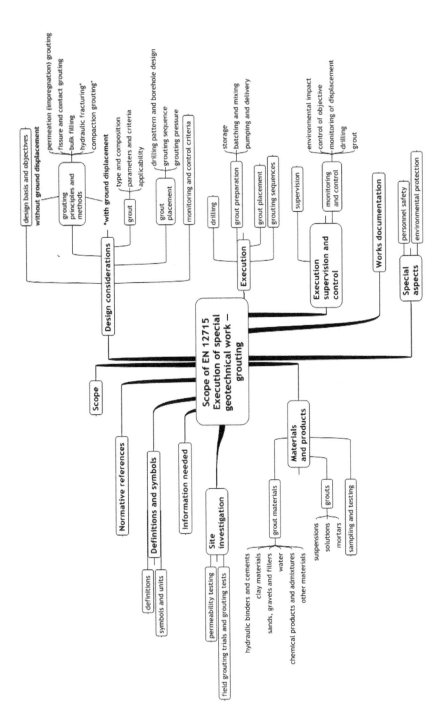

Figure 15.10. Scope of EN 12715, Execution of ... grouting

Non-displacement grouting includes permeation grouting, fissure and contact grouting, and bulk filling, whereas displacement grouting includes hydraulic fracturing and compaction grouting. Although the standard provides a lot of good advice on all aspects of a grouting project, it provides little guidance on the design of the grout mix or details of the method of injection.

EN 12715 was published by CEN in 2000 and re-confirmed for a further five years in 2005.

15.4.2   Jet grouting

Figure 15.11 summarizes the scope of EN 12716,[18] which deals with the execution of jet grouting. Jet grouting is considered distinct from grouting because the process aims to disaggregate the soil or weak rock and mix it with the cement agent. It is usual to produce either a column of grouted ground or a panel, vertical or horizontal.

The main structures formed by jet grouting are walls, slabs, canopies, and blocks. These structures may be created using four main processes: single systems (involving disaggregation and cementation carried out by a single fluid); double air systems (involving disaggregation carried out by cement grout supplemented by an air jet shroud); double water systems (involving disaggregation carried out by a high energy water jet with a cement grout used to cement the soil together); and triple systems (involving disaggregation carried out by high energy water jet supplemented by an air jet shroud with a cement grout used to cement the soil together).

Jet grouting can be used to create foundation piles, underpinning, retaining walls, and low permeability barriers.

As with the execution standard for grouting (see Section 15.4.1), EN 12716 does not provide specific design methods but it does provide useful guidance on what needs to be considered for good design and what should be monitored and recorded during construction.

EN 12716 was published by CEN in 2001 and re-confirmed for a further five years in 2006.

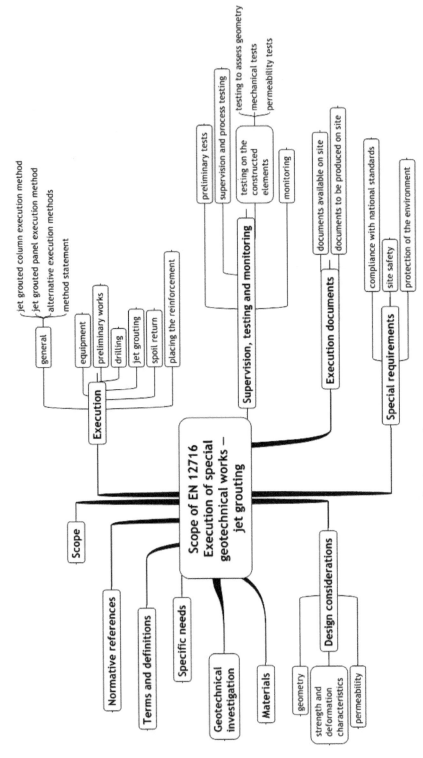

Figure 15.11. Scope of EN 12716, Execution of … jet grouting

15.4.3   Deep mixing

Figure 15.12 summarizes the scope of EN 14679,[19] which deals with the execution of deep mixing.

Deep mixing involves the mixing of in situ materials with a binder plus additives, if required. Mixing is achieved using mechanical tools to a minimum depth of 3m. The columns so formed can be used as load bearing elements, particularly where reinforcement is provided.

The standard provides a range of guidance on what needs to be considered in design. It emphasizes the need to have a good knowledge of the variability of the ground, particularly any harder layers.

As the resultant column is a mixture of the natural soils and binding agent, it is essential that a staged approach to design is adopted. This involves samples of natural soils being mixed in the laboratory to identify suitable design mixes that will achieve required strengths and other properties. Soils vary both vertically and horizontally across sites, making it all the more important that the in situ properties of the improved soil are similar to those of the laboratory samples. Higher strengths are commonly achieved in the laboratory and design mixes must taken this into account.

Two main processes are presented in EN 14679: dry and wet mixing. The main sections of the standard detail the factors that need to be considered in the design and construction of columns by deep mixing. Annex A provides useful information on the principles and practice of dry and wet mixing. This is further enhanced by Annex B, which discusses methods for: settlement reduction; improvement of stability; support of slopes; improvement of bearing capacity; confinement of wastes; containment structures; and reduction of vibrations.

EN 14679 was published by CEN in 2005 and is due for systematic review by TC 288 in 2010. A corrigendum was issued in 2006.

15.4.4   Deep vibration

Figure 15.13 summarizes the scope of EN 14731,[20] which deals with the execution of deep vibration including densification of loose granular soils using vibrators and the provision of vibrated stone columns formed using dry top feed, wet bottom feed, and dry bottom feed.

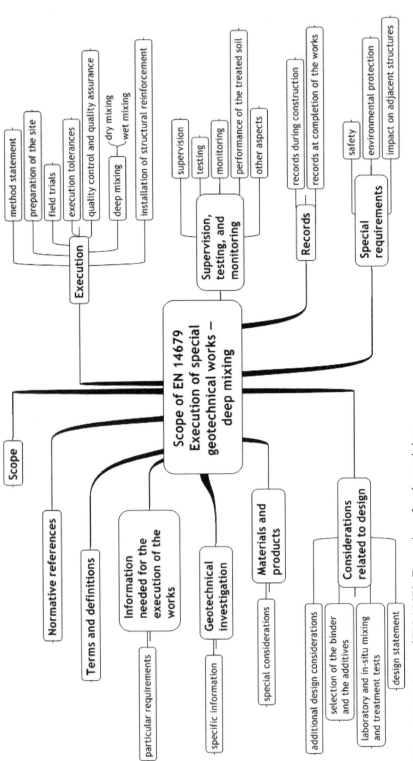

Figure 15.12. Scope of EN 14679, Execution of ... deep mixing

A description of deep vibratory compaction and its applicability is given in Annex A. Annex B describes the process of installing vibrated stone columns and gives recommendations regarding their applicability.

Detailed design methods are not provided but helpful guidance is given on design principles and matters that require consideration when selecting and constructing deep compaction columns.

Emphasis is placed on the use of in situ testing to monitor the improvement of ground properties due to deep vibratory compaction, utilizing cone penetration tests, dilatometer tests, dynamic probing, pressuremeter tests, or standard penetration tests. Similar tests may be used for vibrated stone columns but consideration should also be given to large scale load tests. Plate loading tests should be carried out on individual columns.

EN 14731 was published by CEN in 2005 and is due for systematic review by TC 288 in 2010.

15.4.5   Vertical drainage

Figure 15.14 summarizes the scope of EN 15237,[21] which deals with the execution of vertical drainage including prefabricated vertical drains and sand drains. The standard deals with the requirements for design, drain material, and installation methods for providing vertical drainage for pre-consolidation and reduction of post construction settlements; increasing the rate of consolidation; increase of stability; groundwater lowering; and mitigation of liquefaction effects.

EN 1237 does not cover soil improvement by other methods such as wells, stone columns, or reinforcing elements.

EN 1237 discusses testing of various types of drain: band drains, prefabricated cylindrical drains, and sand drains. For prefabricated band and cylindrical drains, the following are covered: shape and structure measurements; durability; tensile strength and elongation; discharge capacity; filter, its tensile strength and pore size; velocity index; and quality control. There are extensive references to other EN ISOs covering testing of the various elements of drains. This requires a large number of standards to be referred to when specifying prefabricated drains.

It is essential that the consolidation process is monitored to verify that the drains are performing according to expectations. Where necessary, field trials should be carried out to verify the design spacing and depth of drains.

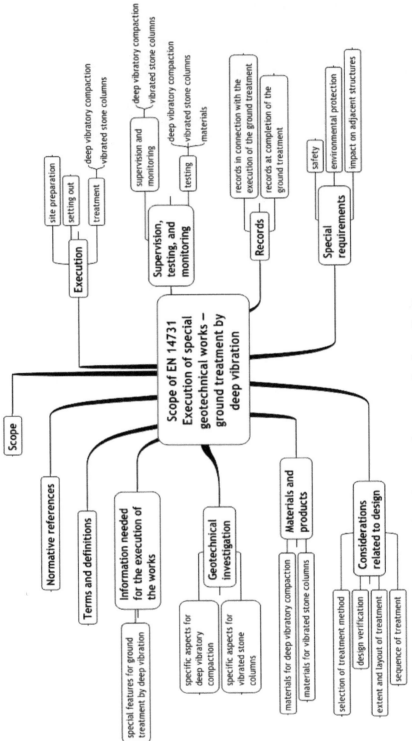

Figure 15.13. Scope of EN 14731, Execution of ... ground treatment by deep vibration

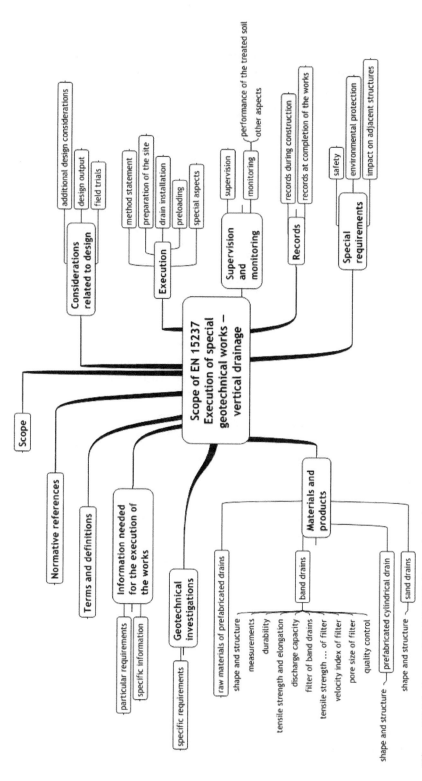

Figure 15.14. Scope of EN 15237, Execution of ... vertical drainage

Annex A of EN 15237 contains useful information on the types of vertical drain and the potential difficulties of installation. It covers methods for determination of band drain discharge capacity and guidance for monitoring. Annex B provides methods for design.

EN 15237 was published by CEN in 2007 and is due for systematic review by TC 288 in 2012.

15.5　　Future developments

As noted in the sub-sections in this chapter, the execution standards are subject to systematic review every five years (as are all European standards). Any confusion caused by conflicts between the execution standards and the provisions of Eurocode 7 and its associated geotechnical testing standards (discussed in Chapter 4) may therefore be remedied within a relatively short time scale, leading to a more coherent and consistent suite of geotechnical standards.

In the UK, de facto standards that already cover the execution of geotechnical works (e.g. SPERW[22] and various CIRIA guides[23]) either already have been or are planned to be updated to complement Eurocode 7 and the execution standards.

15.6　　Summary of key points

The credibility of the execution standards is evident from the following extract compiled from the forewords to the standards themselves:

> 'This [execution] standard has been drafted by a working group comprising delegates from [9-14] countries and against a backdrop of more than [7-30] pre-existing standards and codes of practice both national and international.'[24]

For some countries in Europe, the execution standards represent a valuable source of information and guidance not previously available in their national standards; in other countries, they are regarded as lowest-common-denominator codes that jeopardize good advice already available in national standards.

The UK's Federation of Piling Specialists (FPS) has observed an improvement in the standard of piling and diaphragm walling works – and reduced conflict between contractor and designer on associated construction and detailing issues – since the introduction of ENs 1536 and 1538. However, despite being available for several years, these standards are not widely acknowledged by structural engineers, particularly with regard to reinforcement requirements.[25]

15.7   Notes and references

1. From the foreword to EN 1537. The forewords to the other execution standards carry similar wording.

2. Five of the thirteen execution standards (all of 1999–2001 vintage) use the singular word 'work' instead of 'works' in their titles. Why the title of this suite includes the word 'special' is a mystery.

3. BS EN 1536: 2000, Execution of special geotechnical work – Bored piles, British Standards Institution.

4. BS EN 12699: 2001, Execution of special geotechnical work – Displacement piles, British Standards Institution.

5. BS EN 14199: 2005, Execution of special geotechnical works – Micropiles, British Standards Institution.

6. BS EN 12063: 1999, Execution of special geotechnical work – Sheet pile walls, British Standards Institution.

7. BS EN 1538: 2000, Execution of special geotechnical works – Diaphragm walls, British Standards Institution.

8. BS EN 1537: 2000, Execution of special geotechnical work – Ground anchors, British Standards Institution.

9. Derek Egan (2008, pers. comm.).

10. BS 8081: 1989, Code of practice for ground anchorages, British Standards Institution.

11. BS EN 14475: 2006, Execution of special geotechnical works – Reinforced fill, British Standards Institution.

12. Foreword to BS EN 14475, ibid.

13. BS 8006: 1995, Code of practice for strengthened/reinforced soils and other fills, British Standards Institution.

14. National foreword to BS EN 14475, ibid.

15. See NA.4 of the UK National Annex to BS EN 1997-1, ibid.

16. prEN 14490: 2002, Execution of special geotechnical works – Soil nailing, British Standards Institution.

17. BS EN 12715: 2000, Execution of special geotechnical work – Grouting, British Standards Institution.

18. BS EN 12716: 2001, Execution of special geotechnical works – Jet grouting, British Standards Institution.

19. BS EN 14679: 2005, Execution of special geotechnical works – Deep mixing, British Standards Institution.

20. BS EN 14731: 2005, Execution of special geotechnical works – Ground treatment by deep vibration, British Standards Institution.

21. BS EN 15237: 2007, Execution of special geotechnical works – Vertical drainage, British Standards Institution.

22. Institution of Civil Engineers (2007) *ICE specification for piling and embedded retaining walls* (2nd edition), London: Thomas Telford Publishing.

23. For example, Phear, A., Dew, C., Ozsoy, B., Wharmby, N.J., Judge, J., and Barley, A. D. (2005) *Soil nailing – best practice guidance*, London: CIRIA C637.

24. Compiled from the forewords to ENs 1536, 1537, 1538,12699, 12715, 14199, and 15237.

25. Federation of Piling Specialists (2008, pers. comm.).

Chapter 16

Geotechnical reports

'The report of my death was an exaggeration' – Mark Twain (1835–1910)[1]

16.1   Introduction

Eurocode 7 introduces a plethora of three-letter abbreviations into the geotechnical lexicon, including the limit states STR, GEO, EQU, HYD, and UPL discussed in Chapters 6 and 7. This chapter introduces two more TLAs (three-letter acronyms): 'GIR' for the Ground Investigation Report and 'GDR' for the Geotechnical Design Report.

The Ground Investigation Report is discussed in detail in Section 16.3 and the Geotechnical Design Report in Section 16.4. The reports upon which the GIR depends are discussed in Section 16.2. Finally, Section 16.5 compares the new set of reports defined by Eurocode 7 with those currently specified in existing practice. The relationship between all these reports is illustrated in **Figure 16.1**.

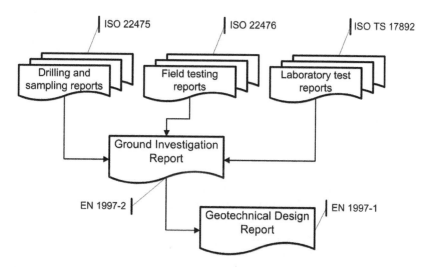

***Figure 16.1**. Summary of geotechnical reporting to Eurocode 7*

16.2   Geotechnical investigation and testing reports

16.2.1   Drilling and sampling reports

Sampling methods and groundwater measurements are covered by EN ISO 22475, details of which are given in Chapter 4.

EN ISO 22475-1 sets out the requirements for reporting the results of drilling, sampling, and groundwater measurements. **Figure 16.2** summarizes the common features of these reports.

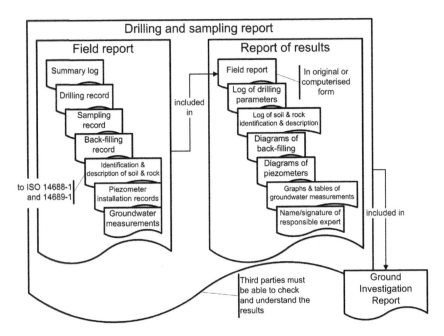

Figure 16.2. Contents of a drilling and sampling report

These reports are similar to those defined in Section 7 of BS 5930[2] and represent typical good practice for reporting borehole data.

16.2.2   Field investigation reports

By the beginning of 2008, international (ISO) standards for geotechnical field testing have been published for the standard penetration test and the dynamic probe only (EN ISOs 22476-2 and -3). Further standards are in development for electrical cone and piezocone penetration tests; Ménard, self-boring, and full displacement pressuremeter tests; flexible and flat dilatometer tests; borehole jack test; field vane test; weight sounding test; Lefranc permeability test; water pressure test in rock; and pumping tests.

Details of these standards are given in Chapter 4. The publication of EN ISOs 22476-2 and -3 has led to the corresponding sections of BS 1377,[3] covering standard penetration and dynamic probe tests, being withdrawn in 2007.

The various parts of EN ISO 22476 set out the requirements for the reports that must be produced during and after the field test. **Figure 16.3** summarizes the common features of these reports. The field report, which is prepared at the test site, must give a summary log of the ground conditions encountered and record the measured values and test results obtained, in such a way that third parties can understand and check the results.

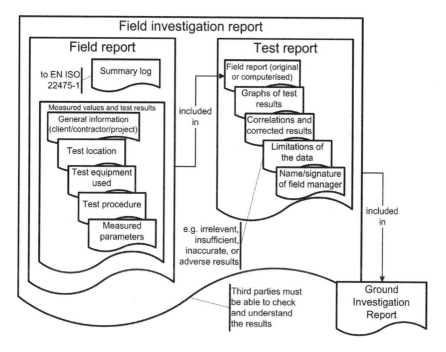

Figure 16.3. Contents of a field investigation report

General information includes the names of the client and contractor; the project name, location, and number; and the name and signature of the test operative. Information on test location includes borehole number, a sketch of the site, and test coordinates relative to a known position. The method of drilling and the make, model, and physical characteristics of the equipment must be included in the field report, together with details of the procedure followed (date of test, reasons for an early end of the test, backfilling of the borehole, etc.). Finally, the actual measurements obtained must be recorded in a format suitable for checking.

The test report includes the field report, either in its original or electronic format, together with graphs showing (uncorrected) test results and, if necessary, corrected results. Information about the corrections applied and limitations of the data (irrelevant, insufficient, inaccurate, or adverse results) must also be recorded. The report must be signed by the field test manager.

16.2.3   Laboratory test reports

International technical specifications for laboratory testing of soil (ISO TS 17892) were published in 2003 and renewed recently for a further three years. They cover the determination of water content, density of fine-grained soils, particle density using the pyncometer, particle size distribution, permeability, and Atterberg limits. Plus the fall cone test, unconfined compression tests on fine-grained soils, unconsolidated undrained and consolidated (drained) triaxial compression tests, and direct shear tests. Further details are given in Chapter 4.

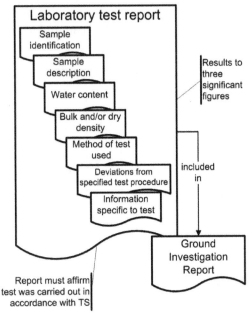

Figure 16.4. Contents of a laboratory test report

The various parts of ISO 17892 set out the requirements for the reports that must be produced during and after the laboratory test. **Figure 16.4** summarizes the common features of these reports.

The laboratory test specifications have not been welcomed in the UK, which currently has a technically superior standard in BS 1377.[4] While the ISO documents retain their Technical Specification (TS) designation, they do not have to be adopted in Member Countries of the Committé Européen de Normalisation (CEN) and the British Standards Institution has decided not to publish them yet, pending a final decision on their status.

16.3   Ground Investigation Report

The requirement for a Ground Investigation Report (GIR) appears in EN 1997-2:

> *The results of a geotechnical investigation shall be compiled in a Ground Investigation Report which shall form part of the Geotechnical Design Report.*   [EN 1997-2 §6.1(1)P]

The contents of the GIR are specified both in EN 1997-1 (as an Application Rule) and in EN 1997-2 (as a Principle):

> *The Ground Investigation Report [should normally]/[shall]† consist of a presentation of all [available]/[appropriate]† geotechnical information including geological features and relevant data; [and] a geotechnical evaluation of the information, stating the assumptions made in the interpretation of the test results.*   [EN 1997-1 §3.4.1(3)] and [EN 1997-2 §6.1(2)P]

In simple terms:

> GIR = Presentation + Evaluation ... of geotechnical information

The contents of the GIR are illustrated in **Figure 16.5** and discussed in the following sub-sections. Physically, the GIR may form a single volume or span several volumes, depending on the size and nature of the investigation.

16.3.1   Presentation

The presentation of geotechnical information is set out in three Principles and one Application Rule in §6.2 of EN 1997-2.

The GIR must provide a factual account of all field and laboratory investigations, presented in accordance with the EN and/or ISO standards used in those investigations. The report must document the methods and procedures used – and results obtained – from desk studies, sampling, field tests, groundwater measurements, and laboratory tests.

The factual account should include a description of the site and its topography, in particular: evidence of groundwater, areas of instability, difficulties during excavation, local experience in the area, and a further eight items given in a checklist.   [EN 1997-2 §6.2(2)]

---

†First wording from EN 1997-1, second from EN 1997-2.

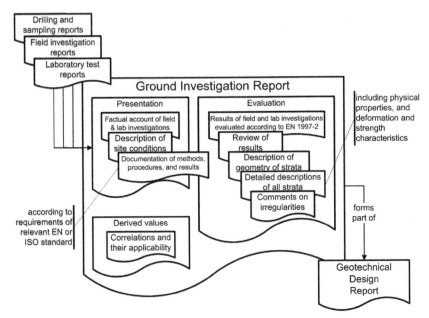

Figure 16.5. *Contents of the Ground Investigation Report*

The level of detail given in Section 6 of EN 1997-2 is less than that provided by Section 7 of BS 5930. It is important to refer to the guidelines provided in BS 5930 when preparing reports intended to meet the requirements of Eurocode 7. It is likely that BS 5930 will be updated to remove any information that conflicts with EN 1997-2.

The results of field and laboratory investigations must be included in the GIR in the format specified in the appropriate European (EN) or International (ISO) standard used to perform the investigation. The requirements of these standards, which are separate from but relied upon by Eurocode 7, are discussed in Section 16.2

16.3.2   Evaluation

The evaluation of geotechnical information is set out in two Principles and six Application Rules in §6.3 of EN 1997-2.

The first Principle requires that the GIR must provide an evaluation and review of the field and laboratory results; detailed description of all strata, including their geometry, physical properties, and strength and deformation characteristics; and comments on irregularities such as cavities and discontinuities.

The second Principle ensures that in situ and laboratory test results are interpreted accounting for various factors including: groundwater, ground type, sampling, handling, transportation, and specimen preparation. Further, it requires that the ground model be revisited in the light of the test results.

The six Application rules provide guidance on presentation of the data, derivation of geotechnical parameters, consideration of anomalies, grouping of similar strata, and deriving boundaries between various geological units across a site.

Importantly, EN 1997-2 requires known limitations of the results to be stated in the GIR. *[EN 1997-2 §6.1(5)P]*

It can be the case that engineers are reluctant to highlight limitations in the data, as it could be construed that contract obligations had not been fully met or their level of competency had been questioned. However, in order that data can be reliably analysed in the future, accurate recording of all the difficulties is essential. For example, when standard penetration tests (SPTs) are carried out below the water table, it is important to record whether the water level in the borehole is maintained above that in the surrounding ground. If this is not the case, then artificially low SPT blow counts could be recorded if the imbalance in water pressure loosens the base of the borehole.

16.3.3 Derived values

Any correlations that have been used to derive geotechnical parameters must be documented in the GIR. Typically this might include such correlations as the relationship between standard penetration test blow count and Young's modulus, consistency index and undrained strength, or plastic limit and California bearing ratio (see Chapter 5).

16.4 Geotechnical Design Report

The contents of the Geotechnical Design Report are specified (as a Principle) in EN 1997-1:

The assumptions, data, methods of calculations and results of the verification of safety and serviceability shall be recorded in the Geotechnical Design Report. *[EN 1997-1 §2.8(1)P]*

In simple terms:

$$GDR = Assumptions + Data + Methods + Verification$$

The contents of the GDR are illustrated in **Figure 16.6** and discussed below. The level of detail included in the GDR depends on the type of design – simple designs may only require a single-sheet report (although these will tend to be limited to Geotechnical Category 1 structures).

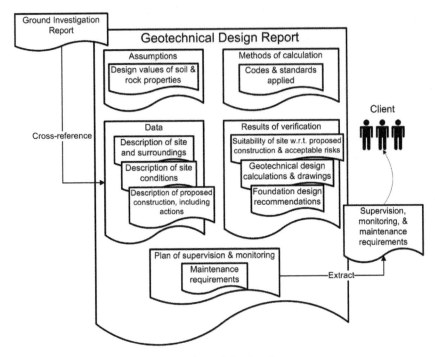

Figure 16.6. Contents of the Geotechnical Design Report

Items which the GDR should normally contain include the design values of soil and rock properties, with justification as necessary (assumptions); descriptions of the site and its surroundings, ground conditions, and the proposed constructions including actions (data); statements about the codes and standards applied (methods of calculation); and statements about the suitability of the site with respect to the proposed construction and acceptable risks, geotechnical design calculations and drawings, and recommendations for foundation design (results of verification). The GDR should refer back to the Ground Investigation Report, where appropriate.

The supervision, monitoring, and maintenance requirements for a project must be included in the GDR and an extract of those requirements must be given to the project owner. This is similar to what we expect when we buy a car from a reputable dealer: we are provided with an owner's manual that specifies the care and attention that our car needs in order to give us a satisfactory performance in the future (happy motoring).

[EN 1997-1 §2.8(4)P] and *[EN 1997-1 §2.8(6)P]*

16.5   Comparison with existing practice

The following sub-sections compare the Ground Investigation and Geotechnical Design Reports with traditional reports on these subjects.

16.5.1   British Standard BS 5930

Current UK practice is specified in Section 7 of BS 5930,[5] which defines the series of reports illustrated in **Figure 16.7**.

Field reports cover all the information that needs to be obtained while working on site, e.g. the recording of in situ tests (such as the standard penetration test, cone penetration test, pressuremeter, etc.) and production of drillers' logs. They also provide outline guidance on the generation of the necessary forms for recording such data.

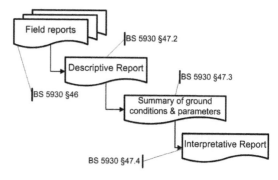

***Figure 16.7**. Reports defined by BS 5930*

The Descriptive Report provides a factual account of all that was carried out in the field. BS 5930 recommends this report is written under the general headings shown in **Figure 16.8**. Detailed guidance is given on suitable content for each section, including examples of good practice in the production of borehole logs (for both cable percussion boring and rotary coring).

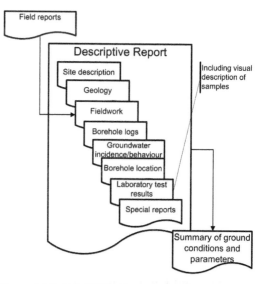

***Figure 16.8**. BS 5930's Descriptive Report*

The summary of ground conditions and parameters is developed from data given in the Descriptive Report. Usually the summary is included either in the Descriptive Report itself or in the Interpretative Report (rather than being presented as a separate report in its own right). Where it appears is normally specified in the contract documents and depends on who takes legal responsibility for corresponding aspects of the work. The contents of the summary, as specified in BS 5930, are illustrated in **Figure 16.9**.

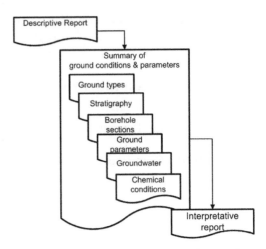

Figure 16.9. BS 5930's 'summary of ground conditions and parameters'

The engineering interpretation must include clear statements about the data on which it is based, the nature of the structures to be built (including their dimensions and loadings), and the ground parameters to be used. The typical content of an Interpretative Report to BS 5930 is shown in **Figure 16.10**.

Outline design information should be given to inform the design and construction of typical geotechnical structures. BS 5930 identifies the following subjects that may require particular attention, if appropriate: spread foundations; piles; retaining walls; basements; ground anchorages; chemical attack; pavement design; slope stability; mining subsidence; tunnels and underground works; safety of neighbouring structures; monitoring of movements; embankments; and drainage. Potential difficulties for

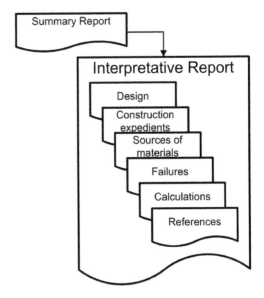

Figure 16.10. BS 5930's Interpretative Report

construction should also be addressed, where relevant, including issues related to: open excavations; underground excavations; groundwater; driven piles, bored piles and ground anchors; grouting; mechanical improvement; and contamination. The report may also include sections on sources of materials for fill and aggregates and failures where these are relevant. Calculations should be included in an appendix.

16.5.2   AGS Guidelines for preparing the Ground Report

The Association of Geotechnical and Geoenvironmental Specialists (AGS) has produced guidance on the preparation of a Ground Report.[6] These guidelines identify the following documents, which together form the Ground Report, and refer to BS 5930 as a key source of detail for their contents:
- desk study
- factual report
- interpretive report
- design report
- validation report.

The Guidelines provide section headings for each report together with outline content for each section.

It is common in UK practice for a project to have a desk study, factual report, and interpretative report prepared according to BS 5930. The design report brings together all the factual and interpretative information together with other design information such as: assumptions; description; design values; calculations; plan of supervision and monitoring; maintenance requirements; and geo-environmental considerations.

AGS recommends that a validation report is written on completion of the works, to provide as-built records of the project's geotechnical elements and to discuss the results of construction monitoring, ongoing maintenance requirements, decommissioning the structure, and re-use of the foundation.

16.5.3   Geotechnical Baseline Reports

The American Society of Civil Engineers recommends[7] the use of Geotechnical Baseline Reports (GBRs) for underground construction in the United States of America.

Geotechnical Baseline Reports, which should be produced by suitably experienced and qualified engineers, recognize the variability and unpredictable nature of the ground. They define the baseline conditions that will affect the successful completion of the construction works. When conditions turn out to be better than the baseline, the contractor benefits; when conditions are demonstrably worse than the baseline, the contractor is suitably reimbursed for the adverse consequences.

Baseline statements need to be carefully prepared to reduce ambiguity and ensure that all parties to a contract are clear about the level of risk they are taking. One of the principal aims of a GBR is to reduce the necessity for the contractor to resort to 'unforeseen ground conditions' clauses in order to recover costs when things go awry.

16.5.4   Report checklists

Various government agencies throughout the world provide checklists of information required for geotechnical reports.[8] For example, the US Federal Highway Administration (FHWA) has produced guidelines and checklists[9] to help ensure that the basic features of a geotechnical report are covered, including:
- site investigation information
- centreline cuts and embankments
- embankments on soft ground
- landslide corrections
- retaining structures
- structure foundations – spread footings
- structure foundations – driven piles
- structure foundations – drilled shafts
- ground improvement techniques
- material sites.

The aim of these checklists is to ensure that all relevant information is provided in the reports that are prepared on behalf of the FHWA.

16.6   Who writes what?

A number of organizations are likely to be involved in the preparation of geotechnical reports:
- client
- design consultant (structural or geotechnical)
- site investigation contractor
- main contractor
- specialist contractor
- design-and-build contractor.

Although this list may not include all parties that might be involved in a construction project, we will use it to illustrate the potential changes that Eurocode 7 will have on the production of geotechnical reports.

The client is unlikely to be involved in the preparation of either the Ground Investigation Report (GIR) or the Geotechnical Design Report (GDR). However, in collaboration with the designer (who may of course be an in-house company), the client will specify which design standards and codes of practice must be used for the project.

Normally, it is the design consultant who determines the required level of geotechnical investigation for a project. If the designer is not a geotechnical specialist, he or she should engage the services of a suitably qualified and experienced geotechnical engineer, especially since (as discussed in Chapter 2) EN 1990 assumes that:

> the choice of the structural system and the design of the structure is made by appropriately qualified and experienced personnel.   [EN 1990 §1.3(2)]

On many small- to medium-sized projects, the GIR will be prepared by the site investigation contractor, while the designer will produce the GDR. If the design consultant is not a geotechnical specialist, the project's success relies even more heavily on the site investigation contractor's expertise in evaluating the field and laboratory test data and selecting suitable derived values of ground parameters.

On large projects and where the design consultant has considerable geotechnical expertise, production of the GIR may become a joint effort between the site investigation contractor (whose role may be limited to presenting the data) and the designer (who would then evaluate that data and determine derived values from it).

It is very unlikely that a site investigation contractor will be in a position to write the GDR, owing to the detailed knowledge of the proposed design that is required by that report. Furthermore, at the time the site investigation contractor is employed on most projects, design loadings (actions) are often not known – another essential ingredient of the GDR.

Main contractors and their sub-contractors are not usually involved in the preparation of geotechnical reports, except under design-and-build contracts where the detailed design is their responsibility.

Specialist contractors (for example, piling companies) often take design responsibility for their works. They will invariably be responsible for producing those parts of the project's overall GDR that relate to their specialist contribution.

Finally, the client will be made aware of monitoring and maintenance requirements of the GDR and will be responsible for taking appropriate measures to comply with those requirements.

16.7   Summary of key points

The Geotechnical Design Report (GDR) defined in EN 1997-1 and the Ground Investigation Report (GIR) defined in EN 1997-2 have many similarities with reports that are currently in use in many parts of the geotechnical world, as summarized in **Figure 16.11**.

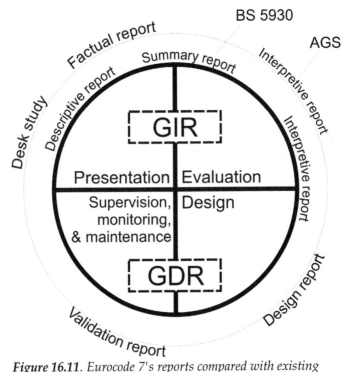

Figure 16.11. *Eurocode 7's reports compared with existing practice*

Put simply, the GIR is equivalent to the AGS factual report (i.e. the descriptive and summary reports defined by BS 5930). The GDR is equivalent to the AGS design report (i.e. interpretive report from BS 5930 plus additional content), but does not assume the contractual significance of a baseline report.

The Geotechnical Design and Ground Investigation Reports take the best of current practice and should enable better control and dissemination of the geotechnical issues associated with any development.

A major shortcoming of the GIR and GDR is their failure to specify the requirements for dealing with contaminated land. Problems associated with contamination are particularly significant for assessing the economics of site development.

16.8   Notes and references

1. Mark Twain (June 2, 1897), *New York Journal*.

2. BS 5930: 1999, Code of practice for site investigations, British Standards Institution, London.

3. BS 1377: 1990, Methods of test for soils for civil engineering purposes, British Standards Institution, London. Clauses 2.2 and 2.3 of Part 9 were withdrawn in 2007.

4. BS 1377: 1990, ibid.

5. BS 5930, ibid.

6. Association of Geotechnical and Geoenvironmental Specialists (2003), *Guidance on the preparation of a Ground Report*.

7. Essex, R. J. (1997) *Geotechnical Baseline Reports for underground construction – guidelines and practices*: American Society of Civil Engineers.

8. For example, Ministry of Defence (1997) *Technical Bulletin 97/39*; British Columbia, Ministry of Transportation and Highways (1998) *Technical Bulletin GM9801*.

9. US Department of Transport, Federal Highway Administration (2003) *Checklist and guidelines for review of geotechnical reports and preliminary plans and specifications*.

Epilogue

'Anyone who listens to my teaching and obeys me is wise like the person who builds a house on solid rock ... but anyone ... who ignores it is foolish like a person who builds his house on sand ...' – Matthew, 7:24–26

The aim of the Eurocodes is to provide a unified approach to structural and geotechnical design that seeks to avoid the pitfalls of founding on rock or sand. But, for the Eurocodes to be effective, their advice must be understood and applied correctly.

Reaction to the Eurocodes

Unfortunately, many engineers' initial reaction to Eurocode 7 is a cross between *The Eurocode Scream* (see **Figure 17.1**) and the natural instinct of an ostrich, which, when frightened, buries its head in the sand. However, when the shock of the new is overcome, views change as the benefits of the Eurocodes become apparent.

The views of many engineers are based on limited knowledge of the Eurocodes and even less experience of using them in practice. **Figure 17.2** summarizes some of the opinions that have been expressed in the civil engineering press regarding the impact of the Structural Eurocodes on civil engineering design.

Figure 17.1. 'The Eurocode Scream' by Jack Offord (after Edvard Munch) – a typical reaction to the complexity of Eurocode 7

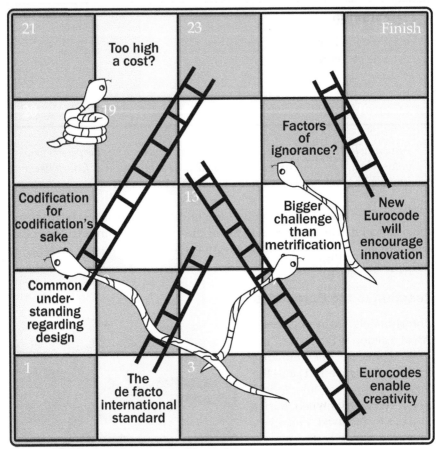

Figure 17.2. *Pros and cons of the Eurocodes*

In our opinion, the Eurocodes provide a unified approach to civil and structural engineering design and bring greater consistency to our treatment of the ground and other structural materials (such as steel and concrete). As engineers become familiar with the numerous documents involved, any antipathy towards them will dissolve.

As evidence of this, we have recently completed a survey of over 400 delegates who have attended our training courses over a six month period. When asked 'What do you think the impact of Eurocode 7 will be on the foundation industry?', their responses were mainly positive (see table below). Several self-confessed sceptics have changed their views about the Eurocodes once their Principles have been explained and their flexibility revealed.

Dissemination

In order for the Eurocodes to become accepted and their Principles to be understood and applied correctly, there needs to be an ongoing programme of education and training. Currently this is taking the form of publications (such as this book), public and in-house training courses, evening lectures, and seminars. These events require funding from industry to pay for tutors to prepare and deliver and for staff time in attendance.

What do you think Eurocode 7's impact will be on the foundation industry?		
Very good	26	6%
Good	227	55%
Neither	98	24%
Bad	30	7%
Very bad	2	< 1%
Unsure	30	7%
Total	413	100%

Source: Geocentrix survey 2007–8

Although the introduction of the Eurocodes has been largely driven by politics, there is no central funding to enable organizations to train their staff and update their procedures. There is a significant burden on organizations to get up to speed with developments and to purchase necessary resources, such as the new codes. The costs are significant and will fall disproportionately on smaller organizations.

Over the next few years a series of publications will become available to explain the application of the Eurocodes. These will include books, open lectures, teaching materials, case studies, and research papers. Each document will provide fresh levels of insight into the subject and will help to uncover any inconsistencies. It is very unlikely that one publication or suite of training events will cater for all needs.

There will be pressure on geotechnical software houses to make their computer programs compatible with EN 1997. This task is made more difficult by ongoing debate about how partial factors should be applied to water pressures, passive earth pressures, etc. (as discussed in this book). There will inevitably be a delay before fully consistent and reliable programs become available.

Looking ahead

Errors and ambiguities in the text of EN 1997 have already been discovered and are being reviewed by a 'Maintenance Group' for future correction. In the first instance, this will lead to publication of a corrigendum, dealing with essential changes to Eurocode 7, followed by a more considered amendment or revision some time after 2010.

As our understanding of the nature of soils and rocks and the complex issues of soil-structure interaction improves, so Eurocode 7 will need further development. The values of partial factors may need to be revised to ensure that adequate levels of reliability are maintained without unnecessary conservatism. The reliance on ultimate limit state calculations may give way to greater consideration of movements and other serviceability conditions.

The introduction of three Design Approaches for limit states GEO and STR is not ideal. The strengths and weaknesses of each Design Approach will become more apparent as engineers become familiar with them. Hopefully, as experience is gained in the use of limit state philosophy in geotechnical engineering, a unified approach accepted across the whole of Europe may emerge. In the meantime, engineers must endeavour to make the most of the common Principles and Application Rules that have been agreed to date.

Conclusion

In the words of Harold Wilson (British Prime Minister, 1964–70 and 1974–76):

'He who rejects change is the architect of decay.'

The last thing an engineer wants to be is an architect!

Appendix 1

Slope stability design charts

This appendix provides charts for designing infinitely long slopes and slopes subject to circular slips according to Design Approach 1.

The charts for infinitely long slopes (**Figure A1.1** to **Figure A1.3**) are based on the equation for the 'characteristic' stability number N_k that is developed in Chapter 9, with partial factors γ_G, γ_c, γ_φ, and γ_{Re} from Design Approach 1. Each chart represents a different r_u value (0, 0.3, or 0.5) and the numbers on the curves (1:1, 1:1.5, etc.) represent the gradient of the slope (i.e. tan β).

The charts for circular slips (**Figure A1.4** to **Figure A1.6**) are based on the method of slices. Each chart represents a different r_u value and the numbers on the curves represent the gradient of the slope (i.e. tan β). Solid lines are for D/H = ∞ and dashed for D/H = 1, where D is the depth to a rigid layer and H the slope height. **Figure A1.6** is for undrained conditions.

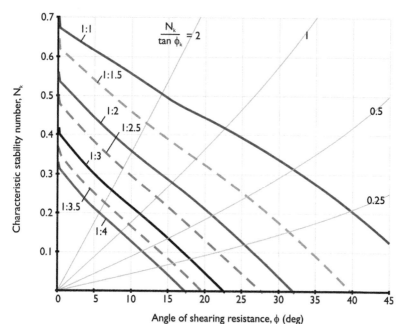

Figure A1.1. Design chart for infinitely long slope with $r_u = 0$

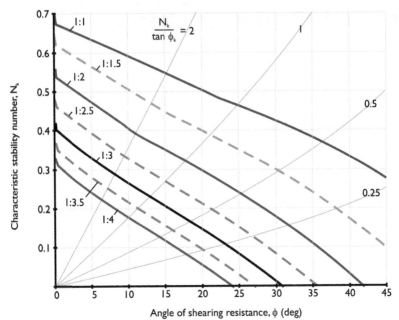

Figure A1.2. Design chart for infinitely long slope with $r_u = 0.3$

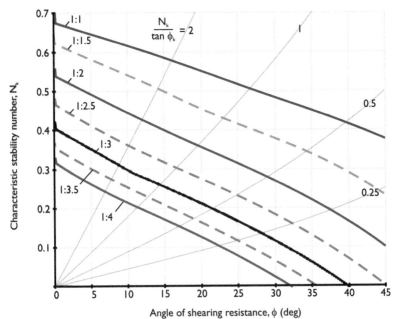

Figure A1.3. Design chart for infinitely long slope with $r_u = 0.5$

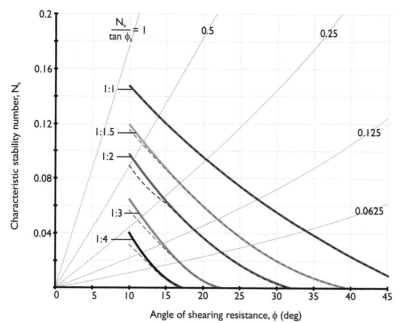

Figure A1.4. Design chart for circular slip in slope with $r_u = 0$

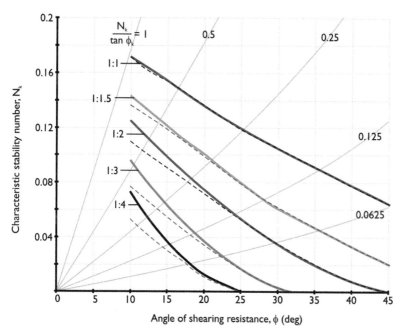

Figure A1.5. Design chart for circular slip in slope with $r_u = 0.3$

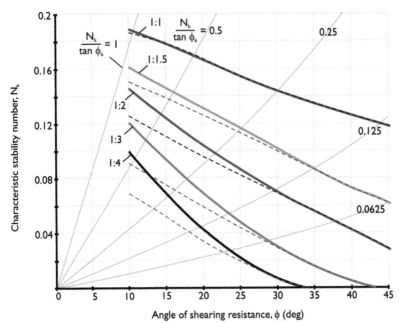

Figure A1.6. Design chart for circular slip with $r_u = 0.5$

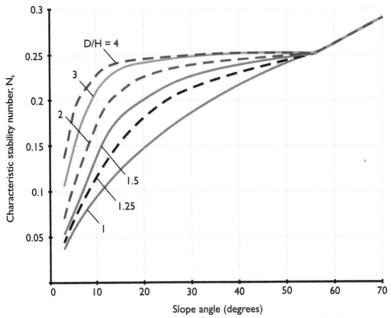

Figure A1.7. Taylor's stability chart for an undrained slope, re-worked to include a partial factor $\gamma_{cu} = 1.4$ as per Eurocode 7

Earth pressure coefficients

EN 1997-1 Annex C provides a numerical procedure for determining active and passive earth pressure coefficients for use in retaining wall design, which is discussed in Chapter 12.

The charts that follow show the variation in $K_{a\gamma}$ and $K_{p\gamma}$ (denoted K_a and K_p on these charts) with angle of shearing resistance φ, for different values of interface friction δ ($0°$, $5°$, $10°$, $15°$, $20°$, $25°$, and $30°$), for vertical walls ($\theta = 0°$). Each figure gives curves for different slope gradients $\tan \beta$ (flat, $\pm1:10$, $\pm1:5$, $\pm1:4$, $\pm1:3$, $\pm1:2.5$, $\pm1:2$, and $\pm1:1.5$), where available. Coefficients $K_{a\gamma}$ and $K_{p\gamma}$ allow the earth pressure acting normal to the wall surface to be calculated (see Chapter 12 for details).

The numerical procedure on which these charts are based imposes certain limitations on the values of interface friction δ and slope angle β that can be used. Specifically, δ must satisfy the condition $\delta \leq \varphi$ and β must satisfy $\beta \geq m_w - m_t$ (see Chapter 12 for the definition of these terms). For this reason, several of the lines shown on the charts are truncated and others lines are missing. If the value of K_a or K_p is not given here, then an alternative procedure must be used to obtain their values (again, see Chapter 12 for possible alternatives).

The practical consequence of these limitations is that the procedure is not valid when $\beta \leq \delta$ for the active case and $-\beta \leq \delta$ for the passive.

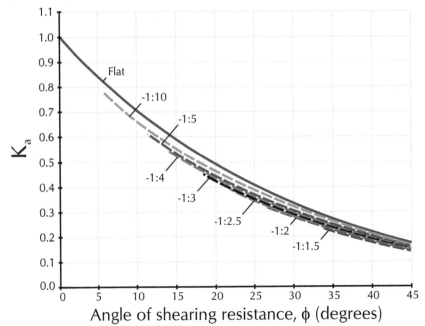

Figure A2.1. Active earth pressure coefficients for $\delta_a = 0°$

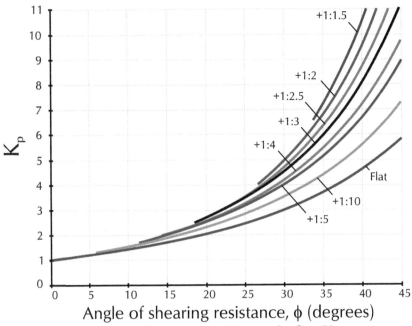

Figure A2.2. Passive earth pressure coefficients for $\delta_p = 0°$

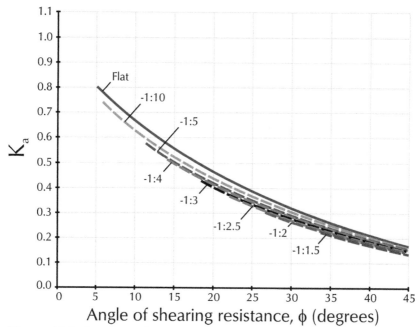

Figure A2.3. Active earth pressure coefficients for $\delta_a = 5°$

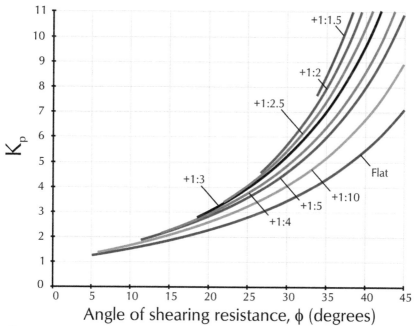

Figure A2.4. Passive earth pressure coefficients for $\delta_p = 5°$

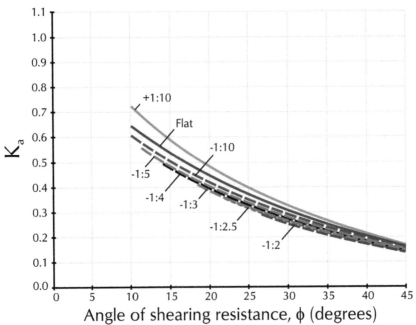

Figure A2.5. Active earth pressure coefficients for $\delta_a = 10°$

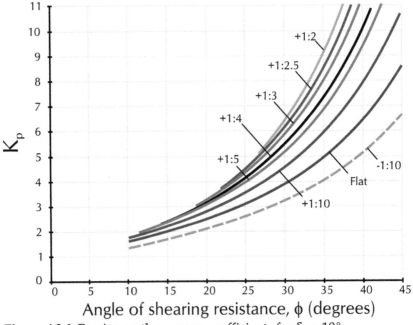

Figure A2.6. Passive earth pressure coefficients for $\delta_p = 10°$

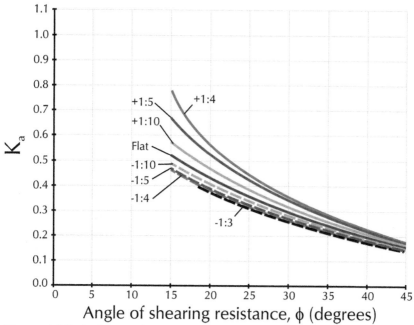

Figure A2.7. Active earth pressure coefficients for $\delta_a = 15°$

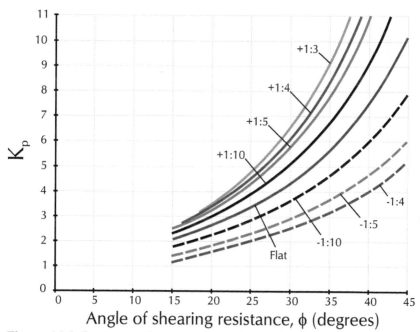

Figure A2.8. Passive earth pressure coefficients for $\delta_p = 15°$

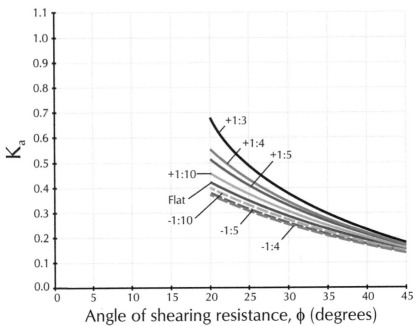

Figure A2.9. Active earth pressure coefficients for $\delta_a = 20°$

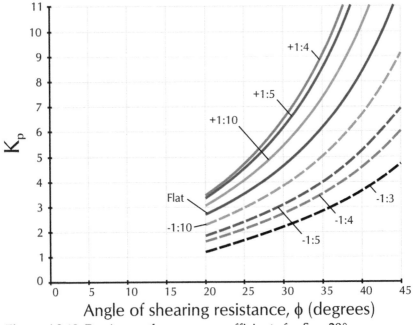

Figure A2.10. Passive earth pressure coefficients for $\delta_p = 20°$

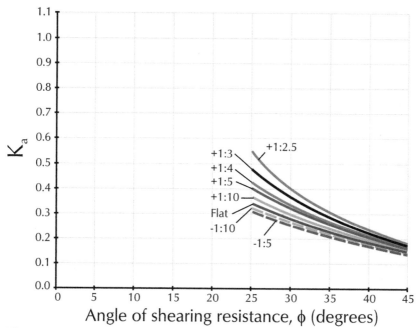

Figure A2.11. Active earth pressure coefficients for $\delta_a = 25°$

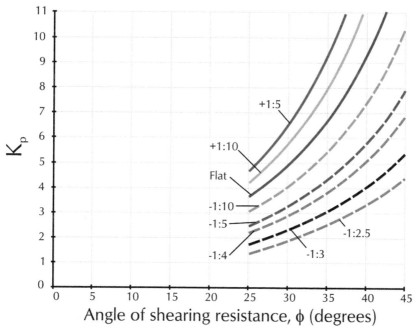

Figure A2.12. Passive earth pressure coefficients for $\delta_p = 25°$

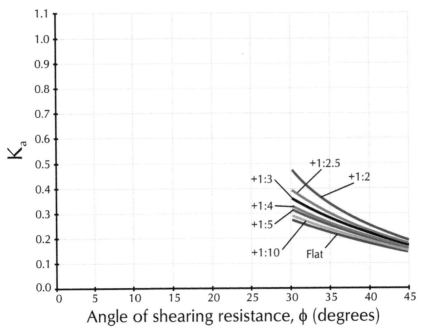

Figure A2.13. Active earth pressure coefficients for $\delta_a = 30°$

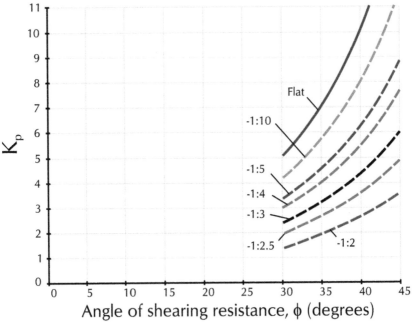

Figure A2.14. Passive earth pressure coefficients for $\delta_p = 30°$

Appendix 3

Notes on the worked examples

The worked examples have been produced using MathCad version 14.0, produced by Parametric Technology Corporation (PTC).

Mathcad requires variables to be defined with unique identifiers and so we have made extensive use of subscripts within the software. Wherever possible, we have employed self-explanatory symbols, but on occasions this has been difficult to do.

Verifications of strength (see Chapter 6) involving Design Approach 1 require two separate calculations to be performed, the so-called Combinations 1 and 2. To reduce the space taken, we present these parallel calculations in matrix format, for example:

$$c_{ud} = \overrightarrow{\left(\frac{c_{uk}}{\gamma_{cu}}\right)} = \binom{42}{30} kPa$$

where the top line of the result (42kPa) is for Combination 1 and the bottom line (30kPa) for Combination 2. Partial factors are introduced into the calculations as follows:

$$\gamma_{cu} = \binom{1.0}{1.4}$$

where 1.0 is the value of γ_{cu} in Combination 1 and 1.4 its value in Combination 2.

The arrow across the top of an equation (see example above) is standard mathematical notation for a matrix cross-product, thereby expanding the previous equation to:

$$c_{uk} = \binom{42}{42} kPa, \gamma_{cu} = \binom{1.0}{1.4}, c_{ud} = \overrightarrow{\left(\frac{c_{uk}}{\gamma_{cu}}\right)} = \overrightarrow{\binom{42/1.0}{42/1.4}} = \binom{42}{30} kPa$$

Without the arrow, Mathcad assumes you want the matrix dot-product, as follows:

$$c_{ud} = \left(\frac{c_{uk}}{\gamma_{cu}}\right) = \frac{42}{1.0} + \frac{42}{1.4} = 72kPa$$

Each worked example concludes with a calculation of the *degree of utilization*, Λ, which is discussed in Chapter 2. This parameter represents the proportion of the available design resistance R_d (or stabilizing design effects of actions $E_{d,stb}$) that is needed to counteract the design effect of actions E_d (or destabilizing design effects of actions $E_{d,dst}$):

$$\Lambda = \frac{E_d}{R_d} \text{ or } \Lambda = \frac{E_{d,dst}}{E_{d,stb}}$$

Provided Λ does not exceed 100%, the design meets the requirements of Eurocode 7, i.e.

$$E_d \leq R_d \text{ or } E_{d,dst} \leq E_{d,stb}$$

(as discussed in Chapter 2).

The worked examples are annotated with numbers ❶, ❷, ❸, etc., where the numbers refer to the notes that accompany each example. These notes attempt to explain key decisions made in the calculations, values that have been taken from Eurocode 7, and any inconsistencies or uncertainties in the application of the standard.

Index

Printed and bound by CPI Group (UK) Ltd, Croydon, CR0 4YY

17/10/2024

01775709-0015